Methane Activation and Utilization in the Petrochemical and Biofuel Industries

Hua Song • Jack Jarvis • Shijun Meng
Hao Xu • Zhaofei Li • Wenping Li

Methane Activation and Utilization in the Petrochemical and Biofuel Industries

Springer

Hua Song
Department of Chemical & Petroleum Engineering
University of Calgary
Calgary, AB, Canada

Jack Jarvis
Department of Chemical & Petroleum Engineering
University of Calgary
Calgary, AB, Canada

Shijun Meng
Department of Chemical & Petroleum Engineering
University of Calgary
Calgary, AB, Canada

Hao Xu
Department of Chemical & Petroleum Engineering
University of Calgary
Calgary, AB, Canada

Zhaofei Li
Department of Chemical & Petroleum Engineering
University of Calgary
Calgary, AB, Canada

Wenping Li
Department of Chemical & Petroleum Engineering
University of Calgary
Calgary, AB, Canada

ISBN 978-3-030-88423-9 ISBN 978-3-030-88424-6 (eBook)
https://doi.org/10.1007/978-3-030-88424-6

This Springer imprint is published by the registered company Springer Nature Switzerland AG
The registered company address is: Gewerbestrasse 11, 6330 Cham, Switzerland

Preface

The oil and gas industry is no stranger to economic downturns, the most recent of which being due to the global economic impact of COVID-19. While recovery is expected in 2021, it will not be a full recovery, and the demand for oil and gas is not expected to return to levels pre-COVID-19. However, despite being threatened by increasing decarbonization strategies from oil and gas companies as they shift focus to low-carbon fuels, and the ever-increasing popularity of renewable forms of energy, natural gas exploration continues to grow and its utilization towards valuable chemicals provides a more elegant solution to the use of this abundant resource.

For the last decade, Dr. Hua Song and his research group, the Green Catalysis Research Group (GCRG), have dedicated research towards catalytic heavy crude oil upgrading, catalytic light olefin upgrading, catalytic coal and biomass upgrading, catalytic paraffin-rich oil upgrading, and other associated fields; all under the presence of natural gas and methane. The group boasts nearly 100 publications in the field of natural gas and methane utilization as well as numerous conference contributions and patented works, making it one of the most significant contributors to this field over the past decade. The inspiration for this book came from the authors' evident passion for the field of catalytic methane utilization, not to mention the ever-increasing demand for energy and a gap that cannot be breached by renewables in the foreseeable future. And was sparked after the publisher approached Dr. Song during the 69th Canadian Chemical Engineering Conference in 2019. During which, he was presenting on "catalytic methane valorization at mild conditions through the coupling effect" as an invited keynote speaker.

This book will take the reader through most aspects of methane utilization in the petrochemical and biofuel industries. The story begins with a historical account of natural gas, including its accidental discovery in ancient China and its initial transportation using bamboo-constructed pipelines. The numerous sources of natural gas are then introduced to the reader along with its production, consumption, processing, and transportation. This detailed introduction (Chap. 1) to natural gas aims to provide all the background necessary before moving on to discussing the topic of methane activation itself in the next chapter.

In Chap. 2, the physical and chemical properties of methane are unravelled, and it becomes clear why methane activation is such a "hot topic" and the reason that its activation has required so much effort and drawn as much interest over the years. This chapter also briefly introduces some non-traditional and emerging methods for activating methane, such as bio-catalysis, plasma activation, photocatalysis, and electrocatalysis. These technologies are promising for methane activation but lack of maturity of other approaches is shown in subsequent chapters. In Chap. 3, the reader receives an introduction to the two "fathers" of catalytic methane activation: dehydroaromatization, and oxidative coupling. Prominent catalysts that promote these reactions are discussed in detail, and the reader is left with an understanding of the active sites involved during these direct methane activation methods. The reader is also introduced to various characterization methods as used in this field of research, for the first time in this book.

The cross-coupling of methane provides a way to reduce input energy requirements when co-fed with higher carbon-number hydrocarbons. This system is covered in Chap. 4 and presented in the form of two case studies: one for the cross-coupling of methane with light straight run naphtha, and the other for the cross-coupling of methane with heavy naphtha. Both co-feeds are often seen as economically unviable in the petrochemical industry and require upgrading via hydrotreating. Of course, these case studies present the feasibility of methane as a hydrogen donor in place of diatomic hydrogen. Chapter 5 demonstrates a similar logic where heavy crude oil, which usually undergoes hydrotreatment with hydrogen, is treated with methane as a hydrogen donor instead. This section provides several studies that prove the feasibility of heavy crude oil upgrading under methane to give lower viscosity, density, and total acid number. The techniques used by researchers and some of the most recent findings are also discussed in detail.

There is one novel technology that is not mentioned in the early sections of this book; nonthermal plasma-assisted catalytic conversion of methane. However, Chap. 6 provides a detailed insight to this technology that can demonstrate room temperature and atmospheric pressure activation of methane, a technology that if commercialized in the future, will be ground-breaking. The final two chapters of this book allow the reader to gain an in-depth understanding of biomass valorization and biomass-derived fuels. In Chap. 7, biomass valorization in the presence of methane is discussed in the form of typical processes, followed by a summary of suitable catalysts, and the two major mechanistic pathways: carbenium and alkyl. Typical characterizations are shared with the reader as well as a summary of the challenges facing commercialization of this technology, including kinetic studies and reactor designs. Chapter 8 gives a detailed description of the current state of biofuel conversion technologies in the presence of methane, using a series of model compounds typically found as major constituents in biofuel. Particular attention is given to carboxylic acids, alcohols/phenols, esters, aldehydes, and ketones.

A surplus of natural gas means that an alternative use beyond heating homes needs to be found. This book will provide a detailed plot of the current technologies as well as ongoing research projects that are utilizing this commodity in a different way. As a topic that is poorly covered outside of the scientific world, this text provides a concise and detailed layout of current progress and what it could mean for the promising future of natural gas.

Calgary, AB, Canada

Hua Song
Jack Jarvis
Shijun Meng
Hao Xu
Zhaofei Li
Wenping Li

Contents

About the Editors

Jack Jarvis started his career in the UK as a Chemistry B.Sc. graduate of Loughborough University. He then operated as a carbon capture chemist, working on a pilot plant utilizing amine solvent technology, which held significant interest from both RWE and Shell Cansolv. Soon after this, Jack recognized the potential in natural gas, especially in North America, and so pursued further studies in the NG field in Canada. For the last 4 years, Jack has dedicated his research towards methane utilization and its coaromatization with light hydrocarbons under the supervision of Dr. Hua Song and GCRG, achieving an M.Sc. at the University of Calgary and currently in the 3rd year of his Ph.D. studies. Department of Chemical and Petroleum Engineering, University of Calgary, Calgary, AB, Canada

Shijun Meng holds a Bachelor's degree in chemical engineering at Tsinghua University (China) in 2017. Currently, he is a 4th year Ph.D. candidate at the University of Calgary. His doctoral research is about nonthermal plasma-assisted catalytic conversion of hydrocarbons. His articles have been published on *Applied Catalysis B: Environmental* and *The Journal of Physical Chemistry Letters.* Department of Chemical and Petroleum Engineering, University of Calgary, Calgary, AB, Canada

Hao Xu got his Bachelor's degree (2014) and Ph.D. degree (2019) in the Department of Chemical Engineering, Tsinghua University, Beijing, China. His research project at Tsinghua University was carbon-based non-mercury metal catalysts for acetylene hydrochlorination. He also worked as a visiting scholar at Rice University, Houston, Texas, USA, in 2017. His topic there was rhodium-based catalysts for aqueous nitrite remediation. He is now a postdoctoral associate under the supervision of Prof. Hua Song in the Department of Chemical and Petroleum Engineering, University of Calgary, Calgary, Alberta, Canada. His research is focused on catalytic heavy oil upgrading using natural gas for clean energy conversion. He has experience in the study of heterogeneous catalysis for over 8 years. Department of Chemical and Petroleum Engineering, University of Calgary, Calgary, AB, Canada

Zhaofei Li obtained his B.Sc. and Ph.D. in Chemistry from Peking University, China, in 2000 and 2005, respectively. He worked as a postdoctoral researcher, research associate, and assistant professor at Ruhr-University Bochum, Tohoku University, and Kyoto University before he joined Petrochemical research institute of PetroChina as a senior engineer. Currently, he is a research associate in the University of Calgary working on heavy oil upgrading and natural gas activation. He has experience in the study of heterogeneous catalysis for over 15 years. Department of Chemical and Petroleum Engineering, University of Calgary, Calgary, AB, Canada

Wenping Li obtained her B.Sc. in Applied Chemistry from Beijing University of Chemical Technology, China, in 2012 and her Ph.D. in Physical Chemistry from Dalian Institute of Chemical Physics (DICP), Chinese Academy of Sciences, in 2019. Prior to joining in the University of Calgary as a postdoctoral associate, she carried out her postdoctoral research at the University of Alberta. Department of Chemical and Petroleum Engineering, University of Calgary, Calgary, AB, Canada

Chapter 1
Methane Sources and Current Uses

1.1 Introduction

Natural gas (NG) is a naturally occurring gas mixture with methane as the primary component. It is one of the three fossil fuels that supply energy for the modern society [1–3]. Normally, it contains varying amounts of other hydrocarbons like ethane, propane, butane, and sometimes a small percentage of water, carbon dioxide, nitrogen, hydrogen sulfide, or a trace amount of carbon monoxide, argon, helium etc.

People have discovered and used natural gas for a very long time. Historically, natural gas was first discovered accidentally in ancient China, as it resulted from the drilling for brines. Also, the first use of this resource was recorded by the ancient Chinese people almost 2500 years ago. They even invented a way to transport gas seeping from the ground in crude pipelines of bamboo, which is exactly the stereotype of pipeline transportation to where it was used to boil salt water to extract salt in southwestern China [1].

Natural gas enjoys tremendous popularity as an energy source, and its global need has grown substantially in the past decades, especially in the twenty-first century due to the two following reasons: natural gas can provide the sustainable energy needed for global development, especially in developing countries as a clean and cheap energy resource. In addition, it can reduce various adverse impacts such as severe air pollution resulting from the widely used energy sources such as coal and petroleum.

Natural gas is a cheap and abundant commodity that can be exploited as a clean fuel in civil and industrial applications as well as the hydrogen and carbon source during the reforming of petrochemical and bio-derived fuels, to form valuable chemicals and precursors for the chemical industry [2]. As a chemical raw material, it can be used to produce acetylene, hydrogen, synthetic ammonia, urea, carbon black, carbon disulfide, methanol, 1,4-butanediol, mono-chloromethane, dichloromethane, chloroform, carbon tetrachloride, and hydrocyanic acid.

H. Song et al., *Methane Activation and Utilization in the Petrochemical and Biofuel Industries*, https://doi.org/10.1007/978-3-030-88424-6_1

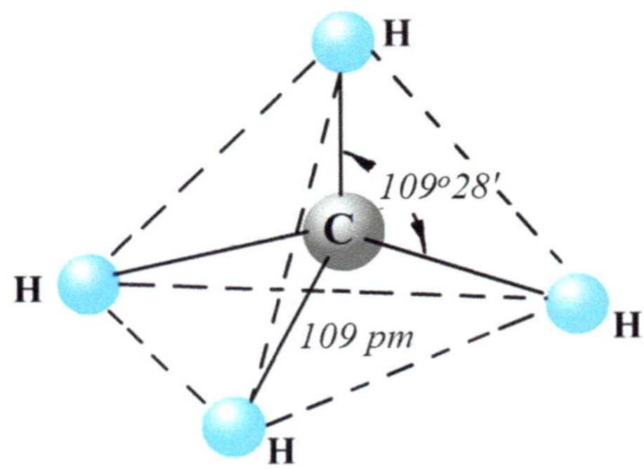

Fig. 1.1 Molecular structure of methane

Methane with its chemical formula of CH_4 is the simplest hydrocarbon. It is composed of one carbon and four hydrogen atoms through sp^3 hybridization of the carbon orbitals to form four identical C-H σ bonds. Therefore, the methane molecule is a regular tetrahedral structure, and the four bonds have identical bond lengths of 109 pm and the same bond angles of 109°28' [3]. Figure 1.1 shows the molecular structure of methane. It has a highly symmetric structure and belongs to the point group of T_d. In addition, CH_4 is a nonpolar molecule due to the fact that it has a symmetric tetrahedral geometrical shape with four identical C-H bonds. The electronegativity of carbon and hydrogen is 2.55 and 2.2, respectively, which causes the partial charges to be almost zero [3].

Therefore, methane is very stable and can widely be found on the earth, considering the fact that the combination energy of C-H covalent bond in methane is as high as 425 kJ/mol [4–6]. The two electrons in the C-H σ bond are very difficult to activate, and the regular visible light cannot excite the electrons in this tetrahedral molecule. Therefore, under standard conditions, methane is a colorless and very stable gas [3]. It has been very challenging to activate methane to produce other chemicals [6]. Nonetheless, researchers have put a huge amount of effort to develop catalyst to lower its activation energy barrier for higher conversion and better selectivity at lower temperatures.

1.2 Advantages and Disadvantages of Methane as a Fuel

It is always complicated to judge whether a fuel is good or not since there are lots of factors influencing the usage of this fuel and the criteria may vary with different human development stages. For example, a series of factors should be first evaluated such as its energy density, price, availability, transportability, toxicity, storability, combustion efficiency, convenience, total storage, durability of supply, and reliability. With the development of modern society, other factors which may cause environmental problems should also be taken into account in evaluating the fuel such as cleanliness and greenhouse effect from itself or its combustion products [1].

Methane has a highly symmetric structure, and the interactions between the neighboring molecules are very weak and make it difficult to liquefy. Therefore, its density is around 0.72 kg/m^3, which is almost about half of the density of air [1–3]. While the density for petroleum-derived liquid fuel such as gasoline and diesel, two of the most utilized fuels, is about 700–850 kg/m^3. It is clearly seen that methane has a very low specific density. Therefore, methane either needs to be converted into compressed natural gas (CNG) or liquefied into (LNG) [2]. CNG usually requires a pressure of around 10–25 MPa and should be stored in high-pressure vessels. While in the latter case, the temperature should reach −161.5 °C for methane to be liquefied. It is clearly seen that a huge amount of energy will be consumed in these two cases to compensate for shortcomings in specific energy density.

Methane is regarded as a clean fuel. It is much better than the other petroleum-based fuels such as gasoline, diesel, kerosene, etc., no need to mention coal, which has a very poor performance with respect to resulting pollutants. The combustion of pure methane with abundant oxygen will produce only carbon dioxide and water. There will be no emissions of NO_x or SO_x (of course the desulfurization process is often needed in the factory to remove possible sulfur-containing species in the natural gas), which are regarded as the causes for acid rain [2]. However, small amounts of SO_x will be emitted even with the very strict regulations to control the sulfur content to a very low value for gasoline and diesel. Moreover, the combustion of coal will not only emit SO_X and NO_X; it will also produce considerable amounts of PM_{10} and $PM_{2.5}$, which are the main cause of haze in developing countries such as China and India. The pollution caused by coal can be greatly mitigated when it used in central stations to generate electricity in power plants by various measures to trap the particles, to remove the SO_x and NO_x in the flue gas. Despite this, it is still not comparable with methane in the power plant.

Methane has almost the same high flexibility of use as electricity. So far, methane can be used in the household space heating, civil cooking, electricity generation, propelling vehicles, etc.

The global warming caused by industrialization has seized the attention of human kind. Therefore, for a specific fuel, its contribution to the generation of greenhouse gases is becoming one crucial factor when judging a fuel. As is well known, CO_2 is a notorious greenhouse gas. However, all carbon-containing fuels will generate some amount of CO_2 to enter the atmosphere. Methane still has the advantage in this point since it has the highest H-to-C molar ratio (4:1) among all hydrocarbons, petroleum-based fuels, and coals. It should be noted that CH_4 itself is a greenhouse gas, and therefore attentions should be paid to lower the amount of methane released to the atmosphere [1].

As a fuel, methane has a good durability of supply. The conventional resources of natural gas still dominate the global market, but it is also seen that a huge amount of methane is produced from coal beds and shales.

Shale gas is natural gas extracted from shale layers. The global shale gas resources are very rich. According to forecasts, the world's shale gas resources are 456 trillion cubic meters, mainly distributed in North America, Central Asia, China, the Middle East and North Africa, Latin America, Russia, and other regions. The potential reserves of shale gas might be larger than the conventional natural gas [7, 8].

In addition, there is one more new type of methane reserve named methane hydrate. Natural gas hydrate is commonly found in deep-sea sediments or permafrost on land. Due to its shallow and wide distribution, huge total volume, and high energy density, it might become the main alternative energy source in the future, and it has received remarkable attention from governments and scientific communities around the world. Future technologies and scientific advancement may lead to the extraction and utilization of this new energy source. It is reported that during the past 40 years, the world gas consumption of natural gas increased around three times, while the global reserve/exploitation ratio has remained quite stable, implying that methane has a high durability of supply [7, 8].

Modern natural gas companies and pipeline transportation have made it very convenient for natural gas to be transported from field to end users, which means it is an excellent energy source for highly populated areas such as large cities [1].

To summarize, natural gas is an excellent energy source for the current and for the future due to the various advantages and benefits it brings to the society. In addition, natural gas is widely envisioned as the primary source of energy in the foreseeable future.

1.3 Natural Gas Sources

Natural gas sources can be divided into two main categories: conventional natural gas and unconventional natural gas [9]. Figure 1.2 shows a schematic geology of natural gas resources. The conventional gas refers to natural gas resources that are

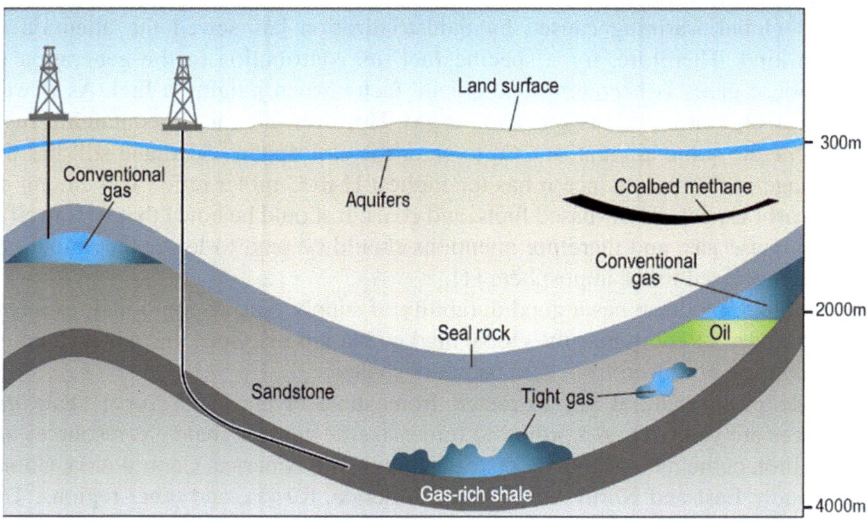

Fig. 1.2 Schematic geology of natural gas resources [9, 10]

normally trapped in porous rocks such as sandstone and can be easily produced from reservoirs using various traditional drilling, pumping, and compression techniques. This is most widely extracted and produced currently. And it will still play a major role in the energy sources [1, 2].

The unconventional natural gas is often found in nonporous rock formations, such as shale [10]. The tight gas is natural gas produced from reservoir rocks with such low permeability that massive hydraulic fracturing is necessary. While the coalbed methane is a form of natural gas which is extracted from coal bed, coalbed methane contains very small amounts of heavier hydrocarbons such as propane or butane and no natural gas condensate. Yet it often contains up to a few percent of carbon dioxide, which needs to be removed before sending to the pipeline transportation [11].

Shale gas is an unconventional natural gas resource extracted from shale layers. Compared with conventional natural gas, shale gas has the characteristics of wide distribution and large thickness in the reservoir. The United States is a relatively successful country in shale gas exploitation, and it was the first to achieve large-scale development. It is first reported that the United States began exploration of shale gas in 1982. After more than 30 years of development, it has gradually matured in shale gas drilling and exploiting technology. With the continuous increase in shale gas production, the United States has surpassed Russia in 2009 and become the world's top natural gas producer. Nowadays, the United States has achieved self-sufficiency in natural gas and has become a natural gas exporter, thanks to the development of shale gas development [10, 11].

Canada is the second country in the world to commercialize shale gas after the United States. Canada and the United States are both located in North America, and their geological structure is very similar. Therefore, the shale gas-related technology in the United States can be easily transferred and applied locally to Canada. The exploration and development of shale gas in the United States has brought about a revolution in the energy industry, and its impact has far exceeded the United States oil and gas industry itself, triggering changes in the world's energy structure and affecting the international geopolitical situation [12]. Currently, shale gas exploit has formed a trend worldwide.

According to the shale gas survey materials published by related companies, among the 32 countries with shale gas resources in the world, the technically recoverable shale gas resources are 187 trillion cubic meters [11].

Although it has recently become very promising and attracted huge attention worldwide, its two main concerns such as the environmental impact and sustainability still might limit its wide utilization in the near future [1].

One newly discovered methane source is the methane clathrate or inflammable ice. It is an ice-like crystalline substance formed by natural gas and water under high-pressure and low-temperature conditions. Because it looks like ice and burns when exposed to fire, it is called "flammable ice." The chemical formula is $CH_4{\cdot}nH_2O$. Natural gas hydrate is commonly found in deep-sea sediments or permafrost on land. The formation of this hydrate clathrate structure requires extreme conditions such as high pressure and low temperature, simultaneously. In the ocean, methane clathrate is mainly stored within 1000 m below the seabed, covering an area of 40 million square kilometers, accounting for about 10% of the total ocean area. In contrast, land-based methane clathrate can be described as very rare, accounting for about 1/100 of ocean reserves [13–15].

Methane clathrate has the characteristics of convenient use, high combustion heat and clean and pollution-free and is recognized as the largest new energy source that has not yet been developed. According to the calculations, 1 m^3 of methane clathrate can release about 164 m^3 of natural gas [16]. Moreover, it is estimated that the total amount of organic carbon contained in combustible ice is equivalent to twice that of known coal, oil, and natural gas in the world. Another advantage of combustible ice is that after burning, only a small amount of carbon dioxide and water will be generated, unlike other conventional fossil energy sources, which will also generate other oxide pollution.

It has similar drawbacks as the shale gas with even more difficulty in the exploitation which limit its usage in industrial practice. In 2010, the cost of extracting natural gas from crystallized natural gas was estimated to be as much as twice the cost of extracting natural gas from conventional sources and even higher from offshore deposits. But some researchers give promising claims about the use of this new material in the foreseeable future [16, 17].

Another type of natural gas source comes from the help of methanogenic bacteria and organic matter. This is one important source of natural gas and it has a very long history. Methanogenic archaea are responsible for almost all kinds of biogas, with huge amounts of methane in our Earth's history. This kind of biogas source includes swamps, marshes, and landfills, as well as agricultural waste materials by the treatment of anaerobic digesters. For example, landfill gas is created by the decomposition of waste in landfill sites. This kind of gas has almost 50% methane in it with others like carbon dioxide, water vapor, nitrogen, and oxygen. The gas can be vented to the atmosphere, flared or burned to produce electricity or heat or be used as cooking gas for some remote rural areas. Biogas can also be more efficiently produced by separating organic materials from waste that otherwise goes to landfills. Like landfill gas, biogas is mostly methane and carbon dioxide, with small amounts of nitrogen, oxygen, and hydrogen [1]. Currently, biogas and landfill gas are already used in some areas, but their use could be greatly expanded. Systems have been established for use in parts of Europe. Using those low-value or even almost no-value waste materials that would cost a considerable quantity of money for further treatment has proved to be profitable via the biogas production. This practice has been very popular in rural communities in southern China, and it has benefited greatly.

As abovementioned, it is clearly seen that natural gas with methane as the most important component has a very broad range of sources, and more new reservoirs and sources are being discovered with the newly developed technologies. More importantly, the unconventional sources should have the chance to be a very important component in our energy consumption structure in the near future.

One main reason of the formation of natural gas is that it was formed by the decaying of plants and animals buried in sedimentary rock layers over a very long period. Normally, natural gas is trapped under layers of rock that prevent it from flowing easily to the surface. But the plant and animal distribution may vary greatly with different geological and ecological conditions.

It is noted that natural gas resources are not evenly distributed globally. The Energy Information Administration estimated that the global natural gas reserves are about 7177 trillion cubic feet. While the biggest reserve is located in the Middle East, which owns around 35% of the total reserves, North America, including the United States and Canada also have huge amounts of natural gas resources [18].

Conventional natural gas production in Canada is mainly from the Western Canada Sedimentary Basin in the provinces of British Columbia, Alberta, and Saskatchewan. There are also some relatively smaller amounts of natural gas resources in Ontario and New Brunswick, while the unconventional natural gas is primarily located in northeast British Columbia and northwest Alberta [19].

It is noted that western Canada, especially the province of Alberta, has a world-class natural resource base, with an estimated 33 trillion cubic feet of remaining established conventional natural gas reserves. Alberta accounts for almost 80% of the natural gas produced in Canada, making Alberta one of the world's largest suppliers of natural gas. Therefore, natural gas and crude oil has been considered as a strategic resource that contributes to economic opportunity and energy security for Alberta. And the Alberta government has spent huge amount of funding to support the development of the natural gas industry in this province, including the exploration, better utilization, and conversion from natural gas to other value-added commodities [19].

It is noteworthy that based on an estimated 2015 world consumption rate of about 3400 km^3 of gas per year, the total estimated remaining economically recoverable reserves of natural gas would last 250 years at current consumption rates. An annual increase in usage of 2–3% could result in currently recoverable reserves lasting significantly less, perhaps as few as 80–100 years. Therefore, it is safe to claim that natural gas can be used as a stable natural source, and there is no need to worry about its complete depletion in the near or distant future.

1.4 Natural Gas Production and Consumption

It is noted that both the natural gas supply and consumption has been increasing in the past years. Especially in the year 2019, both supply and consumption increased, yet the production increased even more rapidly than the consumption increase. Therefore, a decrease of market price was seen in the major markets globally along with the decrease of crude oils [20, 21].

Table 1.1 listed the natural gas production in major areas in the recent 5 years.

It is noted in Table 1.1 that a steady increase of natural gas production is seen globally. North America has the largest increase rate. In the year of 2019, the global production reached 4.1 trillion cubic meters, which increased about 3.4% compared with the value in 2018.

Table 1.1 Natural gas production in recent 5 years. Unit: billion cubic meters

Area	2015	2016	2017	2018	2019
North America	1011	999.7	1006.4	1098.1	1167.3
Central and South America	169.3	174.4 44	168.4	161.5	157.9
Europe	273.5	267.9	2750	273.3	272.5
Asia	804	800.6	847.5	894.9	912.8
Middle East	658.4	669.8	659.7	679.9	700.3
Africa	211.4	217.8	234.3	242.7	249.7
Asia-Pacific	537.9	561.2	603.9	623.1	641.8
In total	3665.5	3691.5	3795.4	3967.0	4102.3

The global LNG production reached 0.421 billion tons/year in 2019, which increased about 10.1% from year 2018. Ten new giant projects have been put into operation in this year, and these are mainly located in the United States, Russia, and Australia. Among them, the United States has the largest increase rate [20, 21].

Table 1.2 lists the natural gas consumption in major regions in the world. It is seen that developed countries consumed a lot more natural gas resources. For instance, North America still has the largest natural consumption percentage, which exceeds one fourth of total global consumption, while Africa has the lowest percentage. In addition, it is seen that the natural gas consumption in 2019 is 3.98 trillion cubic meters, which increased 3.5% compared with the value of 2018. However, this increase rate was lower than the rate from 2018 to 2019 due to the following reasons: the global economic growth in 2019 was not as good as in 2018, the major countries with high consumption didn't increase the consumption as anticipated, the climate change resulted in better weather conditions for the United States, and its increase rate was greatly suppressed [20, 21].

Table 1.3 lists the top ten countries in the natural gas consumption in the past 5 years. It is clearly seen that China has the highest annual increase rate of about 11.3%, while Japan has a negative annual increase rate.

The global natural gas market in 2018 was as active as the previous years. The total international trade amount for natural gas is 253.57 billion USD, which increased about 7.7% compared to the previous year. The traded natural gas amount is 0.776 billion tons. And the average price dropped 5.8% and reached 306.1 USD per ton.

Table 1.4 lists the top ten countries with large natural gas importation. It can be seen that major imported countries are located mainly in Asia. Japan ranked first with a total import amount of 82.9 million tons, which is about 27.4% of the global import amount. China ranked second with a total amount of 53.8 million tons, which is about 17.8% of the global import amount. Korea ranked third with 44.0 million tons, which is about 14.6% of the global import amount. The sum of this three countries is 59.8% of the total global importation [20, 21].

Table 1.5 lists the top ten countries with large natural gas exportation.

Table 1.2 Natural gas consumption in 2019

Area	Consumption in 2019/%
North America	26.6
Central and South America	4.4
Europe	14.0
Eurasia Continent	14.91
Middle East	14.61
Africa	3.9
Asia-Pacific	21.52

Table 1.3 Natural gas consumption in recent 5 years for top ten countries

Area	2015	2016	2017	2018	2019
United States	743.6	750.3	739.5	820.8	848.1
Russia	409.6	420.2	424.8	454.5	485.4
China	185.5	205.8	235.2	277.0	304.0
Iran	191.9	201.4	214.4	229.4	245.5
Canada	102.9	109.5	115.7	121.5	122.2
Saudi Arabic	99.2	105.3	111.4	117.0	122.9
Japan	118.7	116.4	117.1	115.5	115.6
Germany	77.0	84.9	90.2	88.4	88.0
Mexico	78.0	91.8	84.6	86.3	92.9
United Kingdom	71.8	81.0	78.8	81.2	83.5

Unit: billion cubic meters

Table 1.4 Top ten countries with largest LNG import amount

Rank	Area	Amount/million tons	Percentage
1	Japan	82.854	27.4
2	China	53.806	17.8
3	Korea	44.047	14.6
4	India	23.328	7.4
5	Taiwan, China	17.107	5.7
6	Spain	12.077	4.0
7	Egypt	8.037	2.7
8	Pakistan	6.621	2.2
9	Italy	6.373	2.1
10	Thailand	4.442	1.5

It is seen that the major exportation came from countries with huge amounts of natural gas reservoirs like Qatar, Australia, and Malaysia. Among them, Qatar ranked first with a total exportation amount of 76.790 million tons, which was 25.4% of the global exportation market. Australia ranked second with a total amount of 71.074 million tons, which is about 23.5% of the global import amount. Malaysia

Table 1.5 Top ten countries with largest LNG export amount

Rank	Area	Amount/million tons	Percentage
1	Qatar	76.790	25.4
2	Australia	71.074	23.5
3	Malaysia	24.347	8.0
4	United States	22.235	7.3
5	Nigeria	19.680	6.5
6	Russia	16.441	5.4
7	Indonesia	15.614	5.2
8	Trinidad and Tobago	11.630	3.8
9	Omen	9.467	3.1
10	Iran	8.924	2.9

ranked third with 24.347 million tons, which is about 8.0% of the global import amount. The sum of these three countries is 56.9% of the total global importation [20, 21].

In terms of different regions, supply and demand in North America will remain fluctuating in the near future, and LNG exports will grow rapidly. Affected by factors such as the development of alternative energy sources, the demand for natural gas in Europe will grow slowly. Asian natural gas demand will continue to increase under the influence of factors such as the continuous increase in imports from China and other developing countries. Under the background of fluctuating supply and demand in the world natural gas market, Asian LNG prices are expected to fall.

As for the United States, it has abundant natural gas resources, and it changed from net importing to net exporting. The US Energy Information Administration declares that according to their estimations, the United States had about 505 trillion cubic feet of proved wet natural gas resources. In addition, EIA estimated that the United States also had about 2390 trillion cubic feet of unproved dry natural gas resources. Moreover, EIA estimated that through 2050, the majority of the United States' dry natural gas production would be from shale gas and tight gas resources.

1.5 Natural Gas Processing and Transportation

Natural gas processing is a set of industrial processes aiming at the purification of raw natural gas directly from the wells by removing various impurities, contaminants, and other hydrocarbons. The purpose of this practice is to produce the commercial dry natural gas which can meet the criteria regulated by the end users as well as the criteria proposed by the pipeline transportation parties [1, 2].

Normally, the raw natural gas contains different content of impurities such as water, carbon dioxide, ethane, propane, non-hydrocarbon, mercury, and even hydrogen sulfide when it leaves the reservoir. Different fields may have different impurities. This variety brings more difficulty for the transportation. Therefore, it is necessary that the raw natural gas produced from wells must be processed before it can be sent to the pipeline system [2].

The processing of natural gas into liquefied natural gas requires three main steps: the first is the purification treatment to remove all the impurities, carbon dioxide, and sulfur-containing chemicals. The second step is the removal of heavier hydrocarbon molecules, leaving only methane and ethane, and finally, the purified natural gas is continuously refrigerated until the temperature drops to −160 °C. At this moment, natural gas will become a liquid under atmospheric pressure with a much lower volume; the volume of natural gas will be reduced to 1/600 when it becomes a liquid [2].

The most conventional approach is the deep cooling separation to separate ethane, propane, butane, and C_5 and C_{5+}, as well as non-hydrocarbon components in natural gas, so that it can meet the commercial requirements. The product is commercial natural gas, and the by-products are usually ethane, liquefied petroleum gas, and natural gasoline. In terms of economic value, liquefied petroleum gas and natural gasoline should be recovered from natural gas as much as possible. In addition, natural gas liquefaction and helium extraction from natural gas also belong to the scope of natural gas processing [2].

The various storage devices included in this process are called collection tanks. All LNG plants that have been built have more than one collection tank for maintenance and replacement during the production process.

Before being loaded into cryogenic tanks and shipping, LNG will be stored in insulated tanks. Therefore, LNG will be regasified at various receiving terminals owned by LNG buyers before use. That is, natural gas producers will hand over natural gas to companies engaged in liquefaction and then resell it to LNG trading companies after liquefaction.

It can be seen that the LNG construction is a project that requires high investment. The construction and commissioning of an LNG plant generally takes about 4 years, and it usually requires at least a 20-year supply contract to start construction, so as to reduce investment risks and determine the establishment of natural gas production plants, liquefaction processing plants, and a series of economic feasibilities of LNG storage tanks, a receiving terminal, and a regasification plant.

The liquefied natural gas will then be loaded into specially designed tankers. During the 4–30-day voyage, the LNG has been in a refrigerated state, and the distance of the voyage depends on the location of the user's receiving port. Once the carrier reaches the regasification terminal, the LNG is unloaded into a large storage tank.

This fully enclosed system can ensure that the liquefied natural gas is in a refrigeration state until it is converted to a gaseous state. When the LNG is heated and restored to its gaseous state, the natural gas can be transported to household users, power plants, and industrial users through pipelines.

Another very important transportation measure besides the LNG transportation by sea is using natural gas pipelines.

Natural gas pipeline refers to the pipeline that can transport natural gas (including associated gas produced in oil fields) from the field and processing plants to urban gas distribution centers or industrial enterprise users, also known as gas pipelines. Using natural gas pipelines to transport natural gas is a way to transport large amounts of natural gas on land. Natural gas pipelines account for about half of the total pipeline length in the world [22].

Historically, China was the first country to use wooden and bamboo pipes to transport natural gas. In the seventeenth century, Chinese people in Sichuan province established a pipeline system on both flat ground and mountainous areas. Even today, the modern natural gas pipeline systems in China are mostly located in Sichuan Province, the main producer of natural gas [1, 22].

The world's gas pipelines have also experienced a development process similar to that of China. Before the eighteenth century, gas pipelines also used wood and bamboo pipes. Following that, cast iron pipes were used in the late eighteenth century, and steel pipes were used in the 1890s. Gas transmission power began to rely entirely on natural gas wellhead pressure. In 1880, the United States adopted steam-driven compressors. As a result of larger distances, a huge pipeline network system was constructed. Since the 1960s, many transnational pipelines have been built between natural gas-importing and natural gas-exporting countries, such as the 1780-km gas pipeline from the Soviet Union through the former Czech Republic, Slovakia, Austria, and Germany. Long-distance gas pipelines generally use compressors for pressurized transportation. Gas pipeline technology has also developed rapidly in terms of pipe selection, improvement of transmission efficiency, and realization of full line automation. In addition, it also carried out the sequential transportation of gases with different physical properties in the same pipeline and the pipeline transportation test of gaseous and liquid natural gas at a low temperature of −70 °C and a high pressure of 7.5 bar [22].

Although complicated, gas pipelines can be divided into three major types: gas-gathering pipelines, gas transportation pipelines, and gas distribution pipelines according to their uses.

Gas-gathering pipelines are the pipelines from the well of the gas field through the gathering station to the gas treatment plant or the starting gas compressor station and are mainly used to collect untreated natural gas extracted from the formation. Due to the high pressure of the gas well, the pressure of the gas-gathering pipeline is generally above 100 bar.

Gas transportation pipelines are the pipelines from gas processing plants or gas compressor stations to gas distribution centers, large users, or gas storages in major cities. The diameter of the gas pipeline is larger than that of the gas-gathering pipeline and the gas distribution pipeline. The largest gas pipeline has a diameter of 1420 mm. The natural gas is transported under pressure from the starting point compressor station and the compressor station along the line. The gas transmission pressure is 70–80 bar, and the pipeline length can reach thousands of kilometers.

Gas distribution pipelines are the pipelines from the city pressure regulating and metering station to the end users. These operate at low pressures and have multiple branches, dense networks, and small pipe diameters. In addition to the large number of steel pipes, low-pressure gas distribution pipes can also be made of plastic pipes or other materials.

It can be seen that natural gas pipeline transportation has the following features: the gas pipeline system is a continuous closed transportation system. From transportation, storage, to user use, natural gas is under pressure. Due to the small specific gravity of the transported natural gas, the static pressure head is much less affected than the liquid. When the design height difference is less than 200 m, the static pressure head is negligible, and the line is hardly restricted by the longitudinal terrain. There is no water hammer hazard in the liquid pipeline. However, once the pipeline ruptures, it can release a large amount of energy, and there might be chances to cause a fire when it encounters an open flame.

Natural gas pipeline transportation has the advantages of low transportation costs, small land occupation, fast construction, large oil and gas transportation volume, high safety performance, low transportation loss, no-waste gas emissions, low risk of leakage, low environmental pollution, and low impact on harsh weather. In addition, the maintenance is cheap and convenient. Therefore, it is very easy to realize remote centralized monitoring.

1.6 Uses of Natural Gas as Fuel

There are two usages for natural gas. One is the release of energy by combustion of natural gas with air or oxygen to form carbon dioxide and water. During this process, heat can be collected and utilized for various purposes. The other one is the chemical conversion from methane to other value-added commodities such as hydrogen and methanol because a huge amount of carbon and hydrogen elements exist in natural gas [1].

In this part, we will discuss the first usage of natural gas as fuel. Between 6000 BC and 2000 BC, Iran first discovered natural gas seeping from the ground. The natural gas was used for lighting at first by the ancient Persians who worshipped fire to be a torch that never extinguished. China recorded its first use of natural gas at around 900 BC. And the ancient Chinese drilled the first natural gas well in 211 BC, which was about 150 m in depth. And the gas was used as fuel to dry the rock salt in Southwestern China. Until the discovery of natural gas in Britain in 1659, Europeans gradually began to understand how to use it, but it was not widely used. Yet, since 1790, natural gas has become the main fuel for street and house lighting in Europe. In North America, the first commercial application of petroleum products was the application of natural gas in New York in 1821. They delivered natural gas to users through a small-diameter tube for lighting and cooking [1].

Nowadays, natural gas has widespread commercial uses. So far, the most important use of natural gas as an energy source is still to generate heat or steam to drive the large variety of civil and industrial processes.

In the big cities of China, natural gas has gradually replaced the traditional fuel such as wood and coal. It is widely accepted as a cleaner and also cost-effective fuel in everyday life. Even in rural areas or remote places, natural gas has been the preferred choice over coal. With the improvement of people's living standards and the enhancement of environmental protection awareness, the demand for natural gas in most cities has increased significantly. The economic benefit of natural gas as a civil fuel is also greater than that of industrial fuel.

Similar situations were found in North America. Before the 1940s, the wood stove has been very popular in rural areas, while the coal stoves or coal-fired boilers were the choices for urban houses. But after WWII, the fuel was widely changed from fuel oil to natural gas. The furnaces use natural gas to heat the fresh air and distribute warm air to warm up the whole house. Currently, this is the most popular way of heating in North America. In addition, most houses also use natural gas to heat the water. In some houses, natural gas is also used in the cooling equipment such as efficient absorption chillers and steam turbine-driven chillers [1].

Cooking and water heating with natural gas began to make the first inroads in large western cities in the nineteenth century [1]. It can heat up the food to a high temperature very rapidly and cut off heat instantly, which matches the Chinese cooking style very well. Even though it faces the higher risks of open flame, leaks, exhaust gas, etc., it still attracts many people's attention.

Before the 1950s, electricity generation was based on the fuel of coal and heavy oils. But gradually, natural gas has been more popular in the electricity generation power plant due to its affordable prices and steady supply in very large volumes.

In 2018, the globally installed capacity of gas and electricity was 1.8 billion kilowatts, accounting for a quarter of the globally installed capacity of power sources, of which the installed capacity of gas and electricity in North America and the Middle East accounted for a relatively high proportion. North American gas power replaced coal power as the largest power source in the early twenty-first century. In 2018, the installed capacity accounted for 41.2%. The structure of installed power generation capacity in Europe is relatively balanced, with gas power-installed capacity accounting for 27.9%. Power generation in the Middle East is dominated by natural gas, and the installed capacity of natural gas power generation accounted for 66.4% in 2018 [1].

The primary factor restricting the development of the natural gas power generation industry is its high cost of power generation fuel. In recent years, the downward trend in natural gas prices has made it gradually competitive with coal-based power generation. This is critical to the development of gas power in Asia, because Asia has traditionally been the region with the highest natural gas prices in the world and coal in this region, the lowest price. For instance, in the first quarter of 2019, the cost of natural gas-based power plants in Japan and South Korea has been lower than that of coal-based ones.

Under the pressure of climate and environmental pressures, many countries are paying more and more attention to low-carbon development. Compared with Europe and the United States where the proportion of clean electricity is already high, regions such as Asia, which are mostly developing countries, have relatively low natural gas consumption levels and are more aspiring for natural gas demand. Many developing countries have formulated measures to promote the development of natural gas.

On the other hand, natural gas power generation also faces many challenges. In November 2019, the European Investment Bank announced that in response to the challenge of climate change, it will stop investing all fossil energy projects by the end of 2021. According to this new investment policy, most fossil energy projects, including natural gas power generation projects, will be excluded from the scope of financing, which means that the financing of natural gas power generation projects in Europe will be blocked in the future. The logic behind this is not difficult to understand. Most European countries are in the late stage of industrialization, their energy structure, especially electrification, is at a relatively high level, and the proportion of cleanliness is relatively high. It is difficult to make major breakthroughs in natural gas demand. And according to the EU's medium- and long-term development plan, renewable and sustainable energy should be the central topics.

In the "BP World Energy Outlook" (2020 edition) released in September 2020, BP stated that natural gas-based power generation has the following two potential important roles in the low-carbon transition of the energy system: first, it is in the development of rapid economic growth. The growth rate of renewable energy and other nonfossil energy in these countries is not enough to replace coal demand. The use of natural gas can reduce the use of coal. Second, natural gas combined with CCUS technology (carbon capture, utilization, and storage) can achieve zero or near-zero carbon emissions in natural gas-based power plants [7, 8].

It is easily understood that natural gas can be used as a transportation fuel due to the following advantages: it can reduce greenhouse gas emissions by 10–25%. Also, the cost savings of up to 30–40% per kilometer can be achieved. Natural gas-driven vehicles are normally quieter than gasoline- and diesel-driven alternatives, which can result in much less noise pollution in populated areas. Natural gas combustion only produces water and carbon dioxide without NO_x and SO_x in the exhaust gas, which are environmentally benign.

However, natural gas has less energy by volume than diesel fuel, so it must be compressed or liquefied to store enough energy for adequate driving ranges. Therefore, the real fuel for vehicles should be either LNG (liquefied natural gas) or CNG (compressed natural gas).

As a clean energy for sustainable development, LNG has obvious environmental and social benefits. Substituting LNG for gasoline can reduce 90% of sulfur dioxide emissions and 80% of nitrogen oxide emissions, respectively. The environmental benefits are tremendous. It is a high-quality alternative fuel for automobiles. It is foreseeable that cities will gradually replace gasoline or diesel with LNG or natural gas in terms of vehicle fuel. In recent years, it has been valued and promoted by many countries in the world. Russia has accumulated a lot in the use of LNG for automobile transportation, railway transportation, water transportation, and air transportation experience. Nowadays, most British transportation companies use LNG as vehicle fuel.

According to different storage methods of natural gas, natural gas vehicles are roughly divided into CNGV (compressed natural gas vehicles), LNGV (liquefied natural gas vehicles), and ANGV (adsorbed natural gas vehicles).

CNG vehicles refer to vehicles that use compressed natural gas instead of conventional gasoline or diesel as vehicle fuel. The areas with natural gas pipeline networks are mainly developing CNG vehicles.

For the CNG vehicles, the natural gas is stored in a gas cylinder similar to a fuel tank in the vehicle at a pressure of about 25 MPa and used as the fuel. The main process is to boost the pressure of 0.3–0.8 MPa low-pressure natural gas through the natural gas compressor to 25 MPa at the CNG bus station, which is controlled by the sequence control panel and stored in the gas storage cylinder group in the order of high, medium and low pressure, and then by the CNG dispenser to fill the vehicle. The high-pressure gas of the automobile steel cylinder is then decompressed by the pressure-reducing device and supplied to the engine through the gas mixture.

At present, CNG-type vehicles are widely used in China, and LNGV is being spread very rapidly, while ANGV is in the development stages and needs more research to reach real applications. With the massive amount and wide use of LNG, LNG vehicles will gradually be transferred into heavy vehicles and will partially replace small CNG vehicles and buses. CNG (compressed natural gas) refueling stations are also considered at gas stations, which can meet current and future needs.

1.7 Uses of Natural Gas as Raw Material for Chemicals

As abovementioned, methane, as the main component in natural gas, is widely used as a clean fuel. Moreover, it is substituting the traditional fuels very rapidly and in a wide range. In addition, it is also widely regarded as the source of hydrocarbons for petrochemical feedstocks and also a major source for the elements of carbon, hydrogen, sulfur, etc.

Natural gas can also be used as a raw chemical material. The primary processed products using natural gas as raw materials mainly include nearly 20 varieties such as synthetic ammonia, methanol, and carbon black. The important chemical products after secondary or tertiary processing include more than 50 varieties such as formaldehyde, acetic acid, and dimethyl carbonate. Chemical production equipment using natural gas as a raw material has low investment, low energy consumption, less land occupation, fewer personnel, good environmental protection, and low operating costs. The following lists several value-added commodities which are derived from natural gas.

Methane, the most important and abundant component in natural gas, is the most important starting material for the synthesis of ammonia (NH_3), for the production of hydrogen, which is widely used in petrochemical refineries to conduct hydrocracking, hydro-treating, and hydrodesulfurization process to upgrade the fuel, methanol (CH_3OH), and various derivatives [1, 23–42].

In chemical engineering, hydrogen (H_2) is one of the largest and most valuable commodities widely used in synthetic ammonia, petroleum processing, metallurgy, electronics, glass, medicine, aerospace, F-T synthesis, etc. Therefore, steam gasification hydrogen production technology has received greater attention.

Due to its abundant reserves, natural gas has been the main raw material for syngas production and hydrogen production. Although coal reserves are much larger and cheaper, its investment is three times compared with that of a synthesis gas plant using natural gas. When hydrogen is produced from methane, synthesis gas is usually produced first, and pure hydrogen is achieved after transformation and carbon dioxide removal. It is estimated that the production cost of synthesis gas accounts for about 60% to 70% of the entire hydrogen production process. Therefore, the production cost of synthesis gas has an important impact on the overall cost of hydrogen production.

There are several different approaches to achieve hydrogen from methane, such as steam methane reforming, dry reforming methane, partial oxidation of methane, autothermal reforming, methane catalytic decomposition, etc.

Steam reforming of methane has been the traditional hydrogen production method since it first emerged in industrial practice in 1926 [24]. The major reactions are listed as follows [3].

$$CH_4 + \mathrm{H_2O} = CO + 3\mathrm{H_2}\left(\Delta\mathrm{H}^\circ = 206\,\mathrm{kJ/mol}\right)$$

$$CO + \mathrm{H_2O} = CO_2 + \mathrm{H_2}\left(\Delta\mathrm{H}^\circ = -41\,\mathrm{kJ/mol}\right)$$

Conventional methane steam reforming process includes many steps, including feed gas preheating, desulfurization, steam reforming, CO_2 removal, and methanation. The crucial step is of course the steam reforming to form H_2, and in this step, a nickel-based catalyst is applied. The most common water-to-methane ratio is about 3 to 3.5, a reaction residence time of several seconds, a conversion temperature of 820 °C, and a conversion pressure of 1.5 to 3.0 MPa.

Excessive steam is used to prevent the deactivation of the catalyst by carbon deposits. The amount of excess depends on the nature of the natural gas and the catalyst used. In addition, some alkali metals and alkaline earth metals like potassium, calcium, and magnesium are usually loaded into the catalyst structure as additives to accelerate the removal of carbon from the catalyst surface. The typical reaction conversion at the reactor outlet is 90% to 92% based on methane.

During the methane reforming process, coke is inevitable by the following reactions [3].

$$2CO = \mathrm{C} + CO_2\left(\Delta\mathrm{H}^\circ = -172\,\mathrm{kJ/mol}\right)$$

$$CH_4 = \mathrm{C} + 2\mathrm{H_2}\left(\Delta\mathrm{H}^\circ = 75\,\mathrm{kJ/mol}\right)$$

Coke deposited on the surface of the nickel catalyst will damage the structure of the catalyst and cause the pressure drop to increase. In severe cases, the reaction tube will be clogged. Therefore, great efforts should be made to prevent too much coke deposition.

The common industry practices use a high water-to-methane ratio to suppress coke deposition, which will increase process energy consumption and reduce reactor efficiency. Some progress has also been made in the reactor designing. Roy et al. used a fluidized bed to convert methane to steam, adding O_2, CH_4, and H_2O to the fluidized bed at the same time to avoid the intrinsic heat transfer, diffusion resistance, and large axial temperature gradient in fixed bed reactor. The methane conversion increased with the increase of temperature, oxygen flow, and water-to-methane ratio. The oxygen conversion is nearly 100% under all conditions, and the hydrogen production decreases with the increase of pressure and oxygen flow.

Due to thermodynamic limitations, methane steam reforming needs to be carried out at high temperatures (750–920 °C). Lin and his colleagues conducted an integrated study on methane steam conversion and hydrogen purification using a palladium membrane reactor. When the reaction temperature drops to 500 °C, the methane conversion reached 45%. The use of membrane reactors can not only reduce the reaction temperature but also eliminate the subsequent hydrogen purification process, which has a very promising prospect.

Another approach is the production of hydrogen from methane and carbon dioxide as list in the following reaction [3].

$$CH_4 + CO_2 = 2CO + 2\mathrm{H}_2 \left(\Delta \mathrm{H}^\circ = 247\,\mathrm{kJ/mol}\right)$$

Carbon dioxide reforming of natural gas is of great significance not only in CO production but also in the treatment of carbon dioxide. The ratio of H_2/CO in the synthesis gas obtained by steam reforming of methane is 3 to 4, while the synthesis gas obtained by carbon dioxide reforming is 1, which is suitable for F-T synthesis and methanol production. Therefore, the hydrogen production by methane carbon dioxide reforming has attracted wide attention.

It is reported that most of the VII family transition metals have catalytic activity, among which precious metals such as Ru, Rh, and Ir are excellent catalysts for carbon dioxide reforming, with high catalytic activity, strong carbon deposition resistance, and good thermal stability. According to reports, nickel and precious metals are effective catalysts for CO_2 reforming, and Ir/Al_2O_3 is also a catalyst for CO_2 reforming without carbon deposits, and aluminum-supported rhodium catalysts also have catalytic effects.

Recently, the partial oxidation to produce hydrogen from methane has attracted great attentions. The reactions are listed in the following reactions [3].

$$CH_4 + 0.5\mathrm{O}_2 = 2CO + 2\mathrm{H}_2 \left(\Delta \mathrm{H}^\circ = -36\,\mathrm{kJ/mol}\right)$$

$$CH_4 + 2O_2 = CO_2 + 2H_2O\left(\Delta H^\circ = -802\,\text{kJ/mol}\right)$$

As abovementioned, methane steam reforming is a strongly endothermic reaction, which will consume huge amounts of energy. Therefore, huge efforts have been spent to find other methane reforming methods with low energy requirements. It can be seen from the reactions that the partial oxidation of methane is a reaction with a small exothermic effect, and its energy efficiency should be higher than that of steam conversion.

However, it is difficult to control the direct oxidation reaction of methane, and there is a danger of explosion, so its research is limited. The gas phase reaction in the reducing atmosphere will lead to the formation of carbon on the catalyst surface. To this end, people have made unremitting efforts, including adding a small amount of water, reactor design, catalyst design, etc.

It is anticipated that due to increasingly more strict environmental regulations and restrictions on hydrogen production efficiency, the development of hydrogen production from coal will gradually be substituted by using natural gas. Among all those approaches, natural gas steam reforming will still play the most important role in hydrogen production. With the gradual acceleration of the process of alternative fuels, partial oxidation of methane will gradually increase their proportion in the production of synthesis gas with a low hydrogen-to-carbon ratio. Carbon dioxide reforming is more suitable for the cases with higher carbon dioxide content in natural gas or higher gas emission requirements. With the acceleration of the commercialization of fuel cells, the catalytic cracking of methane will be a good choice for hydrogen production. Not only does the hydrogen produced contain no or less carbon monoxide, but also carbon nanomaterials as by-products can be obtained.

If methane can be converted into a chemical that is convenient for storage and transportation, it will promote the development and utilization of methane considerably and avoid the disadvantages resulting from the transportation and storage of methane with high pressure or very low temperature.

As a simple organic liquid, methanol has become a promising chemical for in situ conversion of methane. Methanol is a very important basic chemical; it can be easily transformed into important chemical raw materials and fuels such as olefins and aromatics. Considering the fact that the reserves of natural gas with methane as the main component are huge, if methanol can have high selectively and high efficiency achieved from methane in large-scale industrial production, it will greatly help mankind to get rid of its dependence on oil.

At present, the indirect methanol synthesis method from methane is commonly used in industry [31]. The reaction is carried out at a high temperature of 800–1000 °C, the process flow is long, and the energy consumption and equipment investment are large. At present, there are two major problems in the technology of methane-to-methanol: one is that the C-H bond dissociation energy in methane molecules is very large (425 kJ/mol), and the other is that the selectivity of the target product methanol is difficult to control, and deep oxidation reaction often occurs since methanol is easier to be oxidized than methane, and very often only a large amount of CO_2 can be obtained in this reaction if a careful tuning and development of the catalyst to selectively oxidize methane is not used.

In fact, the direct oxidation of methane to methanol has been studied for decades. In 1993, researchers discovered that in the oleum system, homogeneous Pt catalysts can directly oxidize methane to methanol (Science, 1993, 259, 340–343). After that, people began to explore the use of heterogeneous catalysts to achieve the selective oxidation of methane to methanol. After decades of accumulation, Cu-molecular sieve prepared by ion exchange method is currently the most potential catalyst via the following steps. First, the Cu-doped zeolite catalyst needed to be activated which was generally treated at high temperature in air or oxygen atmosphere, and then it would be cooled to 100–200 °C to react with methane. At this time, methane molecules will be adsorbed on the Cu species, and an O atom is inserted into the C-H bond. After a period of time, water vapor is introduced to desorb the adsorbed methanol to obtain a methanol product. In other words, the direct contact between methane and oxygen is avoided in the above reaction, and the conversion of methane to methanol is realized through a staged reaction.

Traditionally, the conversion of methane to methanol requires steam reforming of methane into synthesis gas, which is then further converted into methanol. Steam reforming is an investment and energy-intensive process. Therefore, new processes are being explored. At present, the methods of direct oxidation of methane to methanol mainly include gas-phase homogeneous oxidation, liquid phase homogeneous catalytic oxidation, gas-solid heterogeneous catalytic oxidation, liquid-solid heterogeneous catalytic oxidation, biological enzyme catalytic oxidation, photocatalytic oxidation and electrochemical catalytic oxidation, etc. [33–42].

The gas-phase homogeneous oxidation of methane usually uses oxygen (or air) as the oxidant and is carried out at 350–500 °C. Since no catalyst is added during the reaction process, the reaction effect is closely related to factors such as reaction temperature, reaction pressure, and reactor material. After optimizations, methane conversion of 5–10% and methanol selectivity of 30–40% could be achieved when the temperature is 450–500 °C and the pressure is 3.0–6.0 MPa.

In the liquid or liquid-solid heterogeneous catalytic oxidation approach with catalysts, a suitable reaction medium can react with methane to form the methyl intermediate groups to avoid deep oxidation of the product. Therefore, the key issue of the liquid phase catalytic oxidation of methane is the introduction of a suitable reaction medium and corresponding catalyst and the use of a suitable oxidant. The reaction media currently studied mainly include strong acid media and mixed media of acetic acid. The oxidants studied are fuming sulfuric acid, oxygen, and hydrogen peroxide.

Currently, the gas-solid heterogeneous catalytic oxidation reaction is drawing great attention since it is easier to recover products than liquid-phase catalysis. The early research catalysts were mostly used in continuous flow reactors. This type of catalyst used O_2 as the oxidant at high temperature and high pressure. Since the methanol produced by the reaction will be further oxidized, the resulting product contains formaldehyde in addition to methanol. The main direction of the research and development of this type of catalyst is to improve the activity and the total selectivity of methanol and formaldehyde [43, 44].

Biological enzyme catalysis has the advantages of high product selectivity and mild reaction conditions, but the cost of obtaining enzymes is relatively high, methanol production capacity is also low, and industrialization is still difficult to overcome economic barriers; photocatalytic oxidation and electro-catalytic oxidation are emerging reaction systems. It has shown good signs and has potential for development, and in-depth research and development should be continued [45–49].

In short, each reaction system currently has certain advantages and disadvantages. By improving the reaction process, reducing the reaction cycle time, and developing cheaper catalysts, the direct oxidation of methane to methanol is expected to replace the traditional two-step process and achieve industrial development.

References

1. V. Smil, *Natural Gas Fuel for the 21st Century* (John Wiley & Sons, Hoboken, NJ, 2015)
2. S. Mokhatan, W.A. Poe, Y.M. John, *Handbook of Natural Gas Transmission and Processing Principles and Practices*, 4th edn. (Gulf Professional Publishing, Burlington, MA, 2019)
3. T. Hua, Y. Wang, J. Bian, J. Chen, *General Chemistry*, 4th edn. (Peking University Press, Beijing, 2013)
4. J.S. Lee, S.T. Oyama, Catal. Rev. Sci. Eng. **30**, 249–280 (1988)
5. J.H. Lunsford, Angew. Chem. Int. Ed. Engl. **34**, 970–980 (1995)
6. G.J. Hutchings, M.S. Scurrell, Chem. Soc. Rev. **18**, 251–283 (1989)
7. BP Energy Outlook 2020 edition (2020)
8. BP Statistical Review of World Energy 2020 (2020)
9. IEA World Energy Outlook 2019 (2019)
10. The geology of natural gas resources, US energy information administration, 12 Feb 2011, www.eia.gov/todayinenergy/detail.php?id=110#
11. U.S. Energy Information Administration, 24 Dec 2020, Eia.gov
12. Explorationandproductionofshaleandtightresources(GovernmentofCanada,2Sept2020),www.nrcan.gc.ca/our-natural-resources/energy-sources-distribution/clean-fossil-fuels/natural-gas/shale-tight-resources-canada/exploration-and-production-shale-and-tight-resources/17677
13. M. Sanchez, C. Santamarina, M. Teymouri, X. Gai, J. Geophys. Res.: Solid Earth **123**(12), 10326–10348 (2018)
14. M. Teymouri, M. Sanchez, C. Santamarina, Mar. Pet. Geol. **120**, 104519 (2020)
15. Z. Chong, S. Yang, P. Babu, P. Linga, X. Li, Appl. Energy **162**, 1633–1652 (2014)
16. A. Macfarlane, China makes 'flammable ice' breakthrough in South China Sea, CNNMoney, 30 Jan 2021, money.cnn.com/2017/05/19/news/china-flammable-ice-sea/index.html
17. R. Anderson, Methane hydrate: Dirty fuel or energy savior, BBC News, 16 April 2014, www.bbc.com/news/business-27021610
18. Natural Gas, U.S. energy information administration, 11 Jan 2021, www.eia.gov/naturalgas/crudeoilreserves/
19. Canada's Natural Gas, Canadian association of petroleum producers, 18 Jan 2021, www.capp.ca/natural-gas/
20. Annual report of petrochemical commodities in China, CNCET, March 2020

21. IEA (2020), Natural gas Information: Overview: Detailed and Comprehensive Annual Data on Natural Gas Supply, Demand and Trade, IEA, Paris., www.iea.org/reports/natural-gas-information-overview. Accessed 16 Jan 2021,
22. J. Cui, Y. Wang, *An Overview of World Pipeline* (Petroleum Industry Press, Beijing, 2009)
23. M. Moirlan, T. Veziroglu, Renew. Sustain. Energy Rev. **3**, 219–231 (1999)
24. P. Grange, Cata. Rev. Sci. Eng. **22**, 135–181 (1980)
25. A. Ozaki, K. Aika, *Catalysis: Science and Technology* (Springer, New York, 1981)
26. C.D. Chang, C.T. Chu, R.F. Socvha, J. Cata. **86**, 289–296 (1984)
27. R.B. Anderson, *The Fischer-Tropsch Synthesis* (Academic Press, New York, 1984)
28. T.V. Choudhary, D.W. Goodman, Catal. Today **77**, 65–78 (2002)
29. M.A. Pena, J.P. Gomze, J.L.G. Fierro, Appl. Cata. A: Gen. **144**, 7–57 (1996)
30. D.E. Ridler, M.W. Twigg, I.M.W. Wwigg, *Catalyst Handbook*, 2nd edn. (Wolfe Publisher, London, 1989)
31. F. B. Noronha, M. Schmal, E. F. Sousa-Aguiar, in *Studies in Surface Science and Catalyst 167: Natural Gas Conversion VIII, Proceedings of the 8th Natural Gas Conversion Symposium*, Elsevier, 2007
32. M. Ahlquis, R.J. Nielsen, P.A. Periana, J. Am. Chem. Soc. **131**, 17110–17115 (2009)
33. M.C. Alvarez-Galvan, N. Mota, M. Ojeda, et al. Catal. Today **171**, 15–23 (2011)
34. A. Holmen, Catal. Today **142**, 2–8 (2009)
35. V.S. Arutyunov, J. Nat. Gas Chem. **13**, 10–22 (2004)
36. K. Fujimoto, in *New Uses of Methane, natural gas conversion II-Proceedings of the Third Natural Gas Conversion Symposium*, 73–84, 1994
37. R.A. Periana, D.J. Taube, S. Gamble, Science **280**, 560–564 (1998)
38. B. Michalkiewicz, K. Kalucki, J.G. Sosnicki, J. Catal. **215**, 14–19 (2003)
39. R. Palkovits, M. Antonietti, P. Kuhn, et al., Angew. Chem. Int. Ed. **48**(37), 6909–6912 (2009)
40. E. Gretz, T.F. Oliver, A.J. Sen, Am. Chem. Soc. **109**, 8109–8111 (1987)
41. X.H. Bao, Z.J. An, X.M. Liu, et al., J. Am. Chem. Soc. **128**, 16028–16029 (2006)
42. W. Chen, J.A. Kocal, T.A. Brandvold, et al., Catal. Today **140**, 157–161 (2009)
43. K.A. Dubkov, V.I. Sobolve, E.P. Talsi, J. Mol. Catal. A **123**, 155–161 (1997)
44. G.I. Panov, A.K. Uriarte, M.A. Rodkin, Catal. Today **41**, 365–385 (1998)
45. M.H. Groothaert, P.J. Smeets, B.F. Sels, J. Am. Chem. Soc. **127**, 1374–1395 (2005)
46. D.K. Pappas, E. Borfecchia, M. Dyballa, J. Am. Chem. Soc. **139**, 14961–14975 (2017)
47. S. Grundner, M.A. Markovits, G. Li, et al., Nat. Commun. **6**, 7546 (2015)
48. J.P. Lange, V.L. Sushkevich, A.J. Knorpp, et al., Ind. Eng. Chem. Res. **58**, 8674–8680 (2019)
49. R. Balasubramanian, S.M. Smith, S. Rawat, Nature **465**, 115–119 (2010)

Chapter 2
Introducing Methane Activation

2.1 Introduction of Methane Activation and Its Challenge

Methane (CH_4) is the simplest and least reactive member of all the paraffin hydrocarbons as well as a member of greenhouse gases whose effect is 25 times of carbon dioxide (CO_2) [1]. The chemical inertness of CH_4 originates from its unique molecular configuration and physical properties (Table 2.1). CH_4 molecule presents a symmetrical tetrahedral configuration with four equivalent C-H bonds because of the sp^3 hybridization of carbon, which brings low electron affinity and polarizability. The first dissociation energy of CH_4 is 439 kJ mol^{-1}, indicating an unusually strong C-H bond compared with other alkanes. CH_4 also demonstrates extremely high ionization potentials with its first ionization potential high up to 13.16 eV, implying the difficulty of removing bound electron of CH_4 and unreactivity in electron transfer reactions. Besides, from the perspective of molecular orbital theory, CH_4 owns a low highest occupied molecular orbital (HOMO) energy level and a high lowest unoccupied molecular orbital (LUMO) energy level, which means either electron loss from HOMO or electron capture of LUMO is a challenge. All these CH_4 properties suggest that both sufficient energy supply and appropriate catalysts are indispensable for methane activation.

Due to the high hydrogen-to-carbon ratio of CH_4, most produced natural gas is consumed as fuel for heating and electricity generation, and only a small fraction is used for chemicals production [7]. To alter the current dependence status on crude oil and considering the uneconomic transportation of abundant methane, it is imperative to convert methane into high-density energy sources or high value-added chemicals. Two main routes could realize this (Fig. 2.1). One is indirect conversion, which is the industrial route for large-scale methane transformation. Methane firstly converts into syngas (a mixture of H_2 and CO) through steam reforming of methane (SRM), and then the acquired syngas is used to produce chemicals mainly including ammonia, methanol, and hydrocarbons in downstream processes. The other is the

H. Song et al., *Methane Activation and Utilization in the Petrochemical and Biofuel Industries*, https://doi.org/10.1007/978-3-030-88424-6_2

Table 2.1 Configuration and physical properties of methane

Property	Value	Ref.
Structure	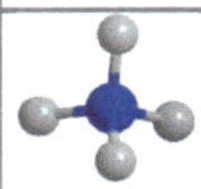	
Electron affinity	−1.9 eV	[2]
Polarizability	2.8×10^{-40} C^2 m^2 J^{-1}	[3]
First dissociation energy	439 kJ mol^{-1}	[4]
First ionization potential	13.16 eV	[5]
Second ionization potential	19.42 eV	[5]
C-H bond length	1.9 Å	[6]

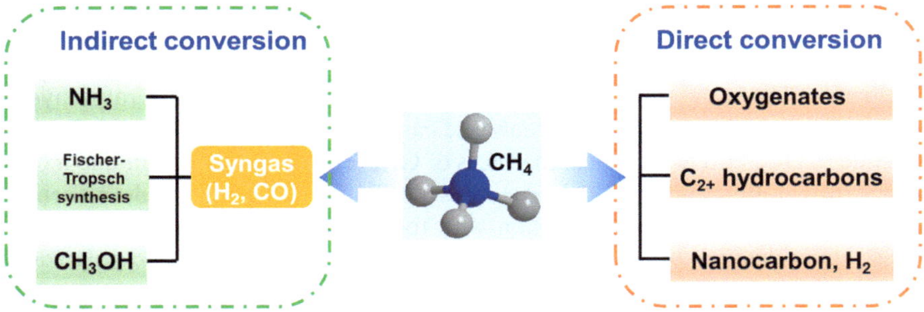

Fig. 2.1 Two main methane conversion routes for valuable fuel and chemical production

direct conversion with main products such as liquid oxygenates (such as methanol) and higher hydrocarbons as well as nanocarbon and hydrogen. Although hydrocyanic acid (HCN), CS_2, and chlorinated methane derivatives also originated from direct methane conversion in industries [8], these chemicals are not in the scope of this book, so no related introduction is presented here. Besides the abovementioned thermo-catalytic methane conversion processes, other energy supply methods, such as light [9], electricity [10], plasma [11], and bioenergy [12], also could drive methane conversion processes.

2.2 Methane Reforming to Syngas

Three reforming methods, SRM, dry reforming of methane (DRM), and partial oxidation of methane (POM), could convert methane into syngases with different H_2/CO molar ratios (Table 2.2). SMR is the most widely adopted method for methane reforming, and the generated syngas exhibits H_2/CO molar ratio of 3 which should be adjusted before used for synthesis. An H_2/CO molar ratio of 2 obtained from

Table 2.2 Chemical equations and Gibbs free energy changes for methane reforming methods

Reaction	Chemical equation	ΔH_{298K} (kJ mol^{-1})
Steam reforming of methane	$CH_4 + H_2O \rightarrow CO + 3H_2$	206
Dry reforming of methane	$CH_4 + CO_2 \rightarrow 2CO + 2H_2$	247
Partial oxidation of methane	$CH_4 + 0.5O_2 \rightarrow CO + 2H_2$	−36

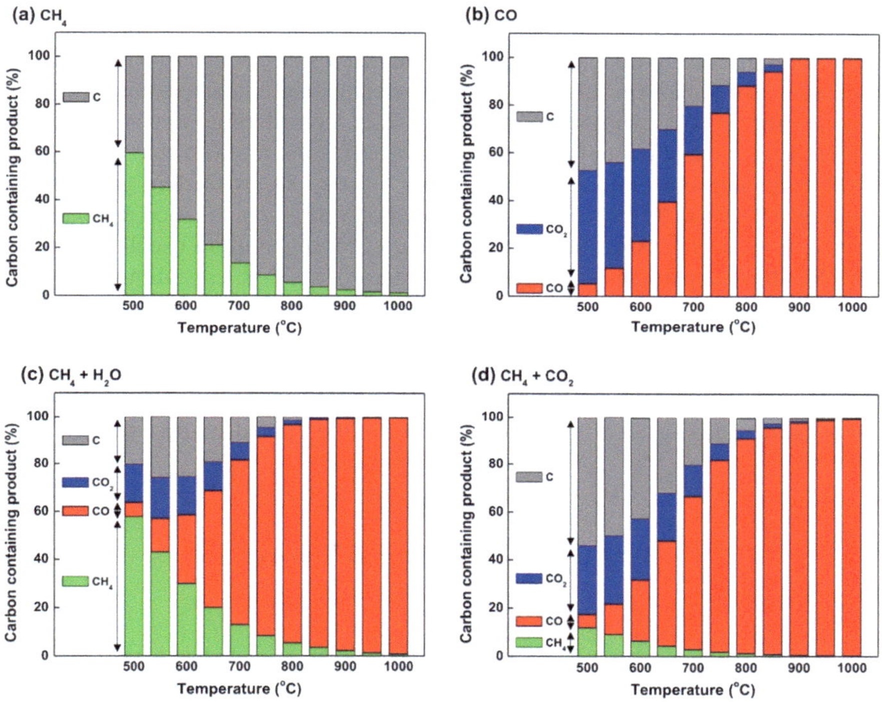

Fig. 2.2 Carbon-containing product for different reactions: (**a**) CH_4 decomposition (100% CH_4), (**b**) CO disproportionation (100% CO), (**c**) SRM (50% CH_4 + 50% H_2O), and (**d**) DRM (50% CH_4 + 50% CO_2). Reprinted with permission from Ref. [18], copyright (2016) Elsevier

POM is appropriate for the methanol synthesis, and DRM produces syngas with an H_2/CO molar ratio of 1 which is suitable for Fischer-Tropsch (F-T) synthesis to produce higher hydrocarbons. Noble metal catalysts (such as Rh and Ru) own excellent catalytic activity and the anti-coking ability for methane reforming, but the expensive prices and limited resources limit their application. In this condition, nickel-based catalysts are the optimal option considering their low cost and catalytic performance comparable to noble metal-based catalysts. The shortage of nickel-based catalysts is the severe coke formation which could be alleviated by using alkaline earth metal oxides as catalyst supports.

SRM is a strongly endothermic reaction and usually operates at high temperature (700–1000 °C) and pressure (3–25 bar) [13] to obtain high performance with steam feeding amount larger than stoichiometry to suppress carbon deposition. However, these regular operation conditions consume a large amount of energy and lead to energy efficiency reduction, which makes anti-coking nickel-based catalysts development necessary. Relative to SRM, DRM exhibits higher selectivity and suitable H_2/CO molar ratio for F-T synthesis. Besides, DRM can efficiently convert biogas and natural gas containing abundant CO_2 while realizing syngas production and greenhouse gases conversion and utilization. Despite all of these, DRM is not considered as a mature industrial process due to the quick catalyst deactivation stemmed from sintering and coke deposition [14]. In the DRM process, two side reactions are the causes of carbon deposition (Fig. 2.2): one is methane decomposition (Eq. 2.1), and the other is CO disproportionation (Eq. 2.2). Considering the conversion and carbon selectivity, thermodynamical calculation results indicate that the optimal DRM operating temperature with 50% CH_4 + 50% CO_2 as feeding gas is in the range of 870–1040 °C [15]. However, the carbon formation situation depends to a great extent on catalysts. For example, Song et al. reported an excellent catalyst with Ni-Mo nanoparticles stabilized at the edges of a single-crystalline MgO support which showed no detectable carbon deposition during an 850-hour continuous DRM operation [16]. POM is a mild exothermic process with quick initiation, suitable syngas for F-T synthesis, and without coke deposition. However, the exothermic property also makes POM hard to control because of local hot spots and temperature runaway occurrences, which leads to the limited studies on POM [17].

$$CH_4 \rightarrow \mathrm{C} + 2\mathrm{H_2} \Delta \mathrm{H}_{298\mathrm{K}} = 75\,\mathrm{kJ\,mol^{-1}} \tag{2.1}$$

$$2CO \rightarrow CO_2 + \mathrm{C} \Delta \mathrm{H}_{298\mathrm{K}} = -172\,\mathrm{kJ\,mol^{-1}} \tag{2.2}$$

2.3 Direct Conversion of Methane to Valuable Chemicals Under Oxidative Conditions

2.3.1 Oxidative Coupling of Methane (OCM) to Higher Hydrocarbons

The C_{2+} products of the OCM process mainly include C_2H_6 and C_2H_4 (Eqs. 2.3 and 2.4), even though C_2H_2 is more thermodynamically favorable. C_2H_4, one of the most important feedstocks,

$$2CH_4 + 0.5\mathrm{O_2} \rightarrow \mathrm{C_2H_6} + \mathrm{H_2O} \Delta \mathrm{H}_{298\mathrm{K}} = -177\,\mathrm{kJ\,mol^{-1}} \tag{2.3}$$

$$2CH_4 + \mathrm{O_2} \rightarrow \mathrm{C_2H_4} + 2\mathrm{H_2O} \Delta \mathrm{H}_{298\mathrm{K}} = -141\,\mathrm{kJ\,mol^{-1}} \tag{2.4}$$

is regarded as an economic indicator of the organic chemical industry. Ethane could also be converted into ethylene via the oxidative dehydrogenation process. It is widely accepted that three steps occurred in OCM: (1) formation of methyl radicals through C-H bond breaking and hydrogen abstraction with the aid of surface oxygen species, (2) the coupling of formed methyl radicals to produce ethane in the gas phase, and (3) ethane dehydrogenated to ethylene. Inevitably, companied with OCM, undesired nonselective oxidation occurs to form CO and CO_2, leading to C_2H_4 selectivity reduction and difficult subsequent product separation. The OCM reaction performance depends on its catalysts. Li-MgO, $Mn/Na_2WO_4/SiO_2$, and Sr/La_2O_3 are the most widely studied catalysts [19]. Despite considerable studies on OCM catalysts, only a few catalysts could meet the criteria of industrial application, i.e., C_{2+} yield >25% and C_{2+} selectivity >80% [20]. To enhance C_{2+} selectivity, mild oxidants such as S_2 was adopted to replace O_2 [21]. However, mild oxidation operating condition means higher methane activation temperature. More information and discussion on OCM catalysts and processes will be presented in Sect. 3.3.

2.3.2 Partial Oxidation of Methane (POM) to Oxygenates

Methanol is a liquid chemical with the merits of economical transportation and direct usage as chemical feedstock as well as fuel for fuel cells. In industries, methanol is produced from methane via a syngas-based route, which consumes a huge amount of energy. In recent years, the selective oxidation of methane to methanol (MTM) has drawn particular attention without energy-intensive syngas production step [22–25]. However, this dream reaction has two tremendous challenges, methane C-H bond activation and a higher reactivity of product than methane, leading to low methane conversion and methanol selectivity.

There are two types of methane C-H bond cleavage mechanisms at low temperatures. As shown in Fig. 2.3, one mechanism involves that electrophilic oxygen atoms including M-O sites on solid catalysts, oxygen-containing radicals, and *high-valent* metal-*oxo* complexes abstract hydrogen from methane to form $\cdot CH_3$ radical. The other mechanism describes that the interaction between CH_4 and coordinatively unsaturated metal atoms directly cuts off methane C-H bond and then forms metal-CH_3 (M-C) σ-bond as an intermediate.

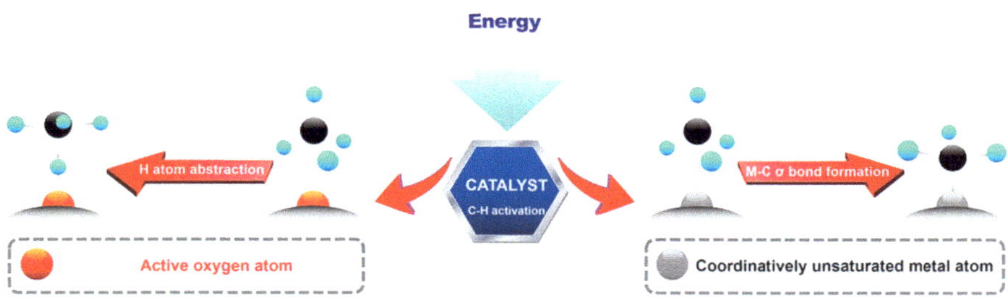

Fig. 2.3 Two kinds of mechanisms of C-H bond cleavage of methane at low-temperature reaction systems. Reprinted with permission from Ref. [26], copyright (2019) Elsevier

The complex catalyst activates methane to form M-C σ-bond in polar media in homogeneous systems. For example, POM catalyzed by (2,2-bipyrimidine)$PtCl_2$ catalyst in fuming sulfuric acid exhibited 72% one-pass methyl bisulphate (CH_3OSO_3H) yield with 81% selectivity [27]. The produced methyl bisulphate converts into methanol via hydrolysis. It is verified that the methyl ester formation is the key to evading product overoxidation and achieving high conversion and selectivity. However, the usage of corrosive medias and expensive molecular catalysts hinders homogeneous system application.

Compared with homogeneous systems, heterogeneous systems are more suitable for industrial applications due to the convenience of product and catalyst recovery. In low-temperature heterogeneous systems, the oxidant activation to produce oxidative species for methane C-H bond cleavage and $\cdot CH_3$ radical formation is the key in most POM processes. With N_2O or O_2 as oxidants at temperatures higher than 450 °C, V- and Mo-based catalysts are active for POM [28, 29]. However, CO_2 and CO were the major products and desired product yields were low. The methane monooxygenase enzymes could manipulate active oxygen species inserted into C-H bond to realize MTM by metal (Fe or Cu)-oxo active structures, which inspires the development of low-temperature MTM catalysts with zeolite-stabilized transition-metal (mainly Fe and Cu) centers in the stepwise or direct reaction process [30–32]. Sushkevich et al. [31] developed a stepwise method for MTM over a copper-containing zeolite (Fig. 2.4) and achieved productivity of 0.202 mol methanol per mole Cu with 97% selectivity. During this process, water functions as a POM oxidant and a promoter for methanol desorption from the active sites. The related mechanism can be described as Cu^{II} oxide active centers in the zeolite activate methane, and then Cu^{I} is reoxidized by water with simultaneous hydrogen formation. The stepwise process realizes different steps operated at their most suitable temperatures, but its space-time-yield is low. Thus, the direct MTM process seems more appealing, despite the low methanol production rate. Recently, Zhu et al. [33] reported a novel catalyst system to improve methanol productivity by in situ hydrogen peroxide generation at 70 °C. This catalyst was composed of AuPd alloy nanoparticles fixed within aluminosilicate zeolite crystals and organosilanes covered on the external surface of the zeolite. The catalyst novelty lies in the silanes

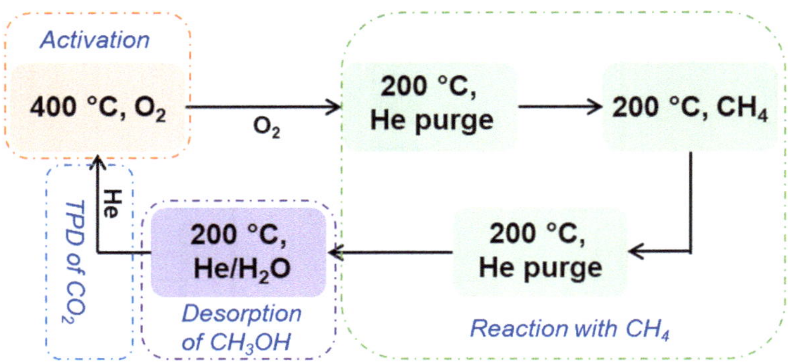

Fig. 2.4 Stepwise process for MTM over a copper-containing zeolite [31]

application, allowing H_2, O_2, and CH_4 to diffuse to the catalyst active sites while hindering the diffusion of in situ generated peroxide, which enhances hydrogen peroxide local concentration to promote methane oxidation. Other advanced materials, including 2D material-confined single-atom catalysts (SACs) [34, 35] and metal-organic frameworks (MOFs) [36, 37], are also promising heterogeneous catalysts for POM. In addition, methane could couple with CO and CO_2 to form acetic acid with high atomic efficiency [22, 38–40].

2.4 Direct Conversion of Methane to Valuable Chemicals Under Nonoxidative Conditions

2.4.1 Nonoxidative Coupling of Methane (NOCM) to C_{2+}

As mentioned above, the existence of oxygen inevitably leads to a decrease of the desired C_{2+} yield due to nonselective oxidation products (CO, CO_2, and H_2O) are thermodynamically more stable than C_{2+}. NOCM could circumvent the deep oxidation because of no oxidant participation (Eqs. 2.5 and 2.6). The NOCM reaction experiences a two-step process [20, 41]: (1) the formation of H_2 and H-deficient carbon species through C-H bonds cleavage as well as the subsequent C-C bond coupling due to the H-deficient carbon species accumulation on the catalyst surface and (2) over-dehydrogenated carbon species saturation by hydrogen to accelerate C-C bond formation, thus cleaning the catalyst surface for the next catalytic cycle. Although operation without oxidant could render NOCM deep oxidation avoidance, it also introduces another shortage of severe coke deposition. Besides, the NOCM process is thermodynamically limited and favors high operating temperature, aggravating the coke problem. Thus, one critical property of NOCM catalysts is the superior anti-coking ability. Guo et al. [42] developed a single iron site embedded in a silica matrix (Fe@SiO_2) catalyst for methane nonoxidative conversion to ethylene and aromatics which delivered an excellent performance with 48.1% methane conversion, 48.4% ethylene selectivity, and no observed deactivation during a 60-hour operation. The adjacent iron site absence suppressed the C-C bond coupling and further oligomerization, thus reducing the coke formation.

$$2CH_4 \rightarrow \mathrm{C_2H_6} + \mathrm{H_2} \Delta \mathrm{H}_{298\mathrm{K}} = 65\,\mathrm{kJ\,mol^{-1}} \tag{2.5}$$

$$2CH_4 \rightarrow \mathrm{C_2H_4} + 2\mathrm{H_2} \Delta \mathrm{H}_{298\mathrm{K}} = 202\,\mathrm{kJ\,mol^{-1}} \tag{2.6}$$

2.4.2 Methane Dehydroaromatization (MDA)

MDR (Eq. 2.7) could convert methane into highly value-added chemicals such as aromatics (benzene and naphthalene) and hydrogen without any oxidants. Aromatics are the main source of

$$6CH_4 \rightarrow \mathrm{C_6H_6} + 9\mathrm{H_2}\Delta\mathrm{H}_{298\mathrm{K}} = 531\,\mathrm{kJ\,mol^{-1}} \tag{2.7}$$

polystyrene and numerous consumer goods. Meanwhile, the produced hydrogen without poisonous CO could be directly applied in the fuel cell industry. However, as shown in Fig. 2.5, MDR is a process limited by thermodynamics, and the methane equilibrium conversion is approximately 12% at 700 °C and almost doubles at 800 °C. The MDA process was firstly realized with Mo/HZSM-5 catalyst in 1993 [43]. Since then, various catalysts based on metal ions (Mo, Fe, Zn, Cr, V, W, etc.) dispersed on zeolites (e.g., ZSM-5, MCM-22, ZSM-8, MCM-49, and mordenite) have been developed and tested to improve MDR performance [44–47]. Detailed catalyst summary and design concept, as well as the suggested mechanism, will be elucidated in Sect. 3.2.

2.4.3 *Methane Cracking to Hydrogen and Carbon Materials*

Two treasured chemicals, hydrogen and nanocarbon, could be efficiently produced in the thermo-catalytic decomposition (TCD) of methane process (Eq. 2.1). Unlike hydrogen from the methane reforming processes, high-purity hydrogen without CO or CO_2 is directly acquired from TCD, and no further separation is needed. Besides, value-added nanocarbon is the only by-product, and the type of formed carbon (e.g., carbon nanotubes, carbon nanofibers, or amorphous carbons) depends on the operating conditions [48–50].

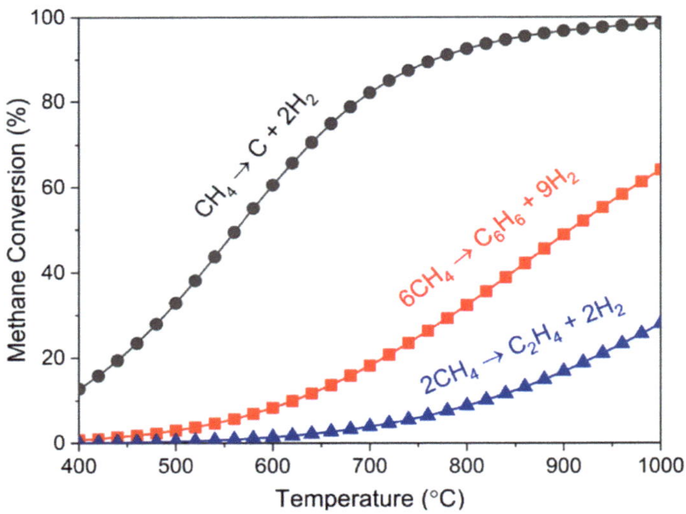

Fig. 2.5 Thermodynamic equilibrium values for methane nonoxidative conversion under 1 bar. Calculations performed with Aspen Plus V10. Reprinted with permission from Ref. [45], copyright (2020) Elsevier

Both metal and carbon materials are active TCD catalysts, and their optimal operating temperature ranges are 700–1100 K and 1100–1300 K, respectively [41]. The most studied TCD metal catalysts are Ni-based and Fe-based catalysts. Unlike most supported catalysts, the metal loading amount of TCD metal catalysts usually exceeds 10% [51]. As metal catalysts show a shortage of coking tendency, carbon materials themselves could be adopted as TCD catalysts for their resistance to deactivation and poisoning. In some cases, the produced carbon acts as a catalyst for methane activation. Defects of carbon materials are suggested as the active sites for the TCD reaction [52, 53].

The particularity of TCD reactions is that carbon is an expected value-added product, but the produced large amounts of carbon solid lead to severe catalyst deactivation by covering active sites, changing the morphology and composition of metal catalysts as well as dramatically reducing the catalyst surface area. A huge breakthrough of TCD catalysts was the development of molten metal alloy catalysts. In the molten alloy system, the produced carbon floats to the surface; thus, it could be skimmed off, having no negative effect on TCD performance. A 95% methane conversion was achieved in a 1.1-m bubble column equipped with a 27% Ni-73% Bi alloy catalyst at 1065 °C, and no activity changes were observed during a 170-hour stability test [54]. Since there is no economic advantage of the TCD process relative to the well-developed SRM, TCD process has not been industrially adopted yet. However, due to TCD which can produce high-quality hydrogen without greenhouse gas emission and further separation necessity, it would play a significant role in hydrogen production in the future.

2.5 Methane Activation and Conversion by Nonthermal Methods

2.5.1 Photocatalytic Methane Conversion

In contrast to thermal catalytic indirect conversion of methane, photocatalytic methane conversion can selectively tune the reaction pathways by modulating the incident light wavelength and photocatalysts redox potentials at room temperature, which could suppress the by-product formation and avoid the catalysts coking and sintering. Photocatalytic methane activation on semiconductor-based photocatalysts is schematically illustrated in Fig. 2.6. Fundamentally, incident light photons with energy equivalent to or exceeding the bandgap (E_g) of the semiconductor can excite electrons from the valence band (VB) to the conduction band (CB), leaving holes in the VB; then the photogenerated electrons and holes separate and migrate to the semiconductor surface and/or the co-catalysts; finally, the surface high-energy electrons and holes, respectively, drive the reduction reactions of protons or oxygen molecules and the oxidation reaction of methane or water molecules. The reactive oxygen species, surface O^- from photogenerated holes, and hydroxyl radicals from

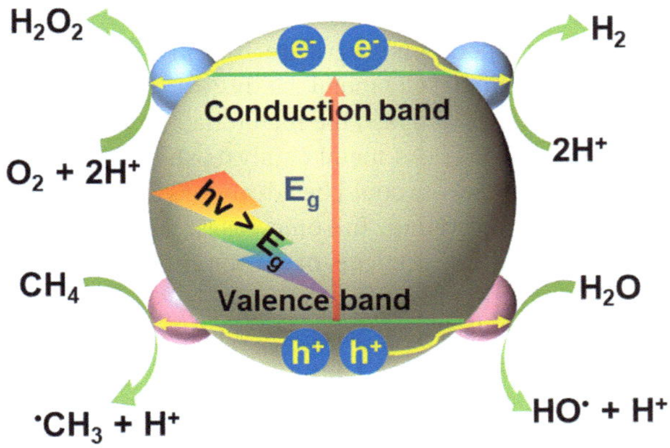

Fig. 2.6 Photocatalytic methane activation. Methane activation over semiconductor photocatalysts in the presence or absence of H_2O and O_2

water photooxidation play pivotal roles in extracting the H atom from methane C-H bond to form methyl radicals ($\cdot CH_3$). The fate of these methyl radicals is determined by the reaction conditions. They can self-couple to ethane and H_2 or react with O_2, H_2O and CO_2 to produce methanol, H_2, and CO.

The photocatalysts for methane conversion are mostly oxides or hybrids of oxides and plasmonic structures, whose valence bands are composed of O 2p orbitals with strong oxidation capability. Using O_2 as the oxidizing agent, V^{6+}/SiO_2-SiO_2 and VO_x/SBA-15 photocatalysts can catalyze the conversion of methane to methanol and formaldehyde under UV light irradiation [55, 56]. Recently, the photocatalyst of ZnO loaded with noble metal co-catalysts (Pt, Pd, Au, or Ag) directly oxidizes methane to methanol and formaldehyde with ~95% selectivity with only O_2 as the oxidant under mild light irradiation at room temperature [57]. Using mild oxidant H_2O, La-doped WO_3 could convert methane to methanol with an alcohol selectivity of 47% [58]. Using H_2O_2 as an oxidant, anatase TiO_2 modified with highly dispersed FeO_x as the co-catalysts realized a methane conversion of 15% (3 h) with 97% alcohol selectivity, where methanol selectivity reached 90% [59].

Over the composite of Zn-zeolite with delocalized electrons in the framework, the methane can be dehydrogenated and coupled to ethane and hydrogen with a stoichiometric ratio of 1:1 under sunlight and ambient temperature and pressure [60]. Impressively, the selectivity of ethane approximates 100%. Pt-loaded Ga-doped hierarchically ordered Ti-Si zeolite materials can also effectively convert methane to ethane with 90% selectivity [61]. Furthermore, the CH_4 conversion reached a high value of ~28% after 32 h of UV light irradiation. ZnO-based photocatalysts are quite active for the dehydrogenated coupling of methane to ethane or ethylene [62, 63]. In addition, nitrides like carbon nitride and GaN can also catalyze methane converted to ethane and even benzene, respectively [64, 65].

2.5.2 *Electrocatalytic Methane Conversion*

Different from conventional thermo-catalytic methane conversion, electrocatalytic methane conversion can be conducted at milder conditions and adjust the reaction activity and selectivity by changing applied potentials. Besides the utilization of intermittent renewable energy via electricity, similar to photocatalytic methane conversion processes, low-temperature operations of electrocatalytic methane conversion also avoids coking and catalyst sintering and could produce some valuable products such as oxygenates [10, 66]. Electron energy in catalytic reactions could be altered by several eV (~3 eV) via applied potentials, generally surmounting the barriers of crucial intermediate steps of C-H activation [26]. However, during most electrocatalytic processes, electric energy often accelerates the reactive oxygen-containing species formation which activates the methane C-H bond via the electrophilic process.

At high temperatures (usually ≥600 °C), electrocatalytic methane conversion can be realized in electrochemical devices such as solid oxide fuel cells (SOFC) and solid oxide electrolysis cells (SOEC) to take place partial oxidation or coupling reactions to form syngas, ethane, or ethylene [67–69]. At relatively low temperatures (300–600 °C), Torabi et al. [70] designed a SOFC with gadolinium-doped ceria (GDC)-Li_2CO_3 electrolyte to convert methane to methanol (Fig. 2.7a) and obtain a 90% methanol selectivity with less than 0.5% CO_2 selectivity. Heo et al. [71] constructed an excellent proton-conducting fuel cell with the composition of H_2, Pt/C|$Sn_{0.9}In_{0.1}P_2O_7$|Pd-Au/C, and CH_4 + O_2 at 50–250 °C. Temperature increase facilities methane conversion although adverse to methanol selectivity

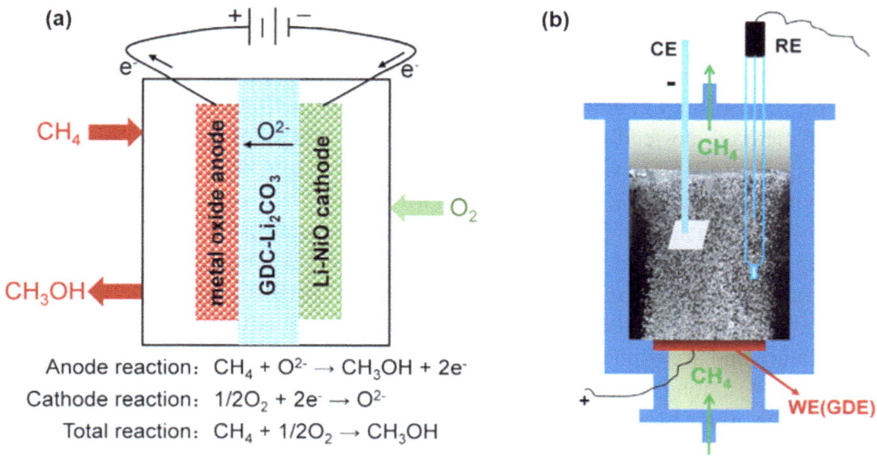

Fig. 2.7 (**a**) Schematic of a SOFC with GDC-Li_2CO_3 electrolyte for CH_4 conversion. (**b**) Schematic of electrolysis cell with liquid electrolyte and GDE for methane conversions. Reprinted with permission from Ref. [66], copyright (2018) Elsevier

enhancement. The optimal methanol production rate was achieved at 250 °C with 6.03% methane selectivity. Apart from solid electrolytes, liquid-phase electrolytes were also widely adopted for electrocatalytic methane conversion [1, 72–74]. The most adopted liquid-phase electrolytes are alkaline or neutral media, where methane dissolves into liquid and is oxidized by solid electrodes with metal oxides or metal hydroxides electrocatalysts. Considering the limited methane solubility in solutions, a gas diffusion electrode (GDE) was adopted to force methane through the electrode, which enlarges gas-liquid-solid interfaces and favors electrocatalytic methane oxidation. Rocha et al. [75] fabricated aTiO_2/RuO_2/PTFE GDC and applied it for electrocatalytic MTM in 0.1 mol L^{-1} Na_2SO_4 electrolyte (Fig. 2.7b). The formation rates of CH_3OH, HCOOH, and HCHO were 103, 112, and 24 mg h^{-1} L^{-1}, respectively, under 2.0 V.

2.5.3 *Plasma-Assisted Methane Conversion*

As the fourth state of matter, plasma contains positive and negative ions, electrons, radicals, and neutral species. Usually, based on gas-phase temperatures, plasmas could be divided into two types, i.e., thermal plasmas such as arc discharge and nonthermal plasmas such as dielectric barrier discharge (DBD), microwave plasma, corona discharge, and spark discharge [76]. In thermal plasmas, all species are in thermodynamic equilibrium and at high temperatures. Although electrons in nonthermal plasma still own high temperatures, the gas-phase temperature of the nonthermal plasma is very low even to room temperature.

Plasma-assisted methane conversion has been suggested as a promising solution to solve most of the issues that existed in thermal processes [79–82]. High-energy plasma electrons collide against neutral methane molecules to produce ions, radicals, and excited species. These active species could rapidly undergo several chemical reactions under atmospheric pressure and mild temperature (Fig. 2.8) [83]. For example, Chawdhury et al. [77] conducted an MTM reaction in a DBD reactor with O_2 and methane co-feeding (Fig. 2.8a). Not only methanol but also by-products including H_2, CO, CO_2, C_2H_6, and HCHO were produced. The acquired methanol selectivity is about 25%; however, it improved to ~35% once glass beads were packed in the reactor, which is attributed to the uniform distribution of the microdischarges and enhanced field strength. Wang et al. [78] realized the synthesis of liquid fuels and chemicals (e.g., methanol, HCHO, ethanol, and acetic acid) from CO_2 and methane in a DBD reactor equipped with a groundwater electrode at room temperature and atmospheric pressure (Fig. 2.8b). The oxygenates' total selectivity was 50–60%, among which the acetic acid (main product) selectivity was high up to 40.2%. The main products of plasma-assisted nonoxidative CH_4 conversion are C_2H_6, C_2H_2, C_2H_4, H_2, and C (Fig. 2.8c). Kado et al. [84] adopted a direct current pulse discharge reactor for methane conversion to acetylene. The acetylene selectivity was larger than 95% with a methane conversion range of 16–52%. Khalifeh et al. [85] studied the TCD of methane in a nanosecond pulsed plasma reactor with Pt-Re/

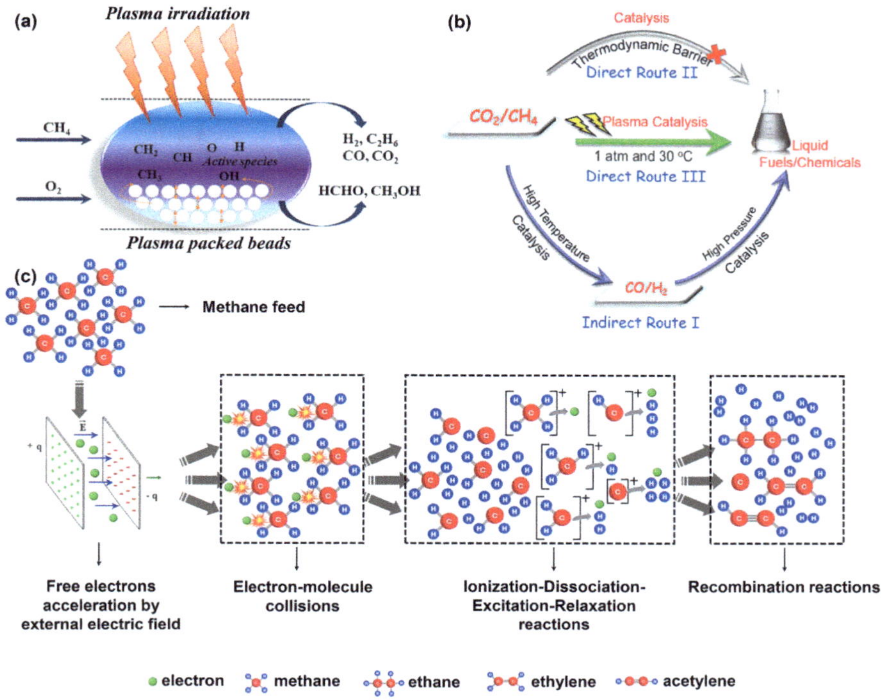

Fig. 2.8 (**a**) Plasma-assisted methane oxidation to methanol and other by-products. Reprinted with permission from Ref. [77], copyright (2018) Elsevier. (**b**) Direct and indirect processes for DRM to liquid fuels and chemicals. Reprinted with permission from Ref. [78] under Creative Commons Attribution. (**c**) Plasma-assisted nonoxidative methane reforming. Reprinted with permission from Ref. [79], copyright (2017) Elsevier

Al_2O_3 and achieved a 100% methane conversion at 9 W discharge power. Detailed information and discussion on plasma-assisted methane conversion will be presented in Chap. 6.

2.5.4 *Biocatalytic Methane Conversion*

At room temperature and atmospheric pressure (20–40 °C, 1 bar), methane monooxygenases (MMOs) biosynthesized by methanotrophs can realize high-selectivity MTM with O_2 as the primary oxidant (Eq. 2.8) [86]. The MMO's ability of cleaving the methane C-H bond originates from the extremely reactive oxygen species formation at the catalytic site. The MMO methane activation mechanism includes directly oxidizing H_2O to ·OH and extracting hydrogen from methane to produce methyl radical. If the produced ·OH is excessive, overoxidation will occur [87]. MMO could be divided into two essentially different types: one is soluble MMO

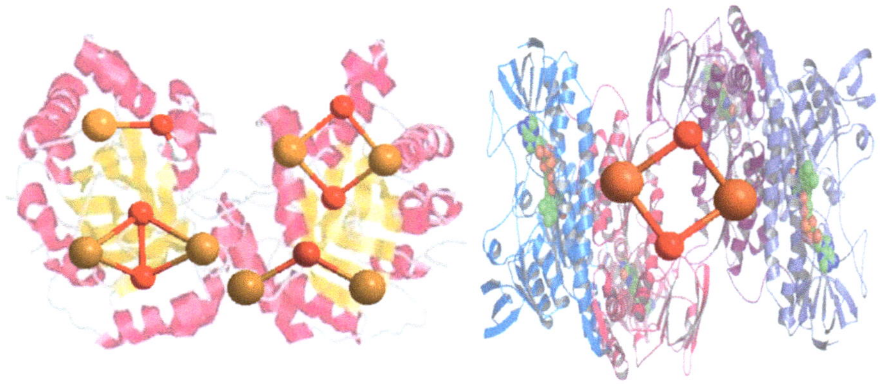

Fig. 2.9 MMO copper and iron active sites for methane activation. Reprinted with permission from Ref. [87], copyright (2020) John Wiley & Sons

(sMMO), and the other is particulate MMO (pMMO) (Fig. 2.9) [88]. sMMO exists in the cytoplasm of methanotrophs, and pMMO is embedded in the cellular membrane. sMMO and pMMO have different structures and active sites. Due to the relative stability of sMMO under abiotic environment, sMMO has been well studied [89, 90]. sMMO enzyme is composed of three proteins, i.e., hydroxylase (MMOH), reductase (MMOR), and component B (MMOB) [91]. Every component is essential for efficient methane oxidation. The active site of sMMO has been identified as a bis(μ-oxo) di-iron core. pMMO is unstable under O_2-containing atmosphere, which brings difficulty for its structural and functional analysis [92]. pMMO enzyme is composed of three subunits, including PmoA (β), PmoB (α), and PmoC (γ), among which PmoB contains mononuclear and dinuclear copper sites, the active site for methane activation [93]. The concentration level of copper ions decides the expressed enzyme type. High copper concentration corresponds to pMMO expression; otherwise, sMMO is expressed. Due to factors such as structural complexity and poor stability, these enzymes are hard to be incorporated into industrial processes [12]. However, as we illustrated in Sect. 2.3.2, the investigation on MMO has brought the rise of high-performance MTM catalysts with zeolite-stabilized TM (mainly Fe and Cu) centers.

$$CH_4 + \mathrm{O}_2 + \mathrm{e}^- + \mathrm{H}^+ \rightarrow CH_3OH + \mathrm{H}_2\mathrm{O} \tag{2.8}$$

2.6 Summary

This chapter is aimed to provide a brief and clear cognition on methane activation via separately introducing methane properties, indirect and direct methane conversion processes, and several novel and promising methane activation technologies.

Although these novel technologies could realize methane conversion to high value-added chemicals at low temperatures and the conversion and utilization of renewable energy, their performance has a big margin for improvement to reach industrialization standards. Thus, this book mainly focuses on thermal catalytic methane conversion processes and their application in petrochemical and biofuel industries.

References

1. C. Oh, J. Kim, Y.J. Hwang, M. Ma, J.H. Park, Electrocatalytic methane oxidation on Co_3O_4-incorporated ZrO_2 nanotube powder. Appl. Catal. B Environ. **283**, 119653 (2021)
2. C.-G. Zhan, J.A. Nichols, D.A. Dixon, Ionization potential, electron affinity, electronegativity, hardness, and electron excitation energy: Molecular properties from density functional theory orbital energies. J. Phys. Chem. A **107**, 4184–4195 (2003)
3. R.D. Amos, An accurate *ab initio* study of the multipole moments and polarizabilities of methane. Mol. Phys. **38**, 33–45 (1979)
4. Y.-R. Luo, *Comprehensive Handbook of Chemical Bond Energies* (CRC Press, Boca Raton, FL, 2007)
5. D.C. Frost, C.A. McDowell, Studies of the ionization of molecules by electron impact - II. Excited states of the molecular ions of methane and the methyl halides. Proc. R. Soc. Lond. Ser. A Math. Phys. Sci. **241**, 194–207 (1957)
6. R.J. Duchovic, W.L. Hase, H. Bernhard Schlegel, M.J. Frisch, K. Raghavachari, Ab initio potential energy curve for CH bond dissociation in methane. Chem. Phys. Lett. **89**, 120–125 (1982)
7. K. Huang, J.B. Miller, G.W. Huber, J.A. Dumesic, C.T. Maravelias, A general framework for the evaluation of direct nonoxidative methane conversion strategies. Joule **2**, 349–365 (2018)
8. T. Baba, A. Miyaji, *Catalysis and the Mechanism of Methane Conversion to Chemicals* (Springer, Singapore, 2020)
9. S. Zhang, J. Bao, B. Wu, L. Zhong, Y. Sun, Research progress on the photocatalytic conversion of methane and methanol. Acta Phys. -Chim. Sin. **35**, 923–939 (2019)
10. A.H. Bagherzadeh Mostaghimi, T.A. Al-Attas, M.G. Kibria, S. Siahrostami, A review on electrocatalytic oxidation of methane to oxygenates. J. Mater. Chem. A **8**, 15575–15590 (2020)
11. T. Nozaki, K. Okazaki, Non-thermal plasma catalysis of methane: Principles, energy efficiency, and applications. Catal. Today **211**, 29–38 (2013)
12. T.J. Lawton, A.C. Rosenzweig, Biocatalysts for methane conversion: Big progress on breaking a small substrate. Curr. Opin. Chem. Biol. **35**, 142–149 (2016)
13. H.-C. Wu, Z. Rui, Y.S. Lin, Hydrogen production with carbon dioxide capture by dual-phase ceramic-carbonate membrane reactor via steam reforming of methane. J. Membr. Sci. **598**, 117780 (2020)
14. J.L. Rogers, M.C. Mangarella, A.D. D'Amico, J.R. Gallagher, M.R. Dutzer, E. Stavitski, J.T. Miller, C. Sievers, Differences in the nature of active sites for methane dry reforming and methane steam reforming over nickel aluminate catalysts. ACS Catal. **6**, 5873–5886 (2016)
15. S. Wang, G.Q. Lu, G.J. Millar, Carbon dioxide reforming of methane to produce synthesis gas over metal-supported catalysts: State of the art. Energy Fuel **10**, 896–904 (1996)
16. Y. Song, E. Ozdemir, S. Ramesh, A. Adishev, S. Subramanian, A. Harale, M. Albuali, B.A. Fadhel, A. Jamal, D. Moon, S.H. Choi, C.T. Yavuz, Dry reforming of methane by stable Ni-Mo nanocatalysts on single-crystalline MgO. Science **367**, 777–781 (2020)
17. J. Tong, W. Yang, H. Suda, K. Haraya, Initiation of oxygen permeation and POM reaction in different mixed conducting ceramic membrane reactors. Catal. Today **118**, 144–150 (2006)

18. W.-J. Jang, D.-W. Jeong, J.-O. Shim, H.-M. Kim, H.-S. Roh, I.H. Son, S.J. Lee, Combined steam and carbon dioxide reforming of methane and side reactions: Thermodynamic equilibrium analysis and experimental application. Appl. Energy **173**, 80–91 (2016)
19. U. Zavyalova, M. Holena, R. Schlögl, M. Baerns, Statistical analysis of past catalytic data on oxidative methane coupling for new insights into the composition of high-performance catalysts. ChemCatChem **3**, 1935–1947 (2011)
20. Y. Xu, X. Bao, L. Lin, Direct conversion of methane under nonoxidative conditions. J. Catal. sis **216**, 386–395 (2003)
21. Q. Zhu, S.L. Wegener, C. Xie, O. Uche, M. Neurock, T.J. Marks, Sulfur as a selective 'soft' oxidant for catalytic methane conversion probed by experiment and theory. Nat. Chem. **5**, 104–109 (2013)
22. J.F. Wu, S.M. Yu, W.D. Wang, Y.X. Fan, S. Bai, C.W. Zhang, Q. Gao, J. Huang, W. Wang, Mechanistic insight into the formation of acetic acid from the direct conversion of methane and carbon dioxide on zinc-modified H-ZSM-5 zeolite. J. Am. Chem. Soc. **135**, 13567–13573 (2013)
23. R. Sharma, H. Poelman, G.B. Marin, V.V. Galvita, Approaches for selective oxidation of methane to methanol. Catalysts **10**, 194 (2020)
24. M.H. Mahyuddin, Y. Shiota, K. Yoshizawa, Methane selective oxidation to methanol by metal-exchanged zeolites: A review of active sites and their reactivity. Cat. Sci. Technol. **9**, 1744–1768 (2019)
25. S. Raynes, M. Shah, R.A. Taylor, Direct conversion of methane to methanol with zeolites: Towards understanding the role of extra-framework d-block metal and zeolite framework type. Dalton Trans. **48**, 10364–10384 (2019)
26. X. Meng, X. Cui, N.P. Rajan, L. Yu, D. Deng, X. Bao, Direct methane conversion under mild condition by thermo-, electro-, or photocatalysis. Chem **5**, 2296–2325 (2019)
27. R.A. Periana, D.J. Taube, S. Gamble, H. Taube, T. Satoh, H. Fujii, Platinum catalysts for the high-yield oxidation of methane to a methanol derivative. Science **280**, 560–564 (1998)
28. K.J. Zhen, M.M. Khan, C.H. Mak, K.B. Lewis, G.A. Somorjai, Partial oxidation of methane with nitrous oxide over V_2O_5-SiO_2 catalyst. J. Catal. **94**, 501–507 (1985)
29. M.M. Khan, G.A. Somorjai, A kinetic-study of partial oxidation of methane with nitrous-oxide on a Molybdena silica catalyst. J. Catal. **91**, 263–271 (1985)
30. P. Tomkins, M. Ranocchiari, J.A. van Bokhoven, Direct conversion of methane to methanol under mild conditions over Cu-zeolites and beyond. Acc. Chem. Res. **50**, 418–425 (2017)
31. V.L. Sushkevich, D. Palagin, M. Ranocchiari, J.A. van Bokhoven, Selective anaerobic oxidation of methane enables direct synthesis of methanol. Science **356**, 523–527 (2017)
32. M.J. da Silva, Synthesis of methanol from methane: Challenges and advances on the multi-step (syngas) and one-step routes (DMTM). Fuel Process. Technol. **145**, 42–61 (2016)
33. Z. Jin, L. Wang, E. Zuidema, K. Mondal, M. Zhang, J. Zhang, C. Wang, X. Meng, H. Yang, C. Mesters, F.S. Xiao, Hydrophobic zeolite modification for in situ peroxide formation in methane oxidation to methanol. Science **367**, 193–197 (2020)
34. X. Cui, H. Li, Y. Wang, Y. Hu, L. Hua, H. Li, X. Han, Q. Liu, F. Yang, L. He, X. Chen, Q. Li, J. Xiao, D. Deng, X. Bao, Room-temperature methane conversion by graphene-confined single iron atoms. Chem **4**, 1902–1910 (2018)
35. Y. Wang, Room-temperature conversion of methane becomes true. Joule **2**, 1399–1401 (2018)
36. T. Ikuno, J. Zheng, A. Vjunov, M. Sanchez-Sanchez, M.A. Ortuno, D.R. Pahls, J.L. Fulton, D.M. Camaioni, Z. Li, D. Ray, B.L. Mehdi, N.D. Browning, O.K. Farha, J.T. Hupp, C.J. Cramer, L. Gagliardi, J.A. Lercher, Methane oxidation to methanol catalyzed by Cu-oxo clusters stabilized in NU-1000 metal-organic framework. J. Am. Chem. Soc. **139**, 10294–10301 (2017)
37. D.Y. Osadchii, A.I. Olivos-Suarez, Á. Szécsényi, G. Li, M.A. Nasalevich, I.A. Dugulan, P.S. Crespo, E.J.M. Hensen, S.L. Veber, M.V. Fedin, G. Sankar, E.A. Pidko, J. Gascon, Isolated Fe sites in metal organic frameworks catalyze the direct conversion of methane to methanol. ACS Catal. **8**, 5542–5548 (2018)

38. J. Shan, M. Li, L.F. Allard, S. Lee, M. Flytzani-Stephanopoulos, Mild oxidation of methane to methanol or acetic acid on supported isolated rhodium catalysts. Nature **551**, 605–608 (2017)
39. Y. Tang, Y. Li, V. Fung, D.E. Jiang, W. Huang, S. Zhang, Y. Iwasawa, T. Sakata, L. Nguyen, X. Zhang, A.I. Frenkel, F. Tao, Single rhodium atoms anchored in micropores for efficient transformation of methane under mild conditions. Nat. Commun. **9**, 1231 (2018)
40. W. Huang, K.C. Xie, J.P. Wang, Z.H. Gao, L.H. Yin, Q.M. Zhu, Possibility of direct conversion of CH_4 and CO_2 to high-value products. J. Catal. **201**, 100–104 (2001)
41. P. Tang, Q. Zhu, Z. Wu, D. Ma, Methane activation: The past and future. Energ. Environ. Sci. **7**, 2580–2591 (2014)
42. X. Guo, G. Fang, G. Li, H. Ma, H. Fan, L. Yu, C. Ma, X. Wu, D. Deng, M. Wei, D. Tan, R. Si, S. Zhang, J. Li, L. Sun, Z. Tang, X. Pan, X. Bao, Direct, nonoxidative conversion of methane to ethylene, aromatics, and hydrogen. Science **344**, 616–619 (2014)
43. L. Wang, L. Tao, M. Xie, G. Xu, J. Huang, Y. Xu, Dehydrogenation and aromatization of methane under non-oxidizing conditions. Catal. Lett. **21**, 35–41 (1993)
44. I. Vollmer, I. Yarulina, F. Kapteijn, J. Gascon, Progress in developing a structure-activity relationship for the direct aromatization of methane. ChemCatChem **11**, 39–52 (2018)
45. U. Menon, M. Rahman, S.J. Khatib, A critical literature review of the advances in methane dehydroaromatization over multifunctional metal-promoted zeolite catalysts. Appl. Catal. A Gen. **608**, 117870 (2020)
46. D. Kiani, S. Sourav, Y. Tang, J. Baltrusaitis, I.E. Wachs, Methane activation by ZSM-5-supported transition metal centers. Chem. Soc. Rev. **50**, 1251–1268 (2021)
47. N. Kosinov, E.J.M. Hensen, Reactivity, selectivity, and stability of zeolite-based catalysts for methane dehydroaromatization. Adv. Mater. **32**, e2002565 (2020)
48. U.P.M. Ashik, W.M.A. Wan Daud, H.F. Abbas, Production of greenhouse gas free hydrogen by thermocatalytic decomposition of methane: A review. Renew. Sust. Energ. Rev. **44**, 221–256 (2015)
49. R. Vander Wal, N.M. Makiesse, Carbons as catalysts in thermo-catalytic hydrocarbon decomposition: A review. C—J. Carbon Res. **6**, 23 (2020)
50. A.F.A. Syed Muhammad, A. Awad, R. Saidur, N. Masiran, A. Salam, B. Abdullah, Recent advances in cleaner hydrogen productions via thermo-catalytic decomposition of methane: Admixture with hydrocarbon. Int. J. Hydrog. Energy **43**, 18713–18734 (2018)
51. U.P.M. Ashik, W.M.A. Wan Daud, J.-I. Hayashi, A review on methane transformation to hydrogen and nanocarbon: Relevance of catalyst characteristics and experimental parameters on yield. Renew. Sust. Energ. Rev. **76**, 743–767 (2017)
52. N. Muradov, F. Smith, A. T-Raissi, Catalytic activity of carbons for methane decomposition reaction. Catal. Today **102–103**, 225–233 (2005)
53. R. Moliner, I. Suelves, M. Lazaro, O. Moreno, Thermocatalytic decomposition of methane over activated carbons: Influence of textural properties and surface chemistry. Int. J. Hydrog. Energy **30**, 293–300 (2005)
54. D.C. Upham, V. Agarwal, A. Khechfe, Z.R. Snodgrass, M.J. Gordon, H. Metiu, E.W. McFarland, Catalytic molten metals for the direct conversion of methane to hydrogen and separable carbon. Science **358**, 917–921 (2017)
55. S. Kaliaguine, B. Shelimov, V. Kazansky, Reactions of methane and ethane with hole centers O^-. J. Catal. **55**, 384–393 (1978)
56. H.H. López, A. Martínez, Selective photo-assisted oxidation of methane into formaldehyde on mesoporous VO_x/SBA-15 catalysts. Catal. Lett. **83**, 37–41 (2002)
57. H. Song, X. Meng, S. Wang, W. Zhou, X. Wang, T. Kako, J. Ye, Direct and selective photocatalytic oxidation of CH_4 to oxygenates with O_2 on cocatalysts/ZnO at room temperature in water. J. Am. Chem. Soc. **141**, 20507–20515 (2019)
58. K. Villa, S. Murcia-Lopez, J.R. Morante, T. Andreu, An insight on the role of La in mesoporous WO_3 for the photocatalytic conversion of methane into methanol. Appl. Catal. B: Environ. **187**, 30–36 (2016)

59. J. Xie, R. Jin, A. Li, Y. Bi, Q. Ruan, Y. Deng, Y. Zhang, S. Yao, G. Sankar, D. Ma, Highly selective oxidation of methane to methanol at ambient conditions by titanium dioxide-supported iron species. Nat. Catal. **1**, 889–896 (2018)
60. L. Li, G.D. Li, C. Yan, X.Y. Mu, X.L. Pan, X.X. Zou, K.X. Wang, J.S. Chen, Efficient sunlight-driven dehydrogenative coupling of methane to ethane over a Zn^{+}-modified zeolite. Angew. Chem. Int. Ed. **50**, 8299–8303 (2011)
61. S. Wu, X. Tan, J. Lei, H. Chen, L. Wang, J. Zhang, Ga-doped and Pt-loaded porous TiO_2-SiO_2 for photocatalytic nonoxidative coupling of methane. J. Am. Chem. Soc. **141**, 6592–6600 (2019)
62. L. Meng, Z. Chen, Z. Ma, S. He, Y. Hou, H.-H. Li, R. Yuan, X.-H. Huang, X. Wang, X. Wang, Gold plasmon-induced photocatalytic dehydrogenative coupling of methane to ethane on polar oxide surfaces. Energy Environ. Sci. **11**, 294–298 (2018)
63. W. Jiang, J. Low, K. Mao, D. Duan, S. Chen, W. Liu, C.-W. Pao, J. Ma, S. Sang, C. Shu, X. Zhan, Z. Qi, H. Zhang, Z. Liu, X. Wu, R. Long, L. Song, Y. Xiong, Pd-modified ZnO-Au enabling alkoxy intermediates formation and dehydrogenation for photocatalytic conversion of methane to ethylene. J. Am. Chem. Soc. **143**, 269–278 (2021)
64. Y. Zhou, L. Zhang, W. Wang, Direct functionalization of methane into ethanol over copper modified polymeric carbon nitride via photocatalysis. Nat. Commun. **10**, 1–8 (2019)
65. L. Li, S. Fan, X. Mu, Z. Mi, C.-J. Li, Photoinduced conversion of methane into benzene over GaN nanowires. J. Am. Chem. Soc. **136**, 7793–7796 (2014)
66. S. Xie, S. Lin, Q. Zhang, Z. Tian, Y. Wang, Selective electrocatalytic conversion of methane to fuels and chemicals. J. Energy Chem. **27**, 1629–1636 (2018)
67. Y.F. Sun, Y.Y. Wu, Y.Q. Zhang, J.H. Li, Y. Luo, Y.X. Shi, B. Hua, J.L. Luo, A bifunctional solid oxide electrolysis cell for simultaneous CO_2 utilization and synthesis gas production. Chem. Commun. **52**, 13687–13690 (2016)
68. C. Zhu, S. Hou, X. Hu, J. Lu, F. Chen, K. Xie, Electrochemical conversion of methane to ethylene in a solid oxide electrolyzer. Nat. Commun. **10**, 1173 (2019)
69. Y. Wang, T. Liu, L. Lei, F. Chen, Methane assisted solid oxide co-electrolysis process for syngas production. J. Power Sources **344**, 119–127 (2017)
70. A. Torabi, J. Barton, C. Willman, H. Ghezel-Ayagh, N. Li, A. Poozhikunnath, R. Maric, O.A. Marina, Developing low-intermediate temperature fuel cells for direct conversion of methane to methanol fuel. ECS Trans. **72**, 193–199 (2016)
71. P. Heo, K. Ito, A. Tomita, T. Hibino, A proton-conducting fuel cell operating with hydrocarbon fuels. Angew. Chem. Int. Ed. **47**, 7841–7844 (2008)
72. Y. Song, Y. Zhao, G. Nan, W. Chen, Z. Guo, S. Li, Z. Tang, W. Wei, Y. Sun, Electrocatalytic oxidation of methane to ethanol via NiO/Ni interface. Appl. Catal. B Environ. **270**, 118888 (2020)
73. M. Jafarian, M.G. Mahjani, H. Heli, F. Gobal, M. Heydarpoor, Electrocatalytic oxidation of methane at nickel hydroxide modified nickel electrode in alkaline solution. Electrochem. Commun. **5**, 184–188 (2003)
74. N. Xu, C.A. Coco, Y. Wang, T. Su, Y. Wang, L. Peng, Y. Zhang, Y. Liu, J. Qiao, X.-D. Zhou, Electro-conversion of methane to alcohols on "capsule-like" binary metal oxide catalysts. Appl. Catal. B Environ. **282**, 119572 (2021)
75. R.S. Rocha, R.M. Reis, M.R.V. Lanza, R. Bertazzoli, Electrosynthesis of methanol from methane: The role of V_2O_5 in the reaction selectivity for methanol of a $TiO_2/RuO_2/V_2O_5$ gas diffusion electrode. Electrochim. Acta **87**, 606–610 (2013)
76. S. Ogo, Y. Sekine, Catalytic reaction assisted by plasma or electric field. Chem. Rec. **17**, 726–738 (2017)
77. P. Chawdhury, D. Ray, C. Subrahmanyam, Single step conversion of methane to methanol assisted by nonthermal plasma. Fuel Process. Technol. **179**, 32–41 (2018)
78. L. Wang, Y. Yi, C. Wu, H. Guo, X. Tu, One-step reforming of CO_2 and CH_4 into high-value liquid chemicals and fuels at room temperature by plasma-driven catalysis. Angew Chem. Int. Ed. **56**, 13679–13683 (2017)

79. M. Scapinello, E. Delikonstantis, G.D. Stefanidis, The panorama of plasma-assisted non-oxidative methane reforming. Chem. Eng. Process.: Process Intensif. **117**, 120–140 (2017)
80. H. Puliyalil, D. Lašič Jurković, V.D.B.C. Dasireddy, B. Likozar, A review of plasma-assisted catalytic conversion of gaseous carbon dioxide and methane into value-added platform chemicals and fuels. RSC Adv. **8**, 27481–27508 (2018)
81. H. Chen, Y. Mu, S. Xu, S. Xu, C. Hardacre, X. Fan, Recent advances in non-thermal plasma (NTP) catalysis towards C1 chemistry. Chin. J. Chem. Eng. **28**, 2010–2021 (2020)
82. J.X. Qian, T.W. Chen, L.R. Enakonda, D.B. Liu, G. Mignani, J.-M. Basset, L. Zhou, Methane decomposition to produce CO_x-free hydrogen and nano-carbon over metal catalysts: A review. Int. J. Hydrog. Energy **45**, 7981–8001 (2020)
83. H.L. Chen, H.M. Lee, S.H. Chen, Y. Chao, M.B. Chang, Review of plasma catalysis on hydrocarbon reforming for hydrogen production-interaction, integration, and prospects. Appl. Catal. B Environ. **85**, 1–9 (2008)
84. S. Kado, Y. Sekine, K. Fujimoto, Direct synthesis of acetylene from methane by direct current pulse discharge. Chem. Commun., 2485–2486 (1999)
85. O. Khalifeh, H. Taghvaei, A. Mosallanejad, M.R. Rahimpour, A. Shariati, Extra pure hydrogen production through methane decomposition using nanosecond pulsed plasma and Pt–Re catalyst. Chem. Eng. J. **294**, 132–145 (2016)
86. S. Sirajuddin, A.C. Rosenzweig, Enzymatic oxidation of methane. Biochemistry **54**, 2283–2294 (2015)
87. L. Sun, Y. Wang, N. Guan, L. Li, Methane activation and utilization: Current status and future challenges. Energ. Technol. **8**, 1900826 (2019)
88. R.S. Hanson, T.E. Hanson, Methanotrophic bacteria. Microbiol. Rev. **60**, 439–471 (1996)
89. B.J. Wallar, J.D. Lipscomb, Dioxygen activation by enzymes containing binuclear non-heme iron clusters. Chem. Rev. **96**, 2625–2658 (1996)
90. M. Merkx, D.A. Kopp, M.H. Sazinsky, J.L. Blazyk, J. Müller, S.J. Lippard, Dioxygen activation and methane hydroxylation by soluble methane monooxygenase: A tale of two irons and three proteins. Angew. Chem. Int. Ed. **40**, 2782–2807 (2001)
91. B.G. Fox, W.A. Froland, J.E. Dege, J.D. Lipscomb, Methane monooxygenase from Methylosinus trichosporium OB3b. J. Biol. Chem. **264**, 10023–10033 (1989)
92. R.L. Lieberman, A.C. Rosenzweig, Biological methane oxidation: Regulation, biochemistry, and active site structure of particulate methane monooxygenase. Crit. Rev. Biochem. Mol. Biol. **39**, 147–164 (2004)
93. R. Balasubramanian, S.M. Smith, S. Rawat, L.A. Yatsunyk, T.L. Stemmler, A.C. Rosenzweig, Oxidation of methane by a biological dicopper centre. Nature **465**, 115–119 (2010)

Chapter 3
Direct Systems: Methane Dehydroaromatization (MDA) and the Oxidative Coupling of Methane (OCM)

3.1 Introduction

Converting methane, the main component of natural gas, to fuels and chemical industry precursors has been of great interest to the petrochemical industry over the years. This is owed to a myriad of factors such as declining oil reserves and the discovery of large reserves of natural gas in the form of shale gas and hydrates [1], leading to predictions of strong natural gas growth in the future [2]. However, current industrial practice is to indirectly convert methane using multiple steps. Synthesis gas (syngas) is a mixture of CO, H_2, and sometimes CO_2, depending on the process. The process to create syngas is expensive and requires intense energy input, and the subsequent steps require an oxygen input to remove CO species to promote desired hydrocarbon production. Figure 3.1 shows a schematic of the indirect processes involved when converting methane to valuable chemical precursors and fuels. These include Fischer-Tropsch synthesis (FTS) [3] and methanol to gasoline (MTG) [4] which are obtained from syngas through the steam reforming process [5] and methanol production from syngas, respectively. These indirect processes are well developed but are hindered by the aspects described above. Highlighted by the dotted black box in Fig. 3.1 are the two systems that will be discussed in this chapter: Ethylene production by direct oxidative routes (a form of OCM) and the direct nonoxidative system, nominally methane dehydroaromatization (MDA). Both have their pros and cons that will be discussed later in the chapter, but for now it should be noted that the positive attributes of these systems have resulted in a surge in research to make them commercially viable. Although not commercially viable at this stage, these two systems provide a desirable alternative to the complex processes currently employed and present the petrochemical industry with a very real possibility of directly utilizing methane on-site, reducing costs and providing higher energy efficiency.

H. Song et al., *Methane Activation and Utilization in the Petrochemical and Biofuel Industries*, https://doi.org/10.1007/978-3-030-88424-6_3

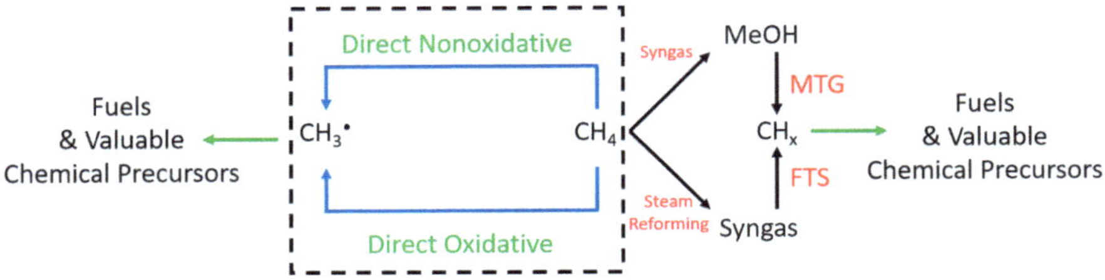

Fig. 3.1 Schematic showing the different routes used for methane conversion to fuels and valuable chemical precursors. Adapted with permission from (Chem. Rev. 2017, 117, 13, 8497–8520). Copyright (2017) American Chemical Society

As discussed in Chap. 2, the activation of methane is a huge challenge, owing to the symmetry of the four C-H bonds, high bond energies, and the lack of electrons in the antibonding orbitals. The MDA reaction when forming benzene is shown as $6CH_4 \rightleftharpoons C_6H_6 + 9H_2$, yielding a Gibbs free energy ($\Delta G^0_{450^\circ C}$) of 267 kJ/mol, indicating a large input of energy is needed to induce reaction. Thus, for the MDA reaction, high temperatures along with a catalyst are still needed to achieve even low methane conversions [6], which results in unselective products and large amounts of naphthalene formation which ultimately coke (deactivate) the catalyst, meaning that frequent regeneration to remove coke is necessary [7]. Despite the use of a catalyst, of which there are many variations, the thermodynamic limitation imposed by the MDA reaction makes it extremely challenging to implement both theoretically and at the commercial scale. Another direct nonoxidative method of methane activation and utilization comes in the form of cross-coupling reactions where methane is co-fed with a higher carbon-number-containing feedstock. However, this process will be discussed in detail in Chap. 4 and has been separated from the two processes discussed here to enable more detail and prevent convolution with the pure MDA process.

The oxidative coupling reaction (OCM) is not limited by thermodynamic constraints and is capable of directly converting methane to ethylene, an in-demand chemical responsible for nearly half of the world's feedstocks in chemical production facilities. Although not limited by thermodynamics ($\Delta G^0_{450^\circ C} = -297$ kJ/mol), the exergonic nature of the summarized and simplified reaction below means that the OCM reaction is kinetically hindered, yielding poor selectivities to ethylene [8]. $4CH_4 + O_2 \rightarrow 2C_2H_6 + 2H_2O$, the ethane then oxidatively dehydrogenates in the presence of oxygen (note that many steps are missing to complete this simplified form of the reaction) to give the following reaction $2C_2H_6 + O_2 \rightarrow 2C_2H_4 + 2H_2O$.

In summary, this chapter will introduce the direct conversion of methane to valuable hydrocarbons (fuel or chemical precursors) by avoiding unwanted intermediate steps such as syngas production. Although a challenging topic, the rewards would mean a more economically viable process for using abundant natural gas as well as a more environmentally friendly system.

3.2 MDA Reaction

3.2.1 *MDA Reaction Over a Monofunctional and Nonmetallic Catalyst*

The conversion of methane to higher hydrocarbons has been conducted in the past through homogeneous catalysis and the use of organometallic media such as Pt chlorides [9], biological activation (possibly more promising when used in-line with biofuels) [10], and even biomimetic catalysts which focus on the conversion of methane to methanol [11]. However, such mediums for the activation of methane have seen little success, and transferring these systems from the lab to an industrial scale is, as of yet, simply not feasible. Furthermore, most of these processes are conducted in the presence of co-feeds, rendering them out of the scope of this chapter where we look at the direct conversion of methane. Thus, this chapter focuses on the use of monofunctional, bifunctional, and heterogeneous catalysis for direct methane activation, which have seen much research in the area and show far more promise for the commercial scale. Before continuing, it is important to note that the role of Brønsted acid sites in the activation of methane has recently been brought into question. In fact, it is now believed that the role of Brønsted acidity is secondary in nature [12] (discussed in the next sub-chapter). However, the importance of shape selectivity still holds true when conducting MDA reactions to produce benzene.

The challenges involved with C-H activation have already been discussed, and so the question becomes how to overcome them? Zeolites have a long and rich history with the petroleum industry [13–15], but it was not until 1993 that HZSM-5, the protonated form of ZSM-5 as invented by Robert J. Argauer in the name of Zeolite Socony Mobil [16], was found to show minor activity for the conversion of methane (1.4%) with outstanding selectivity towards benzene (100%) [17] and pure molecular hydrogen in the gas-phase product. It should be noted that the negative role played by coking and naphthalene formation was not discussed in this work, despite its certain contribution. In fact, the equilibrium conversion of methane to benzene and naphthalene is 12% at 700 °C in equimolar contributions [6], i.e., benzene and naphthalene are the preferred products under these conditions. But because naphthalene likely contributed to solid coking under these conditions, it was not reported as product. Prior to this, the role that zeolites played in fluid catalytic cracking (FCC) was understood to be due to the strong Brønsted acidity, and the same attribution can be applied here when directly activating methane. Brønsted acidity is defined as the ability of a species to donate a proton. In the case of zeolites, nominally ZSM-5, this acidity comes from an irreducible silica matrix with Lewis acidity (the ability to host a proton). ZSM-5 is made up of SiO_2 and Al_2O_3 species, and the Si^{4+} species can be replaced by Al^{3+} cations, creating an additional positive charge in the matrix which is necessary to maintain neutrality. Figure 3.2 illustrates a simple unit and the chemical composition of Brønsted acidity as found in zeolites such as ZSM-5.

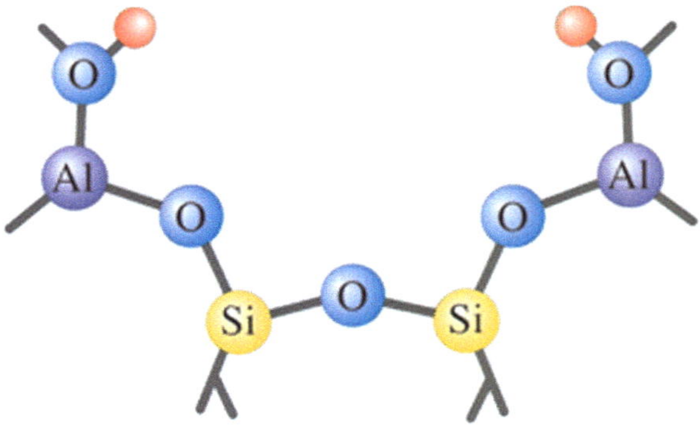

Fig. 3.2 An illustration of Brønsted acidity as found in ZSM-5, the predominant catalyst support used in MDA reactions. Adapted with permission from (Ind. Eng. Chem. Res. 2016, 55, 37, 9895–9906). Copyright (2016) American Chemical Society

This property is also akin to α-quartz material as shown by the density functional theory (DFT) calculations [18], but as suitable dopant Al cations are on the surface of zeolites (as opposed to α-quartz), the amount of Brønsted acid sites available for catalysis (a surface reliant process, whether it be internal or external surface zones) are far higher and thus, more suitable. It should be noted that ZSM-5 is typically synthesized into the ammonium form, depending on the technique used, and is then calcined at high temperatures (>500 °C) to obtain the hydrogen form (HZSM-5). One reason for such low conversions of methane over HZSM-5, even when reacted at high temperatures, is because of the competing hydrogen exchange process which is approximated to have a third of the activation barrier requirements to proceed [19], i.e., a competing reaction is more likely to occur than C-H bond activation. For example, at a moderate acidity, dehydrogenation of methane has an activation barrier of 90.08 kcal/mol as opposed to the hydrogen exchange reaction which has an activation barrier of 33.53 kcal/mol, with this trend being mirrored for other acid strengths of the ZSM-5 support.

P. Blowers and X. Zheng were also able to predict the transition state structures for methane dehydrogenation over a ZSM-5 support via DFT calculations and the cluster approach method, with graphical results shown here in Fig. 3.3.

Although not as competitive as the hydrogen exchange reaction, this transition state is generally believed to be part of the process that eventually leads to aromatization during the overall MDA reaction. Figure 3.3 shows the proton supplied by the Brønsted acidity of the zeolite support, which leads to cleavage of the C-H bond. Note the other oxygen, O(2) acting as the Lewis acceptor of the resulting methyl group. The resulting species is methanium (CH_5^+) which is the transition species observed when conducting computational chemistry. Molecular hydrogen is then lost to yield a methyl cation which may react further with methane species to form a $C_2H_7^+$ intermediate. Finally, this species is decomposed into hydrogen and ethenium, followed by subtraction of H^+ species, likely given to the lattice O, yielding ethylene. It is also suggested, and supported by DFT calculations, that two methyl radicals may form to produce ethane, but this process is so fast at elevated temperatures that ethane species are not observed in the gas product as the ethane molecules dissociate [20]. The ethylene then undergoes further reactions to yield the final

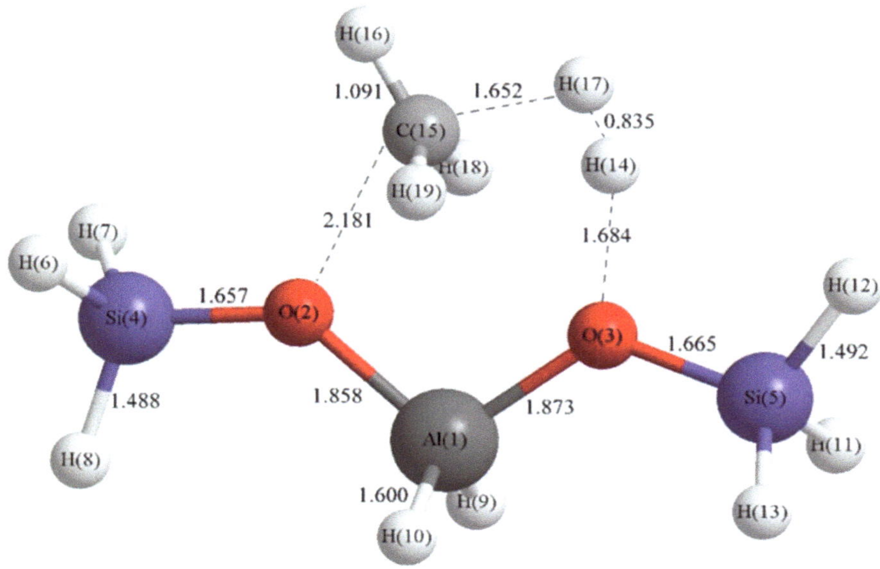

Fig. 3.3 Transition state of methane over the zeolite support for the methane dehydrogenation reaction. Adapted with permission from (Journal of Molecular Catalysis A: Chemical 246 (2006) 1–10). Copyright (2006) Elsevier

product, benzene. These steps are not fully understood or observed and so are not supplied here; however, the ethylene species are generally dehydrogenated and then aromatized after a series of oligomerization [21].

In addition to the benefits of acidity obtained through using zeolites like ZSM-5, the MFI type structure exhibited by ZSM-5 provides a unique morphology which is believed to promote aromatic formation through shape selectivity [22]. Below is shown the detailed structure of ZSM-5 as well as other commonly used supports in the petrochemical industry, to help demonstrate the nature of its potential shape selectivity as one of the many reasons why ZSM-5 is used in aromatization reactions (Fig. 3.4).

ZSM-5 is made up of linked tetrahedra, which combine to make five-membered oxygen rings. Fourteen tetrahedra form a pentasil unit which are linked to form the ten-membered ring porous structure that ZSM-5 is famous for. It is this porous structure that is attributed to the substantial shape selectivity observed for this material. Note that ZSM-5 has a tunable acidity as mentioned earlier, whereas the pure SiO_2 form is referred to as silicalite-1. It is speculated that the Brønsted acid sites catalyze the C-H bond activation, and so no activity is seen over silicalite-1due to the lack thereof. This conclusion holds for a monofunctional, nonmetal-containing catalyst but is not necessarily the case when metals are employed, as will be shown later in this chapter.

This section has outlined the role of a monofunctional catalyst for the direct dehydroaromatization of methane, but because of competing reactions and an

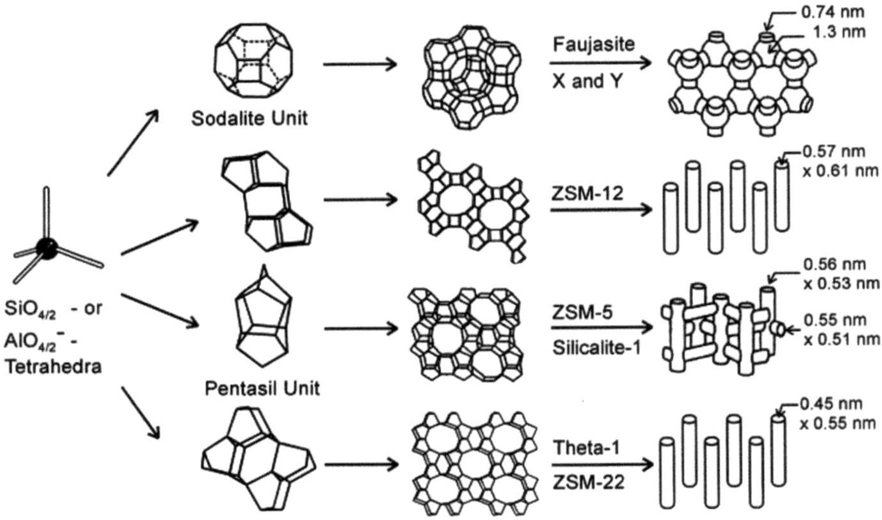

Fig. 3.4 Structures of four zeolites commonly used in the petrochemical industry, including ZSM-5 which is the predominant catalyst used in thermal methane activation. Adapted with permission from (Solid State Ionics 131 (2000) 175–188). Copyright (2000) Elsevier

overwhelming activation barrier, the conversion of methane (1.7% at 700 °C) is not acceptable for commercial application. However, in the same works conducted by Wang et al. [17], it was discovered that metal addition, more notably Mo, could greatly enhance the conversion of methane. The reasons for this are discussed in the next section.

At this point, it is important to note that the role of Brønsted acid sites in the activation of methane has recently been brought into question. In fact, it is now believed that the role of Brønsted acidity is secondary in nature [12] (discussed in the next section). However, the importance of shape selectivity still holds true when conducting MDA reactions to produce benzene.

3.2.2 *The Importance of Acid Sites and Support Topology*

In the works of Wang et al. [17], three types of catalyst were tested: HZSM-5, ZnHZSM-5 (Zn loaded onto HZSM-5), and MoHZSM-5 (Mo loaded onto HZSM-5). While all catalysts exhibited 100% selectivity towards benzene (excluding the contribution of coke as outlined in Sect. 3.2), a twofold increase in methane conversion was observed for the ZnHZSM-5 catalyst (1.4–3%) and a threefold increase for the MoHZSM-5 catalyst (4.4%) at the same acidity ($SiO_2/Al_2O_3 = 25$). Interestingly, the MoHZSM-5 catalyst showed an even more improved methane conversion (7.2%) upon reducing the acidity with an SiO_2/Al_2O_3 equal to 50, a trend not seen for the other catalysts, indicating the significance of acid sites for this reaction, especially when Mo is present. The results from this work led to huge interest in the field of direct methane aromatization, and perhaps surprisingly, MoHZSM-5 acted as a strong base for these studies and remains one of the most active catalysts for the MDA reaction. It is generally accepted that Mo exists as Mo(VI)-oxo species when loaded onto the support and is then reduced or carburized during reaction. Figure 3.5 illustrates the likely structure of these species, the active sites for the MDA reaction.

The MDA reaction was not well understood in 1993 when it was first discovered, but we now know much more about its complexities. This section will demonstrate the most up-to-date works on the MDA reaction and provide a summary of the most current technology. As such, findings between the first discovery of this reaction and the most up-to-date works will be avoided so as not to convolute the subject, and such studies are well documented and readily available [23–25]. It should be noted that exact knowledge of the active sites for the MDA reaction is still lacking. However, there are several important aspects that are known contributors in the bifunctional system: first is the already discussed acid sites and Brønsted acidity.

Although Brønsted acidity may contribute to the MDA reaction to a small degree, the competing hydrogen transfer reaction [26] and experimental data [17, 27] suggest that these sites are only secondary contributors to the reaction. Figure 3.6 shows that at higher SiO_2/Al_2O_3 ratios (lower acidity), the methane conversion and benzene yield decreases, regardless of the Mo loading percentages. Note that 15, 25, and 40 are the indicators of the SiO_2/Al_2O_3 ratios used for these catalysts. While this certainly suggests that a lower acidity hinders methane activation and aromatization, the experimental data alone gives only a fragmented conclusion. Indeed, it is hypothesized that more Brønsted acid sites promote strong anchoring of Mo inside

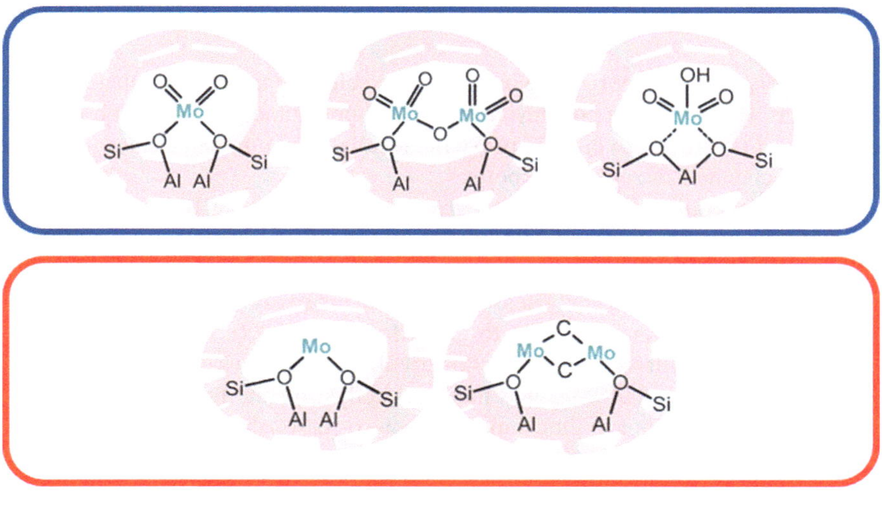

Fig. 3.5 Suggested structures of Mo(VI)-oxo on a silica/alumina support (boxed in blue) and reduced and carburized structures of Mo species on a silica/alumina support (boxed in red). Adapted with permission from (Advanced Materials 32 (2020) 2002565)

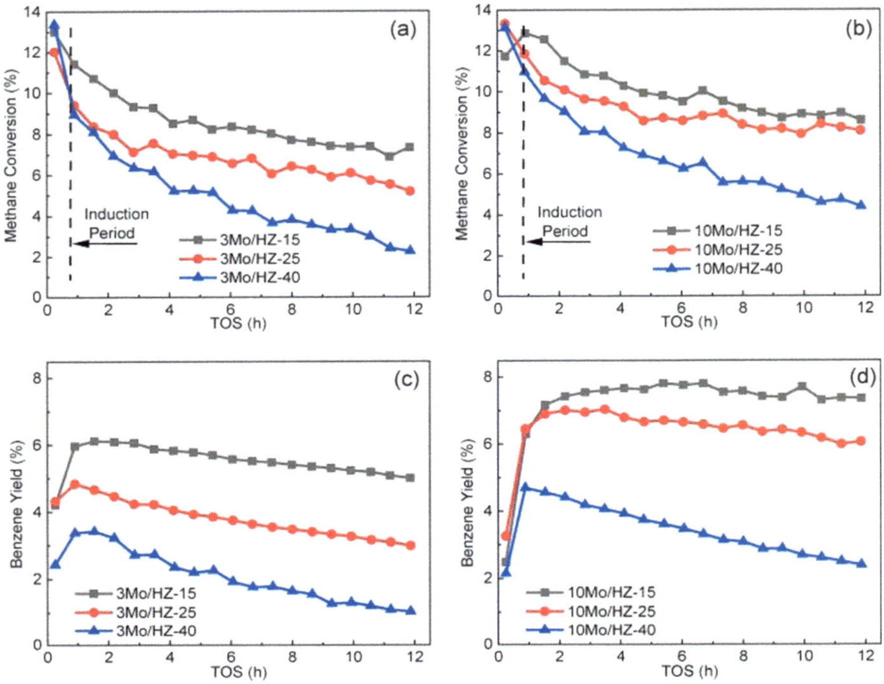

Fig. 3.6 Methane conversion against time on stream (TOS) for (**a**) 3Mo and (**b**) 10Mo catalysts using ZSM-5 with different Si/Al ratios; benzene yield against TOS for (**c**) 3Mo and (**d**) 10Mo catalysts using ZSM-5 with different Si/Al ratios, where 3Mo and 10Mo correspond to 3 wt% and 10 wt% loading of Mo, respectively. Adapted with permission from (Fuel 278 (2020) 118290). Copyright (2020) Elsevier

the channels of the support, allowing the shape selective methane aromatization to take place. However, this conclusion is not full-proof and should not be taken as gospel, as it is difficult to directly prove the localization of metal species inside the pores and channels of zeolitic supports.

Despite this, all experimental data suggests an increase in Brønsted acidity is directly linked to an improved MDA reaction. Indeed, many studies have successfully shown that the MDA reaction can proceed monofunctionally (through active Mo species as opposed to acid sites) [12, 28]. This is easily achievable by the loading of Mo particles on silicalite-1 to give Mo/silicalite-1 (nonacidic). Figure 3.7 demonstrates the hydrocarbon product distribution for a series of catalysts, indicating that although Mo/HZSM-5 was the best catalyst for the formation of benzene from methane, other nonzeolitic supports showed reasonable activity. The activity decreased in the following order: Mo/C < Mo/Al_2O_3 < Mo/SiO_2 < Mo/HZSM-5. This lends yet more credibility to the notion of potential shape selectivity where the channels are necessary to promote benzene formation.

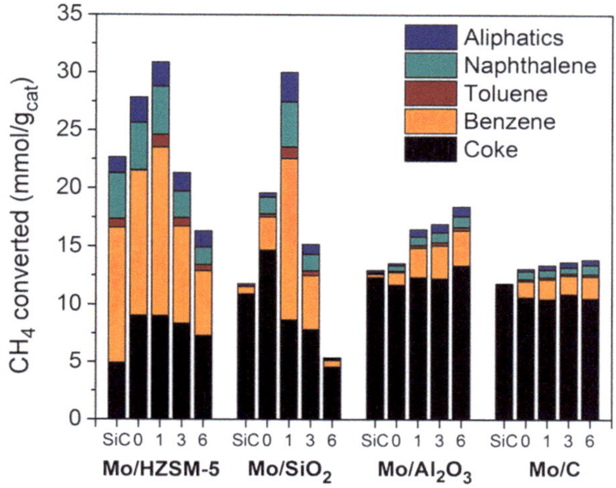

Fig. 3.7 Methane conversion and subsequent product distributions over Mo/HZSM-5, Mo/SiO_2, Mo/Al_2O_3, and Mo/C. Adapted with permission from (ACS Catalysis 7 (2017) 520–529). Copyright (2017) American Chemical Society

The second important aspect of the bifunctional system is the support itself. This has already been alluded to throughout this chapter in the form of shape selectivity and will be better explained at this point. For the case of the MDA reaction, the active sites inside the channels lower the activation energy needed to break the C-H bond (by altering the reaction mechanism) and perform aromatization. However, why is benzene the favored product? One of the reasons is due to steric hinderance inside the channels of the zeolitic supports. ZSM-5's success at performing the MDA reaction and activation of methane to form benzene is thus attributed to its pore diameter of ~5.5 Å [29] (can change due to factors such as calcination, temperature during reaction, and other synthesis processes) close to the diameter of benzene itself (~5.6 Å). This "perfect" pore size is due to the ten-membered ring structures mentioned in the previous section. In fact, there are many catalysts that share this quality, to name a few: MCM-22, MCM-36, TNU-9, ITQ-2, and IM-5.

Naturally, these topologically similar materials are often compared to ZSM-5, and minor improvements have been made to the otherwise optimum Mo/HZSM-5 catalyst. For example, MCM-22 and ZSM-5 loaded with 5% Mo show a reduced methane conversion with respect to Mo/MCM-22 but a lower tendency to form coke, the main deactivation route for these catalysts when undergoing the MDA reaction. Benzene yields are also increased over the MCM-22 support [30]. Figure 3.8 displays this as a function of time on stream, a convention often used in catalysis. Clearly, the rate of deactivation over the ZSM-5 (blue structure in the figure) support is faster than that of MCM-22 (red structure in the figure). The comparison made between Mo_7 (green) and Mo_6 (grey) indicates different precursor salts used when loading Mo to the zeolite supports. The nature of the Mo species will be discussed later in this chapter.

Another study showed that Mo/MCM-22 was also more active and selective than the ITQ-2 support when loaded with Mo because of more naphthalene formation over the latter. This is hypothesized to be because of preferred deposition of

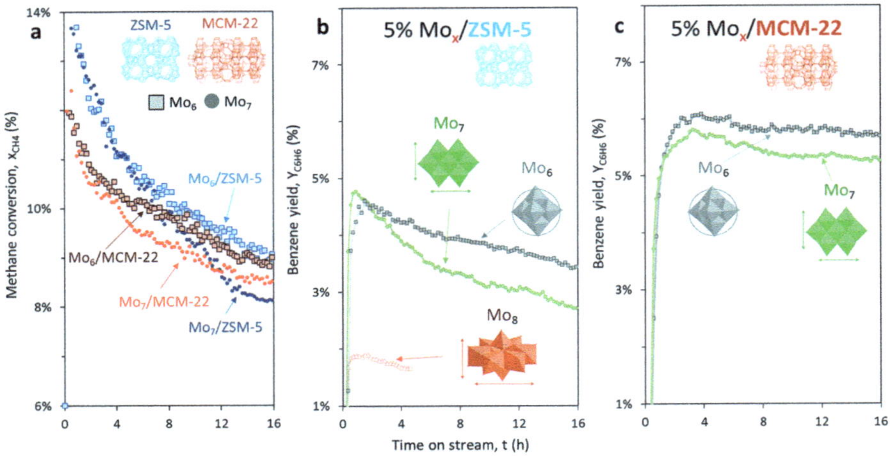

Fig. 3.8 Methane conversion (**a**) and benzene yields for Mo/ZSM-5 (**b**) and Mo/ZSM-5 (**c**) catalysts. Adapted with permission from (Catalysis Science & Technology 9 (2019) 5927–5942)

naphthalene on the external surface (proved by molecular dynamic simulation), and ITQ-2 has a much higher external surface area [31]. In contrast, eight-membered rings (small-pore zeolites) are not as effective for the MDA reaction. Mo/H-SSZ-13, a CHA code zeolite, is made up of eight-membered rings and has been shown to undergo much heavier coking than HZSM-5 when conducting MDA reactions after loading with Mo [32]. Lower benzene formation is also the result of using small-pore zeolites, as discovered in the same study and understandable given the inability of benzene to form/diffuse effectively from these small pores. However, small-pore zeolites do find uses in the petrochemical industry but in olefin selectivity from gaseous feedstocks. While this shows promise for ethylene selectivity from methane, it is outside the scope of the MDA reaction discussed here. There is little, if any, focus on the MDA reaction using large-pore zeolites as such materials are mainly used in the transformation of bulkier species. Besides, the larger pores provide no extra functionality, and so the catalysts are expected to behave no differently to materials such as Mo-loaded Al_2O_3 [33].

3.2.3 Noteworthy Metal Contributions

Thus far, the spotlight has been on molybdenum as the most successful metal to promote the MDA reaction, and while this may be the case, other metals have also seen success for the reaction and can help to understand why Mo is so effective. While it would be impossible to cover all the metals studied for the MDA reaction, some of the more recent metal studies show great promise for the MDA reaction outside of Mo and although poorer performing, help to assess the reason for such positive performances over Mo.

Co can be exchanged onto HZSM-5 using the ion exchange method, where an aqueous solution of a cobalt nitrate salt is mixed with HZSM-5 for a predetermined period, yielding Co/HZSM-5. The resulting catalyst can achieve up to 80% selectivity towards aromatics and higher turnover frequency values than other metal-loaded HZSM-5 catalysts (excluding Mo). The active sites are confirmed by catalyst characterization and indicate Co^{2+} ions anchored to Brønsted acid sites [34]. Figure 3.9 shows the H_2-TPR (temperature-programmed reduction) profiles for the Co/HZSM-5 catalyst prepared by both ion exchange and incipient wetness impregnation (IWI).

The profiles indicate no, or very little, H_2 throughout the broad temperature range for the ion exchange catalyst, indicative of the difficulty to reduce the Co(II) species and their potential stable anchoring at the Brønsted acid site. To support this, UV/Vis spectra obtained by the same group indicated a broad peak between 14,000 and 24,000 cm^{-1}, attributed to Co^{2+} ions, and a band solely observed on the Co/HZSM-5 IWI catalyst between 11,000 and 165,00 wavenumbers [34]. This band is shown to be attributed to the presence of Co^{2+}/Co^{3+} species, a result of the d-d transition band [35]. As these Co^{2+} species on the ion exchanged catalyst are so difficult to reduce, it is understandable that they do not need to be reduced to achieve methane activation, given the successful reaction over pure Co^{2+} species and are present on Al pairs. The reason the Co-loaded HZSM-5 catalyst is unlikely to be used when compared to the superior Mo/HZSM-5 is likely due to the deactivation of these species as they become de-anchored from their Al pairs during reaction. This is shown to occur by using the same H_2 TPR study for spent catalysts, where the ratio of unanchored species to anchored species increased from 7% to 17.4% after only 10 h [34].

Manganese is more than nine times cheaper than cobalt [36] and is a very abundant transition metal, making it a potential suitor for the MDA reaction. Availability and capital cost mean nothing if the metal is not suitable to catalyze the desired reaction. As such, much focus has been spent on testing and elucidating the

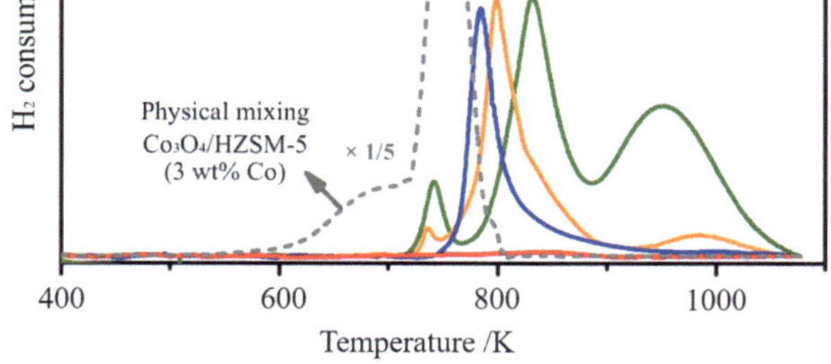

Fig. 3.9 Hydrogen temperature-programmed reduction (TPR) profiles of fresh Co/HZSM-5 catalysts synthesized by IE and IWI. Adapted with permission from (Journal of Catalysis 387 (2020) 102–118). Copyright (2020) Elsevier

contribution of Mn species. It is now generally accepted that a multitude of Mn species exist on the HZSM-5 support after different loading techniques [37, 38]. A more recent study shows that many Mn species exist after the preferred IWI loading and that the nature of Mn-loaded metal particles is dictated by the metal loading weight % [39]. Below 5 wt% Mn loading onto HZSM-5 yields isolated Mn^{2+} ions which are Brønsted acid sites and Mn^{3+} ions which are bound to hydroxyl groups such as Si-OH. Contrastingly, loading above the 5 wt% threshold results in the formation of agglomerated MnO and Mn_2O_3 species. This is confirmed by a strong combination of techniques including X-ray diffraction (XRD), which indicated a reduced crystallinity of the support as Mn loading was increased and NH_3-TPD to confirm the reduction of Brønsted acid sites with increasing Mn loading, indicating the reduction in amount of Mn^{2+} ions. H_2-TPR, X-ray photoelectron spectroscopy (XPS), and diffuse reflectance infrared Fourier transform spectroscopy (DRIFTS) were also employed to help confirm the nature of the Mn species (Fig. 3.10).

The H_2-TPR profiles indicate two types of bands: one broad band at ~390 °C and a sharper band at ~490 °C. The broader band is attributed to Mn^{3+} reduction to form Mn^{2+} and MnO_x species [40], and the sharp band more specifically indicates the reduction of Mn^{3+} present in agglomerated Mn_2O_3 species to Mn^{2+} [37]. The DRIFTS and XPS data assist with the conclusion of the nature of the Mn species on a HZSM-5 support, and the reader is encouraged to review the discussion from the source of this work [39]. The reaction performance data for this system indicated a superior performance on 5 wt% Mn-loaded HZSM-5 with performance being attributed to the critical role played by Mn^{3+} species and the negative influence of Mn agglomerates such as MnO and Mn_2O_3 which block the channel openings of the

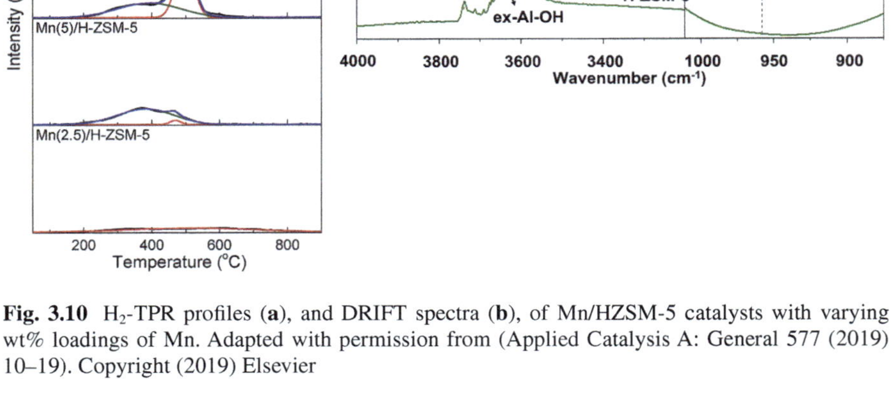

Fig. 3.10 H_2-TPR profiles (**a**), and DRIFT spectra (**b**), of Mn/HZSM-5 catalysts with varying wt% loadings of Mn. Adapted with permission from (Applied Catalysis A: General 577 (2019) 10–19). Copyright (2019) Elsevier

support. We recall that these species were only shown to occur at loadings >5 wt%. The reaction performance data in the form of rate of formation of benzene is given in Fig. 3.11.

Ni has also been investigated with similar findings, where Ni clusters located on the support can cause complete methane decomposition to form coke and deactivate the catalyst and Ni species anchored at the Brønsted acid sites activate methane to give the valuable CH_x species used as precursors to desirable products [41]. In a similar fashion to the Co- and Mn-loaded HZSM-5 catalysts, almost full Brønsted acid site coverage by Ni^{2+} ions which were very hard to reduce can be observed. Conversely, the Ni species loaded by IWI resulted in NiO clusters which tend to form coke.

Fe is another low-cost metal with reasonable activity for the MDA reaction although not as active as the Mo/HZSM-5 system due to high rates of coking. As well as this, the induction time for Fe is much longer than that of the Mo system.

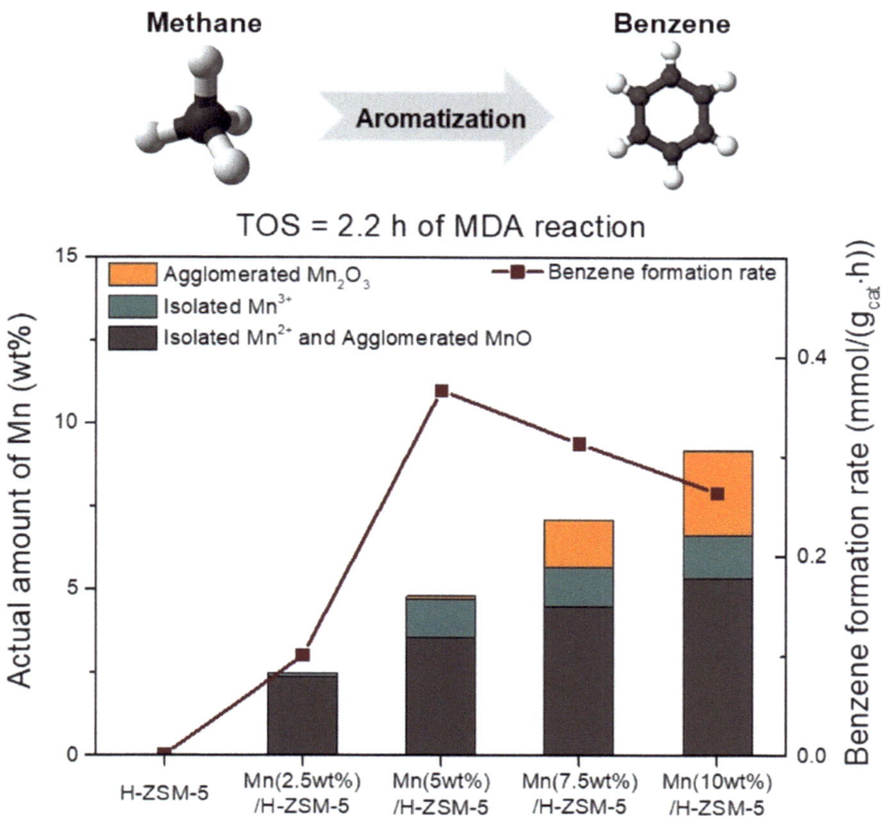

Fig. 3.11 Reaction performance of varying Mn/HZSM-5-loaded catalysts, indicating the amount of different Mn species and the resulting benzene formation rate. Adapted with permission from (Applied Catalysis A: General 577 (2019) 10–19). Copyright (2019) Elsevier

However, unlike the other metals looked at so far, Fe requires a pre-treatment step to carburize; that is to from Fe absorbed with carbon species, making it more active. Indeed, Mo must undergo the same treatment to obtain the best performance and will be looked at in more detail in the next section. For the Fe system, the improved performance is owed to the carburized species' ability to stabilize carbene intermediates through a metal-carbene complex [42]. This action allows the carbene to dimerize to from ethylene, which as we know, can oligomerize to aromatics on the support's acid sites.

3.2.4 On the Nature of Mo Metal Center Contributions

As the most promising catalyst in the MDA reaction, it is prudent to discuss the properties of the Mo/HZSM-5 catalyst in more detail, with a focus on the nature of the Mo species themselves. Using the incipient wetness impregnation technique, the resulting MoO_x species are hosted on external silanol and Brønsted acid sites in the framework, Si-OH and Al-OH^+-Si, respectively. It has been concluded by in situ UV/Vis DRS (diffuse reflection spectroscopy) spectra that the E_g value for a representative 2 wt% Mo loaded on HZSM-5 is 4.8 eV, indicative of isolated MoO_x structures consisting of a single Mo atom. Thus, no Mo chains, bridges, or clusters are formed [43]. Furthermore, in situ Raman spectroscopy, conducted by the same authors on Mo-loaded ZSM-5 with varying Al/Si ratios and varying Mo loading wt%, demonstrates the predominant presence of MoO_x species on framework Brønsted acid sites, with higher Mo loadings showing presence on external silica locations. The significance of this is that users may acknowledge the need for framework Bronsted acid sites for effective Mo loading by impregnation techniques, with external silanol sites only contributing in a secondary nature or when there is no Brønsted acidity left to accommodate the higher Mo loadings.

Mo, like Fe briefly mentioned before, requires specialized treatment before it can be at its most effective for the MDA reaction. It requires carburization. This carburization, or induction period, is the stage of the reaction where Mo is reduced to give its active phase, often simply achieved by flowing methane over the catalyst under reaction conditions. There are several forms of the carburized Mo species that are believed to be active for the MDA reaction, and these include Mo_2C, the same modified by coke, and MoC. As mentioned, these are currently formed through an induction period under reaction conditions with methane, prolonging the reaction time. Indeed, available in situ techniques at the reaction temperatures required are few and far between, and so identifying these carbide Mo species has proven difficult to the field and often only involved theoretical studies [44]. In addition to this, coke formation during the MDA reaction surrounding active centers, and the inability to predict the phase change between induction period and reaction period, makes the analysis of the Mo carbides extremely difficult. To circumvent this problem, Vollmer et al. utilized the knowledge that CO can also be used to effectively form MoC from

MoO_x species [45]. This allows the likely Mo carbides to form but without inducing coking and reaction products that would be seen with CH_4 presence.

To investigate the induction period, active carbide-phase Mo species can be confirmed by CO and methane pulsing comparisons. Pulsing experiments reveal that without CO carburization, benzene is only formed after at least five volumetric pulses of methane; however, with the CO carburization step (2.5% CO in He at 350 °C), benzene formation can be seen immediately [46]. The confirmation of the requirement of a carbidic species such as CO is then further consolidated by attempting to reduce the Mo/HZSM-5 catalyst with hydrogen. The results confirm that a non-carbidic-reducing atmosphere is not enough to remove the induction period due to a delay in the formation of benzene in the same way as a methane-only pulse test. X-ray absorption near-edge spectroscopy (XANES) is a powerful technique, enabling the user to observe changes in the local geometry and the very nature of the species of interest. The same authors were able to conduct operando XANES studies on the untreated and CO-treated catalysts, whose spectra were near enough identical (shown in Fig. 3.12, indicating that the Mo species undergo the same changes regardless of CO treatment). The operando XANES study is also able to show an indicative pre-edge peak shift, which corresponds to gradually carburized Mo after the fifth methane pulse [46]. This technique provides clear evidence that Mo is in a similar chemical environment regardless of CO pre-treatment and supports the theory that CO pre-treatment is only effective for bypassing the induction step and does not affect the ability of MoC species to participate in the MDA reaction.

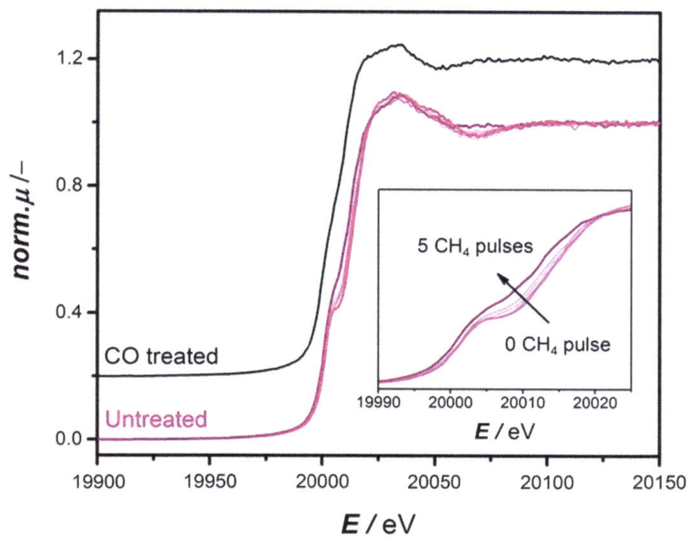

Fig. 3.12 Operando XANES spectra at the Mo K-edge of a Mo/HZ-13 catalyst during methane pulsing at 700 °C and a CO-carburized version of the same. Adapted with permission from Chemical Science, RSC. "On the Dynamic Nature of Mo Sites for Methane Dehydroaromatization"

Isotopic studies in the form of ^{13}C magic-angle spinning nuclear magnetic resonance (MAS NMR) can provide even more information as to the nature of the MoC species. Indeed, there are several major species found on Mo/HZ-13 catalysts: β-Mo_2C (hexagonal close packed) [47] and varying oxycarbidic structures [46]. It has already been mentioned that the advantage of pre-treating with CO enables the observation of the catalyst without inducing aromatization and coking on the catalyst. Thus, a unique opportunity is afforded to observe the interaction of methane with the elucidated Mo carbide active sites in the absence of species that may alter such interactions. Vollmer et al. used $^{13}CH_4$ pulsing experiments on a pre-carburized catalyst using ^{12}CO, enabling them to track ^{12}C from the Mo carbide sites into the products [46]. The conclusion from works such as this yielded the following: when using ^{12}C for carburization and ^{13}C for methane pulsing (Fig. 3.13), the first pulse sees that the 83/84 ratio reaches a value of 0.67 and then decreases to 0.28 over the subsequent eight pulses of methane. The higher value of 83/84 during the first few pulses can confidently be attributed to the $^{12}C^{13}C_5H_6$ presence. The increased abundance of masses 82 to 78 then further supports the incorporation of ^{12}C into benzene [46].

Generally, the molybdenum oxide becomes a molybdenum carbide through the following generalized route: $MoO_3 \rightarrow MoO_xC \rightarrow Mo_2C$. Given what has been learnt so far in this chapter, and the numerous works conducted over the past 3 decades, it is possible to elucidate some mechanisms for the MDA reaction. And this is given in the next section.

3.2.5 *Suggested MDA Reaction Mechanisms*

While it is generally accepted that the MDA reaction system is made up of three major stages: induction, dehydroaromatization, and deactivation; There are only two main reaction mechanisms that have been seriously considered by the field. These two

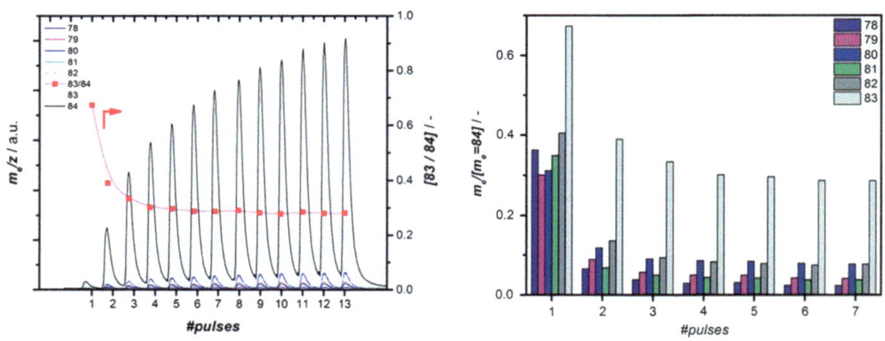

Fig. 3.13 Mass/charge ratio evolution for fragmentation of benzene with pulsing of $^{13}CH_4$ and the area under each pulse for 78–83 mass/charge ratios. Adapted with permission from Chemical Science, RSC. "On the Dynamic Nature of Mo Sites for Methane Dehydroaromatization"

mechanisms are shown in Fig. 3.14 as the bifunctional mechanism and hydrocarbon pool mechanism. The bifunctional mechanism suggests C-H bond activation and C-C coupling over the Mo carbide sites, where C_2 intermediates then occur on the Brønsted acid sites, although as mentioned earlier in the chapter, the role of Brønsted acid sites has more recently been suggested as secondary and not directly related to aromatization. It is still debated as to whether ethylene or acetylene is the key intermediate for this reaction [28]. The hydrocarbon pool mechanism is perhaps more suited to the notion that Brønsted acid sites play a secondary role in directing Mo species to the preferred location inside the pores for shape selectivity towards benzene. In this mechanism, C-H bonds are still activated on Mo carbides, but the hydrocarbon intermediates, confined in the zeolite pores, are transformed to aromatic products.

In summary, the MDA reaction is hindered by thermodynamic constraints which have been partially overcome by the synthesis of Mo carbides loaded onto shape selective solid acid supports (predominantly ZSM-5). Although great strides have been made as to the elucidation of the relevant mechanisms, more understanding is needed before this technology can be commercially viable.

3.3 OCM Reaction

3.3.1 Catalyst Properties for OCM

The oxidative coupling of methane does not possess thermodynamic limitations, but poor selectivity is the product of such a kinetically unfavorable reaction in the presence of oxygen. This results from the preferred oxidation of products to give

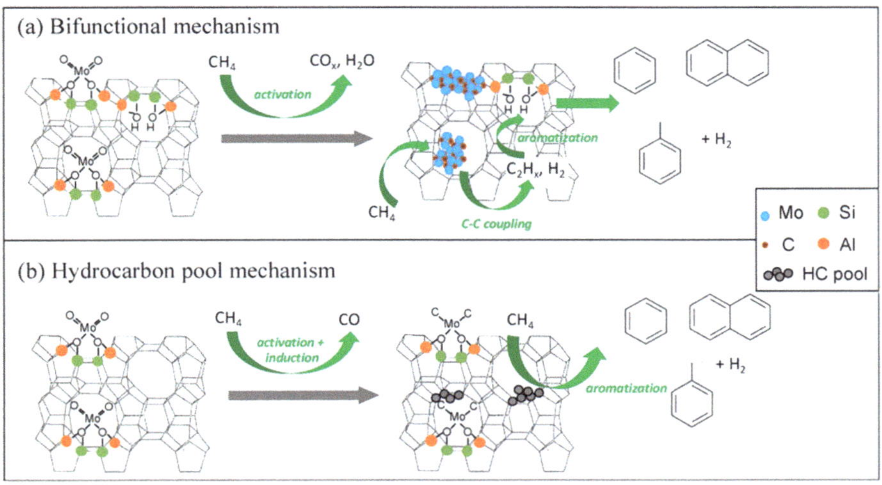

Fig. 3.14 The two proposed mechanisms for the MDA reaction: Bifunctional mechanism (**a**), and Hydrocarbon pool mechanism (**b**). Adapted with permission from (Applied Catalysis A: General 608 (2020) 117870). Copyright (2020) Elsevier

components such as CO_2. Ethylene is the most desirable product from this reaction due to its reasonable rate of formation and because of ethylene's use in the synthesis of many petrochemical products [48]. Oxide catalysts are typically used, such as groups I and II and transition metal oxides. And it is the reactive oxygen species on the surface of these catalysts that are heavily involved in the activation of the C-H bond. Understandably, the varying materials that give rise to these oxygen species has resulted in much dispute over the exact nature of the active sites themselves. The major catalysts used in the OCM reaction are more specifically transition metal oxides, lanthanide oxides, or group I and group II oxide catalysts.

Perhaps the most successful catalysts for the OCM reaction are the transition metal oxide catalysts which come with poor selectivity. Thus, they require modifications to improve their properties. Such improvements have yielded the M-W-Mn/SiO_2 catalysts as the most promising for the OCM reaction (M = Na or K). It is this catalyst that will be discussed in detail in this chapter.

Li et al. first demonstrated the Na_2WO_4-Mn/SiO_2 catalyst in 1992, and it has since been dedicated to in many studies. The results for methane conversion immediately show the ease to which methane is converted, with conversions reaching nearly 40% and reasonable selectivities of ~70% towards C_2 products (ethane and ethylene) [49]. When synthesizing this material, one may utilize various techniques that are shown in Table 3.1, along with the identified bulk phases and typical precursors used to achieve the synthesis.

IWI is most frequently used, owing to its simplicity and its less economically demanding nature. When confirming the identity of these materials, it has recently become possible to do so under reaction conditions as opposed to pre and post. In situ Raman spectra of $Mn_xO_y/Na_2WO_4/SiO_2$ were obtained and were able to show extensive surface mobility at reaction conditions (>650 °C) [50]. Figure 3.15a shows in situ Raman spectra conducted by the same group on the $Mn_xO_y/Na_2WO_4/SiO_2$ catalyst which clearly shows the restructuring of supported Na_2WO_4 (orange band) at temperatures above 600 °C. This is assisted by the melting of this phase at even higher temperatures and thus, enables the following reaction: $Mn^{2+}Mn^{3+}{}_6SiO_{12} + 7Na_2WO_4 \leftrightarrow 7MnWO_4 + 7Na_2O + SiO_2 + 1.5O_2$. Referring again to the Raman spectra, we can see a reduction of Mn^{3+} (Mn_7SiO_{12}) to Mn^{2+} ($MnWO_4$), followed by molecular oxygen release (points 3 and 4 on the TGA-DTA-MS spectra (Figure 3.15b). It is this process that likely provides mobile oxygen from the lattice when the system is under steady-state operation.

Despite the ease of synthesis for a IWI catalyst, the POSS technique is capable of producing smaller crystal sizes of Mn_2O_3 (<20 nm) as opposed to the larger crystal sizes produced by IWI (40 nm) [51]. It was also found that the sol-gel synthesis technique yielded a more homogeneous active structure distribution [52] as shown by Fig. 3.16 which indicates a very homogenous distribution of Na- and Mn-containing phases when energy-dispersive X-ray (EDX) spectrometry is employed.

However, despite the merits of these alternative synthesis methods, long-term stability studies are lacking. This is not the case for IWI-prepared catalysts where most of the research has been focused.

Table 3.1 Synthesis methods and the corresponding crystalline bulk phases identified with XRD for the supported $Mn_xO_y/Na_2WO_4/SiO_2$ catalysts. Adapted with permission from (ACS Catalysis 9, 7 (2019) 5912–5928). Copyright (2019) ACS Publications

Technique	Precursors and solvents	Major bulk phases
Incipient wetness impregnation (IWI)	$Mn(NO_3)_2$ and Na_2WO_4 in H_2O	Mn_2O_3, Na_2WO_4, WO_3, $MnWO_4$, SiO_2 (α-cristobalite)
Mixed slurry (MS)	$Mn(NO_3)_2$ and Na_2WO_4 in H_2O	Mn_2O_3, Na_2WO_4, WO_3, $MnWO_4$, SiO_2 (α-cristobalite)
Sol-gel (SG)	$(C_2H_5)_4SiO_4$ in C_2H_5OH and HNO_3	Mn_2O_3, Na_2WO_4, $MnMn_6SiO_{12}$, SiO_2 (α-cristobalite), SiO_2 (tridymite), SiO_2 (quartz)
Physical mixing (PM)	$Mn(NO_3)_2$ and Na_2WO_4	Na_2WO_4, Mn_2O_3, SiO_2 (α-cristobalite)
Polyhedral oligomeric silsesquioxanes (POSS)	$(C_4H_9)_7Si_7O_9(OH)O_2Mn$ and $(C_4H_9)_7Si_7O_{12}W$ and $(C_4H_9)_7Si_7O_9(OH)_2ONa)$ in hexane-toluene	Mn_2O_3, Na_2WO_4, SiO_2 (α-cristobalite and tridymite)
Fluidized bed (FB)	$Mn(NO_3)_2{\cdot}4H_2O$ and of $Na_2WO_4{\cdot}2H_2O$ in H_2O	Mn_2O_3, Na_2WO_4, SiO_2 (α-cristobalite), $MnWO_4$ (only in spent catalyst)
Solution combustion (SC)	$Mn(NO_3)_2$ and Na_2WO_4 in $C_8H_{20}O_4Si$ was the precursor for Si, as well as fuel	Na_2WO_4, Mn_2O_3, SiO_2 (α-cristobalite)
Flame spray pyrolysis (FS)	W-ethoxide, Mn-2-ethylhexanoate, Na-2-ethylhexanoate and hexamethyldisiloxane in xylene	Na_2WO_4, SiO_2 (α-cristobalite), $MnWO_4$ (only in Na-free catalyst) Mn_2O_3 (only in spent catalyst)

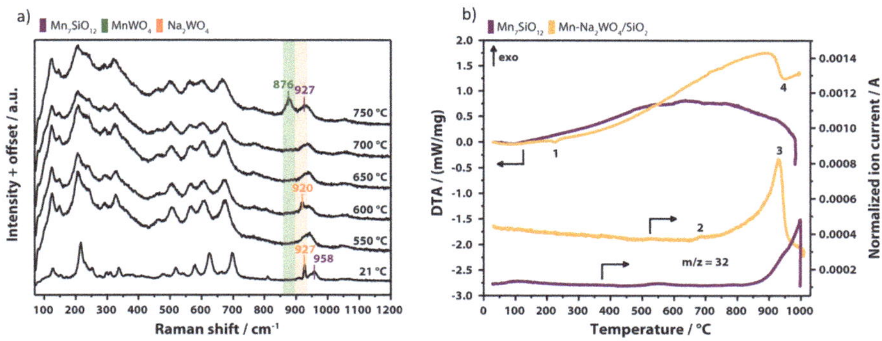

Fig. 3.15 In situ Raman spectra of a $Mn_xO_y/Na_2WO_4/SiO_2$ catalyst (**a**) and TGA-DTA-MS of the same and a reference sample of Mn_7SiO_{12} (**b**). Adapted with permission from (Angewandte Chemie 132, 35 (2020) 15031–15036). Copyright (2020) Wiley-VCH

As the OCM reaction is conducted at high temperatures (800–900 °C), it is difficult to analyze the actual structures and nature of active sites during reaction. However, some active site roles have been speculated and will be summarized here. Wang et al. acknowledged early on that Na-O-Mn species are the direct active site by comparing $Mn/Na_2WO_4/SiO_2$, $MnNa_2WO_4/MgO$, and $NaMnO_4/MgO$ catalysts

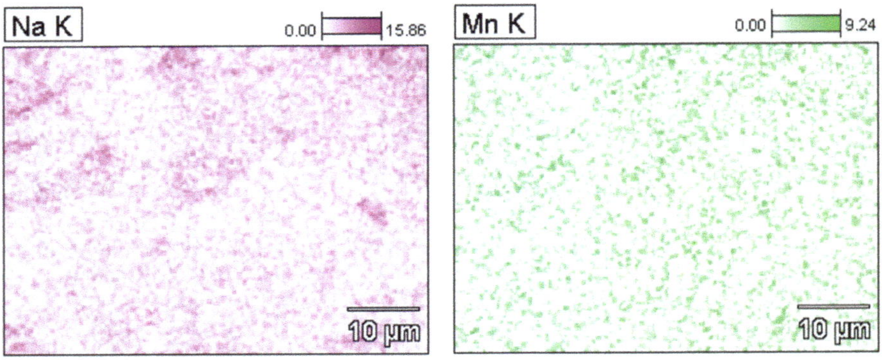

Fig. 3.16 EDX analysis of the sol-gel synthesized $Mn_xO_y/Na_2WO_4/SiO_2$ showing Mn and Na dispersion homogeneity. Adapted with permission from (Catalysis Today 236 (2014) 12–22). Copyright (2014) Elsevier

and realizing that similar catalytic behaviors resulted from duplicate active sites [53]. They were also able to observe through CH_4 pulse reaction sequences that these active species are unstable under high-temperature reaction conditions without gas-phase O_2 presence. Around the same time, a study was made into the role of distorted WO_4 species within the same catalyst. Using Raman spectroscopy, the authors were able to identify the role of Na^+ ions in creating a distorted WO_4 tetrahedron which increases C_2 hydrocarbon formation [54]. Seven years later, a study revealed that Na assists with transporting Mn to the surface of the catalyst, an important aspect given the surface reliant nature of heterogenous catalytic activity [55]. It is also suggested in the same work that Na, W, and Mn have a synergistic effect and that the $Na_2W_2O_7$ crystal phase can be active for the OCM reaction and that Na and O mostly occupy the catalyst surface, with Mn and W species occupying a secondary sub-layer. Studies showing more of the same have since been conducted, but a more recent study demonstrates that TiO_2 doping is capable of achieving similar methane conversions (26%) and reasonable C_2 hydrocarbon selectivities (76%) at a lower temperature (720 °C) than the standard 800–900 °C. A significant improvement on the excessive energy requirements that would be needed to conduct this reaction on a commercial scale. This is attributed to the ability of the $MnTiO_3$ species to trigger the $Mn^{2+} \leftrightarrow M\ n^{3+}$ cycle which is pivotal in activating O_2 and simultaneously converting methane to C_2 and C_3 hydrocarbons [56]. Nonetheless, the exact nature of this catalyst is not fully understood, although progress is being made and the same applies to the catalytic mechanisms, which will be discussed in the next section.

3.3.2 *Mechanistic Pathways for the OCM Reaction*

There are two mechanisms of note when it comes to the OCM reaction. The first, which was initially proposed by Professor Jack H. Lunsford is shown in Fig. 3.17. This mechanism suggests molecular oxygen adsorption (dissociative) at the outer catalyst surface Na sites to give surface oxygen atoms which are responsible for cleaving the C-H bond in methane, forming methyl radicals (CH_3•). The confirmation of these species was achieved by matrix isolation electron spin resonance experiments by various authors where the combination of these radicals was also shown to form ethane [57].

The second mechanism was suggested around the same time as Lunsford et al. but instead, by the original discoverers of the catalyst itself, Li et al. [54]. This mechanism is widely cited and has received more recent support through further studies [58]. Depicted here in Fig. 3.18, we can see a synergetic effect is proposed, where the methane molecule is activated by the W^{6+} species arising from WO_4 tetrahedrons, which provide lattice surface oxygens. This generates methyl radicals, similar to the first mechanism shown, leading to the formation of W^{5+} species, electron transfer to Mn^{3+} species which are also present, leading to eventual regeneration of the W^{6+} sites and the reducing cycle discussed earlier where Mn^{2+} is formed. Recall that the Mn^{3+} is regenerated after the Mn^{2+} species contribute to the activation

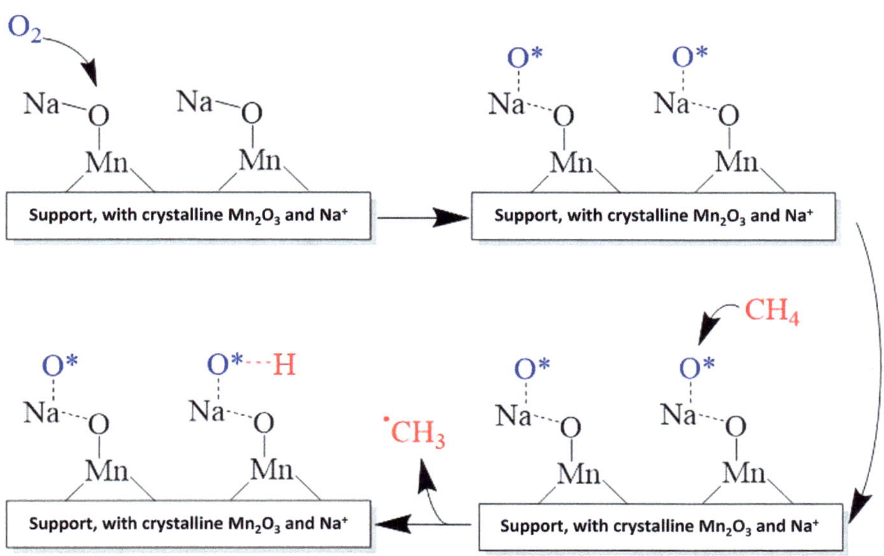

Fig. 3.17 OCM reaction mechanism as proposed by Lunsford over a $Mn/Na_2WO_4/SiO_2$ catalyst. Adapted with permission from (ACS Catalysis 9, 7 (2019) 5912–5928). Copyright (2019) ACS Publications

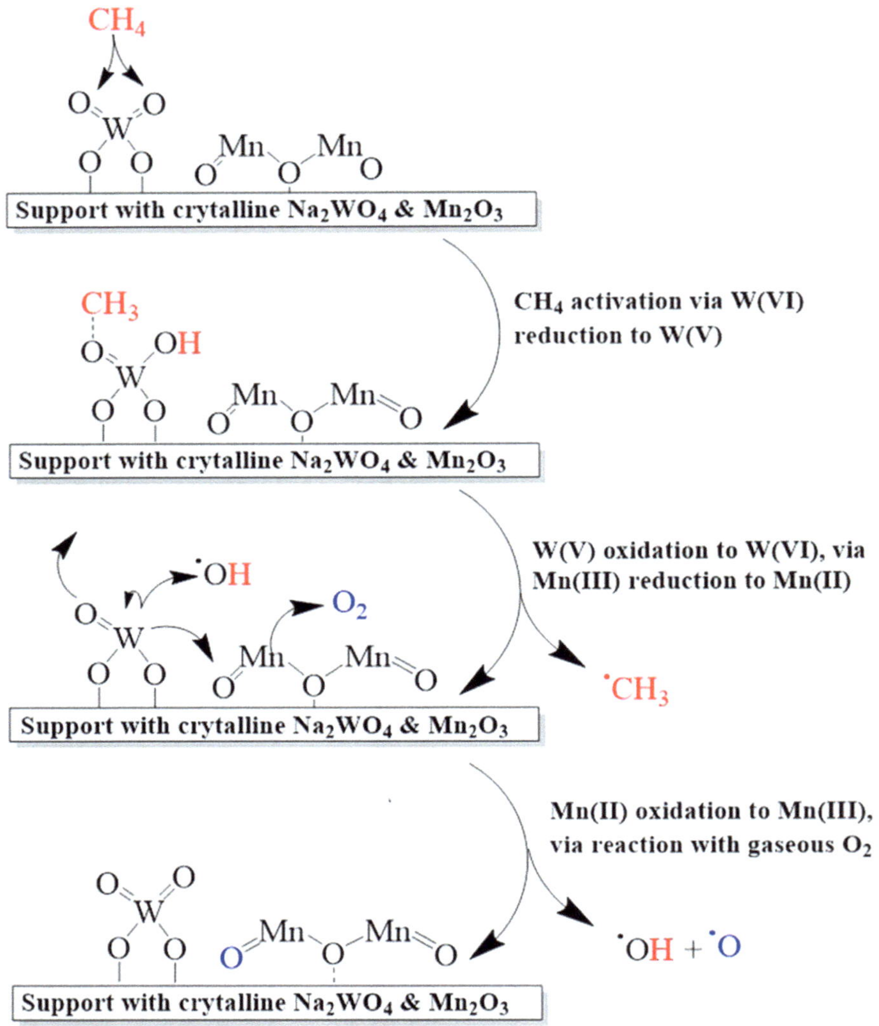

Fig. 3.18 OCM reaction mechanism proposed by Li et al. over a $Mn/Na_2WO_4/SiO_2$ catalyst. Adapted with permission from (ACS Catalysis 9, 7 (2019) 5912–5928). Copyright (2019) ACS Publications

of molecular oxygens in the gas phase. Despite this, the second mechanism suggested lacks precise and direct evidence for the catalytically active surface sites and their respective structures and so requires further work before being taken as gospel [59].

The reader will note the heterogeneous nature of these mechanisms, but it should be taken into account that at elevated temperatures (>600 °C), the mechanism is also occurring homogeneously in the gas phase. This understandable occurrence means that methane conversions can more easily go towards gas-phase products but with less selectivity in the presence of a catalyst. These gas-phase reactions can even go as far as to replicate steam reforming methane, yielding CO_2 and even CO. Moving back towards the heterogenous aspects of the OCM reaction, another recent study elucidated that there are two types of lattice oxygen species present on the solid surface of the catalyst, each providing different product species [60]. These findings indicate both strong and weakly adsorbed lattice oxygen species that are responsible for full oxidation (CO_2) and the more desirable formation of C_2 hydrocarbons, respectively.

3.4 Summary

To summarize this chapter, the MDA and OCM reactions provide not only valuable solutions to the utilization of natural gas (predominantly methane) but also more understanding, and the support to scale such processes up is still required before their commercialization is possible. The nonoxidative MDA reaction shows great promise for the very selective production of valuable chemicals without the hinderance of unselective and undesirable products. This reaction is also much easier to control than its oxidative counterpart. However, suitable methane conversion remains its "Achilles' heel," and until significant improvements are made, commercialization of this technology remains to be seen. Conversely, the OCM system provides users with a suitable level of methane conversion, but a lack of control means more understanding is needed and perhaps lower temperature solutions to avoid the more unselective gas-phase reactions. Many publications over the past few decades are testament to the potential of these technologies, and indeed, great strides have been made as a result of these efforts. The continued drive to understand and utilize natural gas is ongoing, and more readily available natural gas through sources such as shale means that improvement of these technologies is not only desirable but necessary.

Chapter 4 will take us through an alternative and more feasible way to utilize methane in the petrochemical industry, the cross-coupling of methane with nonoxidative hydrocarbons.

References

1. R.A. Kerr, Natural gas from shale bursts onto the scene. Science (80-.) **328**(5986), 1624–1626 (2010). https://doi.org/10.1126/science.328.5986.1624
2. S.H. Mohr, J. Wang, G. Ellem, J. Ward, D. Giurco, Projection of world fossil fuels by country. Fuel **141**, 120–135 (2015). https://doi.org/10.1016/j.fuel.2014.10.030
3. M.E. Dry, The Fischer-Tropsch process: 1950-2000. Catal. Today **71**(3–4), 227–241 (2002). https://doi.org/10.1016/S0920-5861(01)00453-9

4. D.J. Wilhelm, D.R. Simbeck, A.D. Karp, R.L. Dickenson, Syngas production for gas-to-liquids applications: Technologies, issues and outlook. Fuel Process. Technol. **71**(1–3), 139–148 (2001). https://doi.org/10.1016/S0378-3820(01)00140-0
5. J.R. Anderson, M. Boudart, Catalytic Steam Reforming, in *Catalysis Science & Technology*, (Springer, Berlin, 1984), pp. 1–117
6. Y. Xu, X. Bao, L. Lin, Direct conversion of methane under nonoxidative conditions. J. Catal. **216**(1–2), 386–395 (2003). https://doi.org/10.1016/S0021-9517(02)00124-0
7. N. Kosinov, E.A. Uslamin, L. Meng, A. Parastaev, Y. Liu, E.J.M. Hensen, Reversible nature of coke formation on Mo/ZSM-5 methane dehydroaromatization catalysts. Angew. Chem. Int. Ed. **58**(21), 7068–7072 (2019). https://doi.org/10.1002/anie.201902730
8. V.R. Choudhary, S.A.R. Mulla, V.H. Rane, Surface basicity and acidity of alkaline earth-promoted La2O3 catalysts and their performance in oxidative coupling of methane. J. Chem. Technol. Biotechnol. **72**(2), 125–130 (1998). https://doi.org/10.1002/(SICI)1097-4660(199806)72:2<125::AID-JCTB880>3.0.CO;2-3
9. A.E. Shilov, G.B. Shul'pin, Activation and catalytic reactions of alkanes in solutions of metal complexes. Russ. Chem. Rev. **56**(5), 442–464 (1987). https://doi.org/10.1070/rc1987v056n05abeh003282
10. C.A. Haynes, R. Gonzalez, Rethinking biological activation of methane and conversion to liquid fuels. Nat. Chem. Biol. **10**(5), 331–339 (2014). https://doi.org/10.1038/nchembio.1509
11. C.H. Yeh, S.S.F. Yu, S.I. Chan, J.C. Jiang, Quantum chemical studies of methane oxidation to methanol on a biomimetic Tricopper complex: Mechanistic insights. ChemistrySelect **3**(18), 5113–5122 (2018). https://doi.org/10.1002/slct.201800550
12. N. Kosinov, F.J.A.G. Coumans, E.A. Uslamin, A.S.G. Wijpkema, B. Mezari, E.J.M. Hensen, Methane dehydroaromatization by Mo/HZSM-5: Mono- or bifunctional catalysis? ACS Catal. **7**(1), 520–529 (2017). https://doi.org/10.1021/acscatal.6b02497
13. W. Vermeiren, J.P. Gilson, Impact of zeolites on the petroleum and petrochemical industry. Top. Catal. **52**(9), 1131–1161 (2009). https://doi.org/10.1007/s11244-009-9271-8
14. T. Degnan Jr., Applications of zeolites in petroleum refining. *Top. Catal.* **13**, 349–356 (2000). https://doi.org/10.1023/A%3A1009054905137
15. P. Nachtigall, J. Cejka, R.E. Morris, *Zeolites in Catalysis, Thomas Graham House in Cambridge*, (RSC, 2017)
16. R. J. Argauer, "Crystalline Zeolite ZSM-5 and Method of Preparing the Same," 1972
17. L. Wang, L. Tao, M. Xie, G. Xu, J. Huang, Y. Xu, Dehydrogenation and aromatization of methane under non-oxidizing conditions. Catal. Lett. **21**(1–2), 35–41 (1993). https://doi.org/10.1007/BF00767368
18. V. Agarwal, H. Metiu, Hydrogen abstraction energies and ammonia binding to BEA, ZSM-5, and α-quartz doped with Al, Sc, B, or Ga. J. Phys. Chem. C **119**(28), 16106–16114 (2015). https://doi.org/10.1021/acs.jpcc.5b04171
19. X. Zheng, P. Blowers, A computational study of methane catalytic reactions on zeolites. J. Mol. Catal. A Chem. **246**(1–2), 1–10 (2006). https://doi.org/10.1016/j.molcata.2005.10.009
20. S. Wu, X. Tan, J. Lei, H. Chen, L. Wang, J. Zhang, Ga-doped and Pt-loaded porous TiO2-SiO2 for photocatalytic nonoxidative coupling of methane. J. Am. Chem. Soc. **141**(16), 6592–6600 (2019). https://doi.org/10.1021/jacs.8b13858
21. I. Vollmer, E. Abou-Hamad, J. Gascon, F. Kapteijn, Aromatization of ethylene – Main intermediate for MDA? ChemCatChem **12**(2), 544–549 (2020). https://doi.org/10.1002/cctc.201901655
22. J.S. Jarvis et al., Inhibiting the Dealkylation of basic Arenes during n-alkane direct aromatization reactions and understanding the C6Ring closure mechanism. ACS Catal. **10**(15), 8428–8443 (2020). https://doi.org/10.1021/acscatal.0c02361
23. S. Ma, X. Guo, L. Zhao, S. Scott, X. Bao, Recent progress in methane dehydroaromatization: From laboratory curiosities to promising technology. J. Energ. Chem. **22**(1), 1–20 (2013). https://doi.org/10.1016/S2095-4956(13)60001-7

24. Y. Xu, L. Lin, Recent advances in methane dehydro-aromatization over transition metal ion-modified zeolite catalysts under non-oxidative conditions. Appl. Catal. A Gen. **188**(1–2), 53–67 (1999). https://doi.org/10.1016/S0926-860X(99)00210-0
25. S. Majhi, P. Mohanty, H. Wang, K.K. Pant, Direct conversion of natural gas to higher hydrocarbons: A review. J. Energ. Chem. **22**(4), 543–554 (2013). https://doi.org/10.1016/S2095-4956(13)60071-6
26. W. Ding, S. Li, G.D. Meitzner, E. Iglesia, Methane conversion to aromatics on Mo/H-ZSM5: Structure of molybdenum species in working catalysts. J. Phys. Chem. B **105**(2), 506–513 (2001). https://doi.org/10.1021/jp0030692
27. M. Rahman, A. Infantes-Molina, A.S. Hoffman, S.R. Bare, K.L. Emerson, S.J. Khatib, Effect of Si/Al ratio of ZSM-5 support on structure and activity of Mo species in methane dehydroaromatization. *Fuel* **278** (2020). https://doi.org/10.1016/j.fuel.2020.118290
28. N.K. Razdan, A. Kumar, B.L. Foley, A. Bhan, Influence of ethylene and acetylene on the rate and reversibility of methane dehydroaromatization on Mo/H-ZSM-5 catalysts. J. Catal. **381**, 261–270 (2020). https://doi.org/10.1016/j.jcat.2019.11.004
29. W. Widayat, A.N. Annisa, Synthesis and characterization of ZSM-5 catalyst at different temperatures. IOP Conf. Ser. Mater. Sci. Eng. **214**(1), 1–8 (2017). https://doi.org/10.1088/1757-899X/214/1/012032
30. I. Julian et al., Polyoxometalates as alternative Mo precursors for methane dehydroaromatization on Mo/ZSM-5 and Mo/MCM-22 catalysts. Cat. Sci. Technol. **9**(21), 5927–5942 (2019). https://doi.org/10.1039/c9cy01490j
31. A. Martínez, E. Peris, G. Sastre, Dehydroaromatization of methane under non-oxidative conditions over bifunctional Mo/ITQ-2 catalysts. Catal. Today **107–108**, 676–684 (2005). https://doi.org/10.1016/j.cattod.2005.07.051
32. P. Collier et al., Understanding the deactivation phenomena of small-pore mo/h-ssz-13 during methane dehydroaromatisation. Molecules, 1–15 (2020)
33. N. Kosinov et al., Stable Mo/HZSM-5 methane dehydroaromatization catalysts optimized for high-temperature calcination-regeneration. J. Catal. **346**, 125–133 (2017). https://doi.org/10.1016/j.jcat.2016.12.006
34. Y. Xu, M. Chen, T. Wang, B. Liu, F. Jiang, X. Liu, Probing cobalt localization on HZSM-5 for efficient methane dehydroaromatization catalysts. J. Catal. **387**, 102–118 (2020). https://doi.org/10.1016/j.jcat.2020.04.021
35. Kuterasiński et al., Spectroscopic studies of MFI and USY zeolite layers over stainless steel 316L wire gauze meshes. *Spectrochim. Acta Part A Mol. Biomol. Spectrosc.* **230** (2020). https://doi.org/10.1016/j.saa.2020.118060
36. T. Zell, R. Langer, Iron-catalyzed hydrogenation and dehydrogenation reactions with relevance to reversible hydrogen storage applications. Recycl. Catal. **2**(1), 87–109 (2016). https://doi.org/10.1515/recat-2015-0010
37. G. Lv, F. Bin, C. Song, K. Wang, J. Song, Promoting effect of zirconium doping on Mn/ZSM-5 for the selective catalytic reduction of NO with NH3. Fuel **107**(x), 217–224 (2013). https://doi.org/10.1016/j.fuel.2013.01.050
38. A. De Lucas, J.L. Valverde, F. Dorado, A. Romero, I. Asencio, Influence of the ion exchanged metal (Cu, Co, Ni and Mn) on the selective catalytic reduction of NO X over mordenite and ZSM-5. J. Mol. Catal. A Chem. **225**(1), 47–58 (2005). https://doi.org/10.1016/j.molcata.2004.08.036
39. T.H. Lim, D.H. Kim, Characteristics of Mn/H-ZSM-5 catalysts for methane dehydroaromatization. *Appl. Catal. A Gen.* **577**, 10–19 (2019). https://doi.org/10.1016/j.apcata.2019.03.008
40. Q. Sun, W.M.H. Sachtler, Mn/MFI catalyzed reduction of NOx with alkanes. Appl. Catal. B Environ. **42**(4), 393–401 (2003). https://doi.org/10.1016/S0926-3373(02)00278-3
41. Y. Xu, M. Chen, B. Liu, F. Jiang, X. Liu, CH4 conversion over Ni/HZSM-5 catalyst in the absence of oxygen: Decomposition or dehydroaromatization? Chem. Commun. **56**(32), 4396–4399 (2020). https://doi.org/10.1039/d0cc01345e

42. P. Tan, Active phase, catalytic activity, and induction period of Fe/zeolite material in non-oxidative aromatization of methane. J. Catal. **338**, 21–29 (2016). https://doi.org/10.1016/j.jcat.2016.01.027
43. J. Gao, Y. Zheng, J.-M. Jehng, Y. Tang, I.E. Wachs, S.G. Podkolzin, Identification of molybdenum oxide nanostructures on zeolites for natural gas conversion. Science (80-.) **348**(6235), 686–690 (2015)
44. D. Zhou, S. Zuo, S. Xing, Methane dehydrogenation and coupling to ethylene over a Mo/HZSM-5 catalyst: A density functional theory study. J. Phys. Chem. C **116**(6), 4060–4070 (2012). https://doi.org/10.1021/jp209037c
45. J. Dang, G. Zhang, L. Wang, K. Chou, P.C. Pistorius, Study on reduction of MoO2 powders with CO to produce Mo2C. J. Am. Ceram. Soc. **99**(3), 819–824 (2016). https://doi.org/10.1111/jace.14042
46. I. Vollmer et al., On the dynamic nature of Mo sites for methane dehydroaromatization. Chem. Sci. **9**(21), 4801–4807 (2018). https://doi.org/10.1039/c8sc01263f
47. T.C. Xiao et al., Preparation of molybdenum carbides using butane and their catalytic performance. Chem. Mater. **12**(12), 3896–3905 (2000). https://doi.org/10.1021/cm001157t
48. A. Penteado, E. Esche, D. Salerno, H.R. Godini, G. Wozny, Design and assessment of a membrane and absorption based carbon dioxide removal process for oxidative coupling of methane. Ind. Eng. Chem. Res. **55**(27), 7473–7483 (2016). https://doi.org/10.1021/acs.iecr.5b04910
49. J. Wang et al., Comparative study on oxidation of methane to ethane and ethylene over Na2WO4-Mn/SiO2 catalysts prepared by different methods. J. Mol. Catal. A Chem. **245**(1–2), 272–277 (2006). https://doi.org/10.1016/j.molcata.2005.09.038
50. M.J. Werny, Y. Wang, F. Girgsdies, R. Schlögl, A. Trunschke, Fluctuating storage of the active phase in a Mn-Na 2 WO 4 /SiO 2 catalyst for the oxidative coupling of methane. Angew. Chem. **132**(35), 15031–15036 (2020). https://doi.org/10.1002/ange.202004778
51. I.Z. Ismagilov et al., Effect of preparation mode on the properties of Mn-Na-W/Sio2 catalysts for oxidative coupling of methane: Conventional methods vs. POSS nanotechnology. *Eurasian Chem. J.* **18**(2), 93–110 (2016). https://doi.org/10.18321/ectj430
52. H.R. Godini et al., Sol-gel method for synthesis of Mn-Na2WO4/SiO 2 catalyst for methane oxidative coupling. *Catal. Today* **236**(PART A), 12–22 (2014). https://doi.org/10.1016/j.cattod.2014.01.005
53. D. Wang, M.P. Rosynek, J.H. Lunsford, Oxidative coupling of methane over oxide-supported sodium-manganese catalysts. J. Catal. **155**(1), 390–402 (1995). https://doi.org/10.1006/jcat.1993.1273
54. J. Wu, S. Li, Role of distorted WO4 in the oxidative coupling of methane on tungsten oxide supported catalysts. J. Phys. Chem. **99**(13), 4566–4568 (1995). https://doi.org/10.1021/j100013a030
55. S.F. Ji et al., The relationship between the structure and the performance of Na-W-Mn/SiO2 catalysts for the oxidative coupling of methane. Appl. Catal. A Gen. **225**(1–2), 271–284 (2002). https://doi.org/10.1016/S0926-860X(01)00864-X
56. P. Wang, G. Zhao, Y. Wang, Y. Lu, MnTiO3-driven low-temperature oxidative coupling of methane over TiO2-doped Mn2O3-Na2WO4/SiO2 catalyst. *Sci. Adv.* **3**(6), e1603180 (2017). https://doi.org/10.3390/catal9040363
57. S. Pak, P. Qiu, J.H. Lunsford, Elementary reactions in the oxidative coupling of methane over $Mn/Na_2WO_4/SiO_2$ and $Mn/Na_2WO_4/MgO$ catalysts. *J. Catal.* **179**(1), 222–230 (1998). http://ac.els-cdn.com/S002195179892228X/1-s2.0-S002195179892228X-main.pdf?_tid=8a808ff6-6054-11e5-95d8-00000aab0f6b&acdnat=1442835372_00fbd4886436a976338f149501f8056f
58. P. Wang, G. Zhao, Y. Liu, Y. Lu, TiO2-doped Mn2O3-Na2WO4/SiO2 catalyst for oxidative coupling of methane: Solution combustion synthesis and MnTiO3-dependent low-temperature activity improvement. *Appl. Catal. A Gen.* **544**, 77–83 (2017). https://doi.org/10.1016/j.apcata.2017.07.012

59. D. Kiani, S. Sourav, J. Baltrusaitis, I.E. Wachs, Oxidative coupling of methane (OCM) by SiO2-supported tungsten oxide catalysts promoted with Mn and Na. ACS Catal. **9**(7), 5912–5928 (2019). https://doi.org/10.1021/acscatal.9b01585
60. Y. Gordienko et al., Oxygen availability and catalytic performance of NaWMn/SiO2 mixed oxide and its components in oxidative coupling of methane. Catal. Today **278**, 127–134 (2016). https://doi.org/10.1016/j.cattod.2016.04.021

Chapter 4
The Cross-Coupling of Methane with Nonoxidative Hydrocarbons

4.1 Introduction

Catalytic naphtha reforming is employed worldwide to meet varying legislations interested in increasing the octane number of low-octane hydrocarbon-containing streams. Current processes require an element of hydrotreating, where the hydrogen is obtained from the steam reformation of natural gas, an expensive process. This chapter shows the technical feasibility of employing methane (the predominant component of natural gas) as a direct hydrogen source by co-feeding it with various naphtha streams at moderate temperatures and pressures using heterogenous catalysis.

Naphtha oil in this context refers to the light oil fraction obtained from the refinement of crude oil. These streams can contain any number of olefinic, paraffinic, aromatic, and naphthenic C_5-C_{12} hydrocarbons with varying distributions depending on the refinery and the crude oil source. However, n-paraffins are usually the main constituents [1]. The octane number of a fuel indicates the anti-knocking properties of a fuel, a factor that effectively determines the ability of a fuel to resist compression before it is ignited in the engine. N-paraffins have very poor octane numbers, and so it is understandable these components hold little value and are upgraded to either their isomers or, preferably, aromatics. The most desirable of these higher-octane-upgraded chemicals are benzene, toluene, and xylene (BTX) not only for their higher-octane values to prevent engine knocking but also for their use in the chemical industry as precursors for polymer synthesis. In fact, naphtha reforming is responsible for approximately 40% of the worlds production of high-octane gasoline [2], an apt demonstration of the focus of naphtha reforming and the need to refine reforming processes.

The first real commercial outfit designed for the reforming of naphtha was in 1949 by UOP using a Pt/Al_2O_3 catalyst modified with a Cl-containing organic [3]. Although selectivity's towards i-paraffins were appreciable, the catalyst was

H. Song et al., *Methane Activation and Utilization in the Petrochemical and Biofuel Industries*, https://doi.org/10.1007/978-3-030-88424-6_4

non-regenerable and often posed corrosion problems. Since then, researchers have placed a heavy focus on improving the process with three main categories as outlined in Fig. 4.1.

From this alone, it is clear to see how catalyst formulation is the most important factor when reforming naphtha feedstocks. The year 1969 saw the formulation of Zeolite Socony Mobil-5 (ZSM-5) [5], the catalyst featured in the previous chapter for the MDA reaction and which has an MFI-type framework. The morphology of ZSM-5 makes it very successful at promoting aromatization through the shape of its pores as well as an increased surface area with micro- and mesoporosity.

Arguably one of most exciting basic attributes of this support is the ability to fine-tune the acidity. Made up of SiO_2 and Al_2O_3 species, the Si^{4+} species can be replaced by Al^{3+} cations to create an additional positive charge which is necessary to keep the material neutral. Acid-catalyzed reactions (explained later) can therefore be promoted or depressed. In recent years, there has been a lot of interest focusing on the methods of zeolite synthesis, using different organic templates to obtain the correct morphology [6], bases such as KOH to form hierarchical structures [7], and even techniques used to alter the size of zeolite crystals formed. Arguably one of the most effective ways to alter the performance of the support is the addition of a metal or various metals to provide further attributes to the catalyst. Depending on the metal used, the performance can be increased through enhanced dehydrogenation which can then be further promoted by the addition of a second metal to form intermetallic particles that alter the electronic properties of the metals in question.

Catalytic naphtha reforming has a rich history, and research is still growing in this area as the world attempts to better understand the processes involved when isomerizing and aromatizing n-paraffins. The next subsection will outline the motivation of this research, considering the reasoning behind using methane as a co-feed.

During the naphtha reforming process, hydrogen partial pressure is achieved to promote hydrocracking where the lager carbon number components are broken down. Hydrogen is mostly obtained through reforming, where natural gas (whose major component is methane) is steam reformed at extreme temperatures (~800 °C)

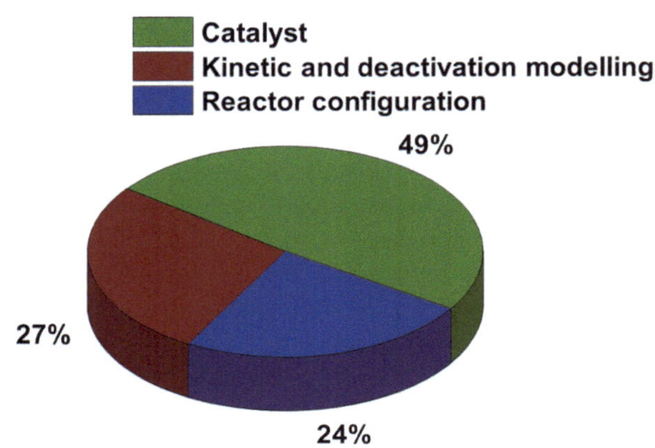

Fig. 4.1 The percentage of completed scientific studies of different categories focused on naphtha reforming from 1949 to 2013 [4]

and produces CO_2 as a major greenhouse gas contributor [8]. Understandably, it would be a great boon to the industry if this step were eliminated, both from an economical and environmental standpoints. As shown in Fig. 4.2, natural gas saw a very successful year in 2017 with increases in global production by 4%, nearly double the average growth rate for a period of 10 years (2.2%). This comes as a product of increased demand giving natural gases avenue in many areas including power generation, domestic heating and fuel, transportation fuel, fertilizer, and many manufacturing processes.

Therefore, it is proposed in the scientific field that methane can be used as a direct donor of hydrogen due to its recent abundance and availability through technological developments which have enhanced the exploration and recovery of shale gas. Currently, this is not reasonable at the commercial scale; however, since the discovery of the feasibility of low-temperature nonoxidative activation of methane in the presence of olefins and higher carbon number paraffins [10], the field has boomed. Thus, this research has been motivated by the prospect of reforming various naphtha feeds in the presence of methane as an alternative source of hydrogen, with the major target products being BTX for their use as gasoline additives and valuable industrial-scale precursors in the chemical industry.

Many years of research have been dedicated to the effective activation of methane so that it can be used for applications beyond residential heating but with limited success. Thus, the development of a technology capable of converting methane into valuable chemical feeds is highly desirable. The dehydrogenation and further aromatization of methane under nonoxidative conditions was initially observed over a HZSM-5 support with transition metal additions [11, 12]. However, the direct formation of aromatics such as benzene is not thermodynamically favorable at low temperatures.

$$6CH_4 \rightarrow C_6H_6 + 9H_2$$

$$\Delta H_{(25^\circ C)} = 531 \frac{kJ}{mol}$$

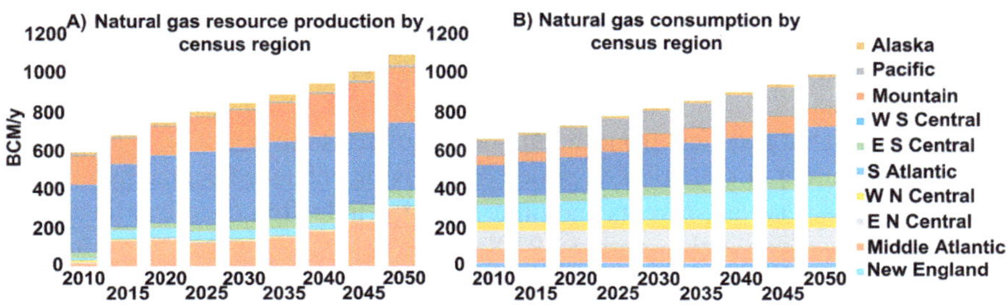

Fig. 4.2 Natural gas production by region (billion cubic meters per annum) and natural gas consumption by region (billion cubic meters per annum), demonstrating the growth of natural gas demand [9]

$$\Delta G_{(25^{\circ}C)} = 443 \frac{kJ}{mol}$$

As a result of this, high reaction temperatures are required which lead to fast catalyst deactivation because of coking. However, Choudhary et al. were able to show that co-feeding methane with hydrocarbons [13], such as those found in light and heavy naphtha, enables much lower temperatures for effective methane conversion and therefore, less coking and catalyst deactivation. A series of subsequent publications have confirmed this finding and are the driving force behind the use of methane as an alternative hydrogen source and in the cross-coupling of methane with hydrocarbons.

4.2 Co-Feeding with Light Naphtha

Light straight run (LSR) naphtha produced by petroleum refineries contains an abundant mixture of many C_5–C_7 (predominantly C_5) hydrocarbons [14] which require conversion to more valuable products for the purpose of increasing the octane number of gasoline fuels and/or producing valuable chemical feedstocks. Thus, branched paraffins are becoming increasingly important given their increased research octane number (RON) compared to normal paraffins, especially i-pentane and i-hexane (~80) compared to n-pentane and n-hexane (~64) [15]. Isomerized paraffins are not only important for the enhancement of a feedstocks octane number but are also more environmentally friendly than olefins when used in gasoline [16]. Thus, it is important to minimize olefin formation during the upgrading of such LSR feedstocks.

$AlCl_3$ and metal on Al_2O_3 catalysts are not used anymore due to their poor performance [17], and Pt-Cl/Al_2O_3 encounters reactor corrosion issues [18]. Pt/Mordenite avoids these issues, and it has even been reported that Pt/beta results in more activity for these isomerization reactions [19]. It is widely accepted that bifunctional mechanisms proceed through the following reactions: dehydrogenation of alkanes on Pt, protonation of produced olefins on Brønsted acid sites to form carbenium ions, rearrangement, β-scission, and de-protonation and hydrogenation over Pt to result in the final isomerized forms of normal paraffins, as well as cracked gaseous products.

Other desirable products from the upgrading of LSR are benzene, toluene, and the xylenes (hereafter called BTX) which can provide enhanced octane numbers to fuels as well as serve as solvents and other valuable industrial chemical feeds. Not only does the aromatization provide BTX but also hydrogen as a by-product, an increasingly valuable commodity for its use in hydro-consuming processes such as fuel cells and as a future energy carrier [20]. Thus, the use of a successful aromatization catalyst which can facilitate both isomerization and aromatization could potentially enhance the value of LSR through the production of i-paraffins, BTX, and hydrogen as a gaseous by-product. One such catalyst is ZSM-5, a heterogenous catalyst capable of such reactions and, furthermore, upon selective metal loading, can increase the selectivity of naphtha feeds towards BTX in particular [21].

Deactivation of these catalysts during naphtha reforming occurs predominantly through coking, an issue which is resolved in industry by applying a hydrogen atmosphere which effectively reduces the environment to minimize carbon deposition [22]. Hydrogen also serves as a source to assist with hydrocracking during these upgrading reactions, but it comes at a price due to its natural unavailability. A predominant method of acquiring hydrogen is through the steam reforming of natural gas which consists mainly of methane [23]. This process is conducted at high temperatures, 900 °C, and produces large amounts of CO_2 during the recovery of more hydrogen through the water-gas shift reaction ($CO + H_2O \leftrightarrow CO_2 + H_2$) which is an undesirable by-product acting as a contributor to the greenhouse effect. Thus, we turn our attention to CH_4, an abundant, naturally occurring and cheap source of hydrogen if effectively activated. Although the prospect of such an attainable hydrogen source seems promising, methane's notorious stability, attributed to its stable tetrahedral C-H bonds, as well as its electronic structure obtained from sp^3 hybridization, has proven difficult to overcome [24]. Nonetheless, due to the large volume of research focusing on the conversion of methane and despite the understandably more reactive components formed, nonoxidative activation of methane at low temperatures has been achieved in the presence of olefins and higher carbon number paraffins as co-feeds. Thus, a strong desire among the scientific community continues to drive towards the ability to overcome the challenge of activating CH_4 through novel catalysis so that it can contribute to the desirable liquid products while simultaneously providing the hydrogen needed to inhibit catalyst deactivation.

Many studies have been conducted on single metal loading onto ZSM-5 with regard to Zn [1, 25–28] and Pt [29–31] as well as some attention to their performance as an alloy. The Zn function of HZSM-5 has been shown to promote the formation of aromatics [32] by providing dehydrogenation sites on the catalyst surface which lead to olefin formation, followed by aromatic production. Pt, on the other hand, although more active, is less selective and results in the formation of undesirable aromatics ($>C_9$) because of its preferred reaction pathway of hydrogenolysis which sees C-C bonds cleaved by H_2, enabling the formation of higher aromatics. Thus, there has been recent interest in the effect of Pt-Zn alloys on the aromatization of hydrocarbons, in this case LSR, with the intent that the alloy can promote the direct isomerization over less acidic, porous HZSM-5 so that the positive morphological characteristics can still be taken advantage of. The idea is that the less selective effect of sole Pt can be nullified using a Zn-Pt alloy while simultaneously omitting the need to form gas products, which may remain unconverted if sole Zn is used, and we rely on its strong oligomerization tendencies.

4.2.1 *Experimental Studies of Light Paraffin Cross-Coupling with Methane*

Note that there are many variants of the catalysts shown here that are effective for the application of cross-coupling of methane with hydrocarbons, and the ones shown here serve as a suitable case study to avoid convolution.

LSR naphtha can be replicated for model studies, one such replica is shown here in Fig. 4.3 and was designed to recreate a typical feedstock found in the petrochemical industry, and the catalyst is designed to convert the n-paraffins to their isomers and/or valuable aromatics (nominally BTX).

Figure 4.4 shows the selectivity, conversion, and BTX yield results for these experiments using metal-loading variations of HZSM-5 (280:1), including no metal loading for comparison and the effect of methane and nitrogen environments. Although sole Pt or sole Zn loading onto HZSM-5 have both shown promise for the aromatization of paraffins in recent years, the improvements are often "small" stepping-stones with regard to improved performance. The results shown in Fig. 4.4 immediately shows us the marked improvement when both Pt and Zn are loaded, in the form of yield and selectivity. Upon initial inspection, the use of both metals under a methane environment improves the selectivity drastically as well as the BTX yield. Not surprisingly, a lack of metal loading, i.e., HZSM-5 (280:1), shows the worse performance, with sole Pt and sole Zn, showing slight improvements. As expected, sole Pt, with the exception of no metal loading, shows the worse selectivity, likely a result of its strong capacity for hydrogenolysis.

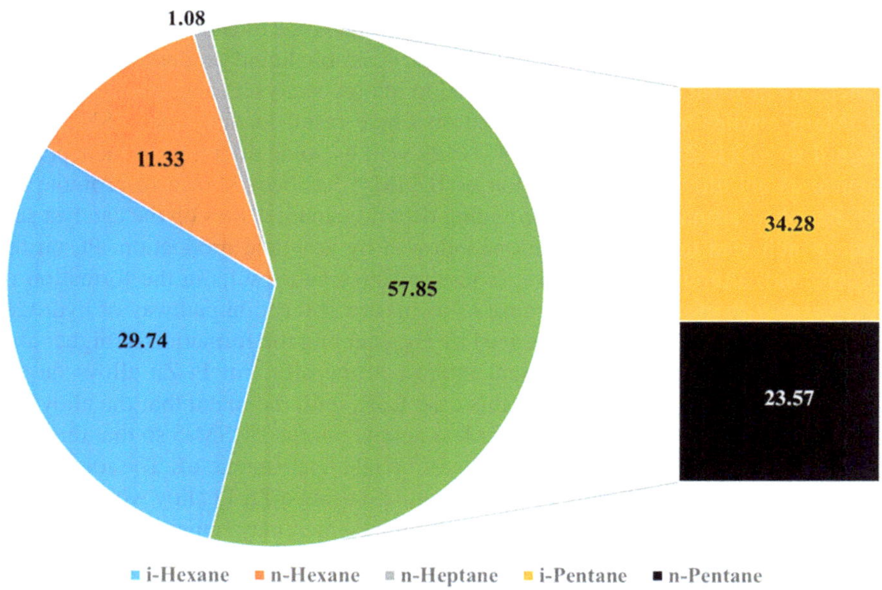

Fig. 4.3 Example distribution of components found in LSR naphtha [33]

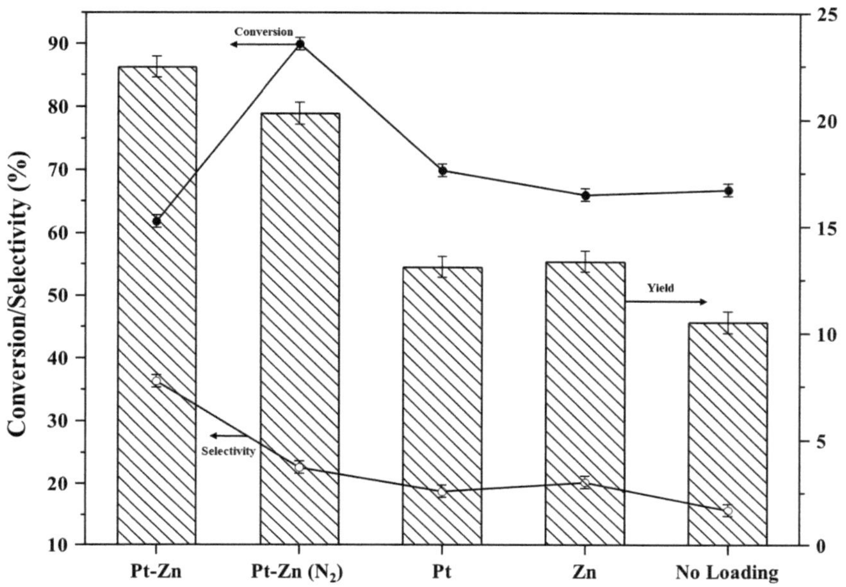

Fig. 4.4 Graph displaying the performance of various HZSM-5 (280:1) catalysts with regard to BTEX selectivity, conversion, and BTEX yield, where BTEX selectivity is calculated as total selectivity, including gas products [33]

Liquid selectivities may also provide valuable information as shown in Figs. 4.5 and 4.6. As expected, sole Pt gives the poorest selectivity, barely surpassing that of conventional ZSM-5 with no metal loading, with sole Zn's promotion of oligomerization of gaseous C_3 species as well as hydrogenation capacity accounting for an increase in total liquid yield and increased formation of BTX. The increased formation of BTX when only Zn is employed when compared to sole Pt employment is likely a result of increased oligomerization of gaseous olefinic species, C_2 in particular.

Figure 4.7 shows gas product compositions and agrees with this assumption by demonstrating a significant decrease in ethylene content when comparing sole Zn to sole Pt loading, likely a result of a lack of Zn presence for effective oligomerization. Gas products also show a higher mol % of hydrogen when Pt-Zn are loaded onto HZSM-5, which agrees with the higher liquid selectivity witnessed during the formation of BTEX than seen for other catalysts.

As well as witnessing the effects of metal loading on the catalyst, the presence of methane as an atmosphere compared to nitrogen is also demonstrated in previous figures. Despite reducing feedstock conversion, BTX selectivity, yield, and even total liquid yield understandably increase, leading to a much more desirable range of products being in their liquid form as opposed to undesirable gaseous species. This is displayed in Fig. 4.4 with the liquid yield of Pt-Zn/HZSM-5 under a nitrogen environment being much lower than that when CH_4 is employed, suggesting a

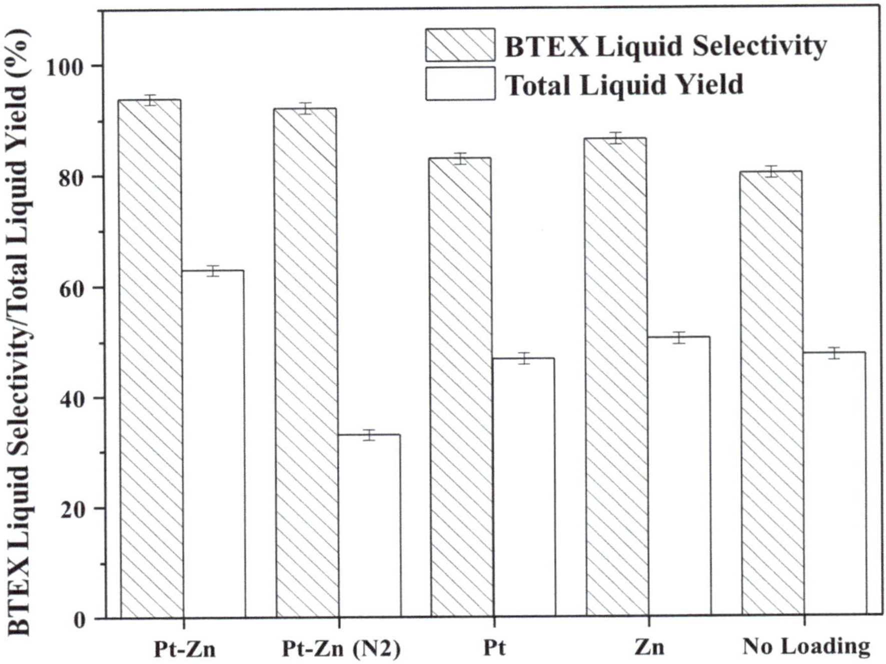

Fig. 4.5 A graph displaying the performance of various HZSM-5 (280:1) catalysts with regard to BTEX liquid product selectivity and total liquid yield [33]

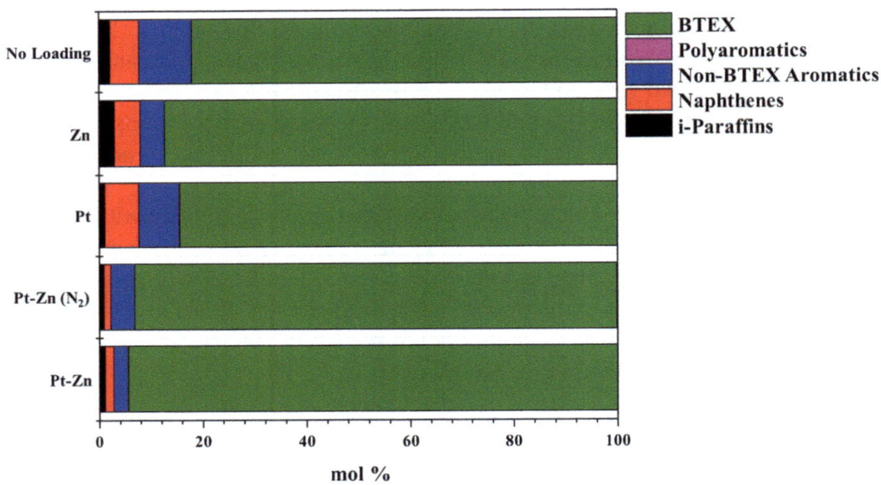

Fig. 4.6 A graph displaying the liquid product distribution of various HZSM-5 (280:1) catalysts [33]

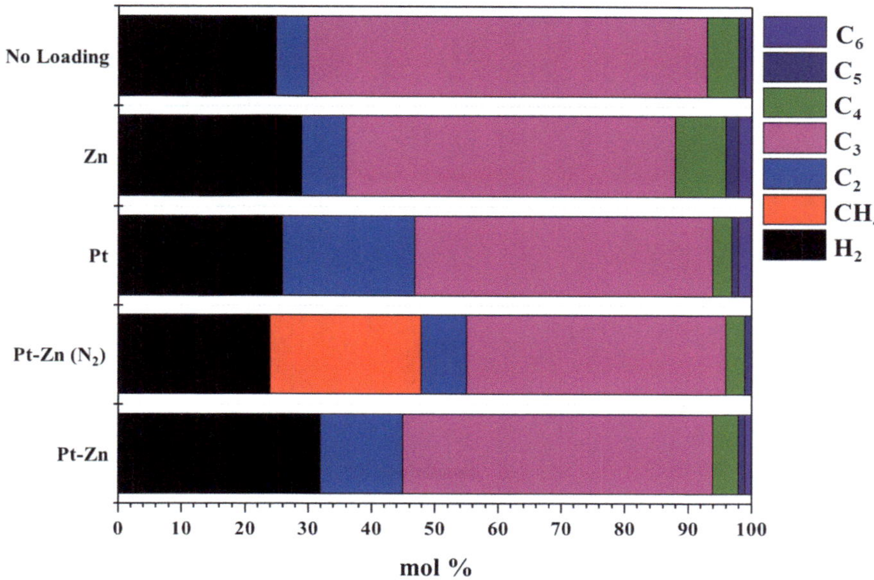

Fig. 4.7 Gaseous product distribution for varying HZSM-5 (280:1) catalysts on a fixed bed reaction system with LSR naphtha and methane or nitrogen [33]

positive effect of methane's reductive atmosphere as well as potential CH_4 incorporation into the liquid products.

In summary, the Pt-Zn synergy on a catalyst has shown a marked improvement in performance in all aspects, regardless of environment, suggesting a potential synergetic effect and/or a highly superior alloy that potentially takes the most positive attributes of the metals employed. This could be through Pt acting as a dispersion promoter or even a result of Pt effectively assisting dehydrogenation to olefins, with Zn contributing to oligomerization, followed by aromatization, thanks to the morphology of ZSM-5 and the presence of dehydrogenating metals.

Characterizations are an important aspect of this field as it is necessary to show the catalytic properties both before and after the reactions to ascertain reaction mechanisms. Figure 4.8 shows TEM images of spent Pt-Zn/HZSM-5 under a methane environment (a) and a nitrogen environment (b), where particle agglomeration can be seen. An average particle diameter of 1.4 nm and 2.1 nm under methane and nitrogen, respectively, was found, suggesting that the presence of methane inhibits particle agglomeration, thus providing another explanation for the enhanced performance under a methane environment due to a better distribution of active sites throughout the reaction. Lattice fringing is seen (c), supporting XRD data which shows a maintained crystalline structure during synthesis. Figure 4.9 demonstrates the presence and dispersion of Zn and Pt on fresh Pt-Zn/HZSM through HAADF elemental mapping.

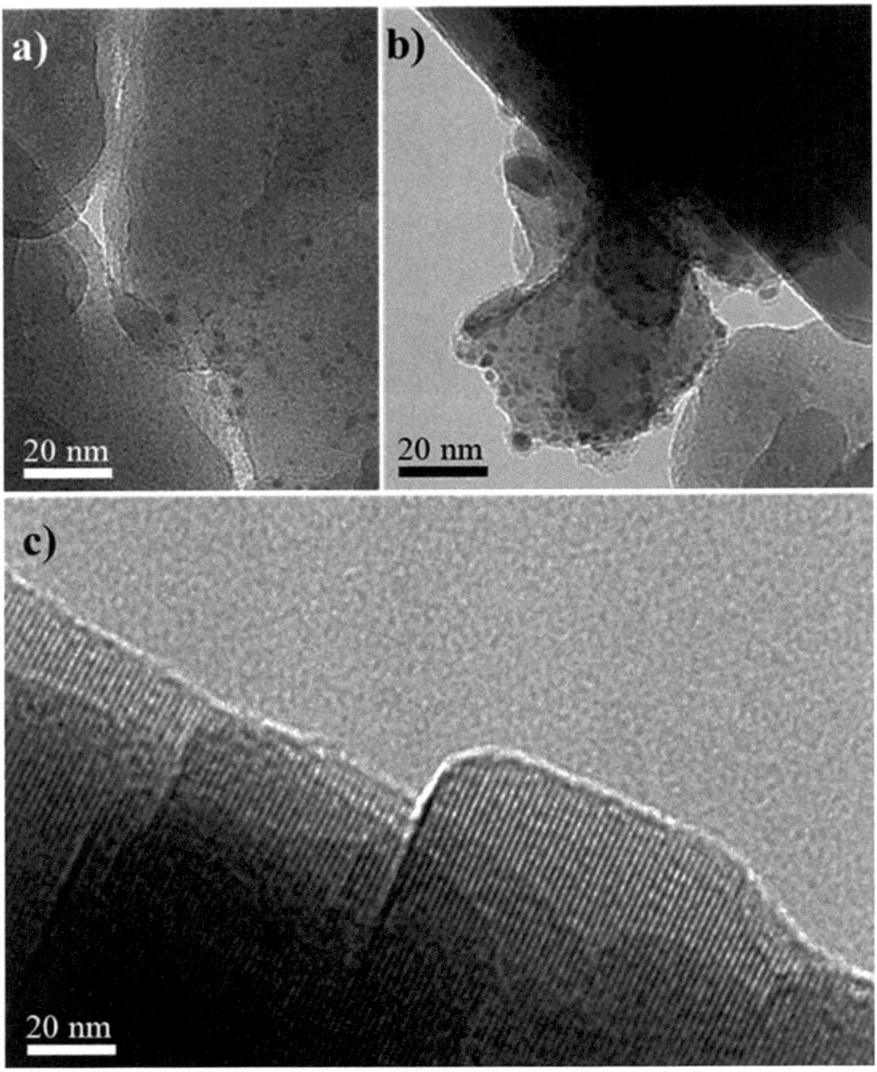

Fig. 4.8 TEM images of (**a**) Pt-Zn/HZSM-5 post reaction under a methane environment, (**b**) Pt-Zn/HZSM-5 post reaction under a nitrogen environment, and (**c**) pristine Pt-Zn/HZSM-5 lattice fringing [33]

XRD patterns are shown in Fig. 4.10, and all cases exemplify the MFI crystal structure before and after the reaction. However, upon closer inspection, it is evident that HZSM-5 with no metal loading experiences no significant change after reaction, likely a result of no metal presence to migrate and cause crystal deformations during the reaction. Sole Pt-loaded catalysts experience an intensity reduction in the 010 crystal plane post reaction, accompanied by an increase in intensity of 501, 151,

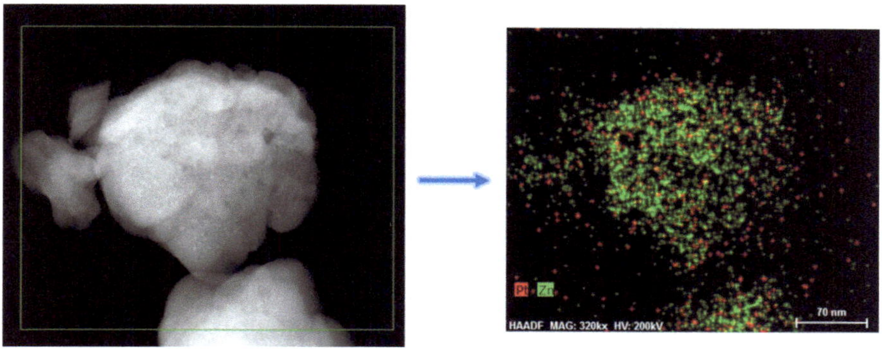

Fig. 4.9 HAADF STEM images of pristine Pt-Zn/ZSM-5 (280:1) with elemental mapping: red corresponds to Pt particle locations, with green corresponding to Zn particle locations

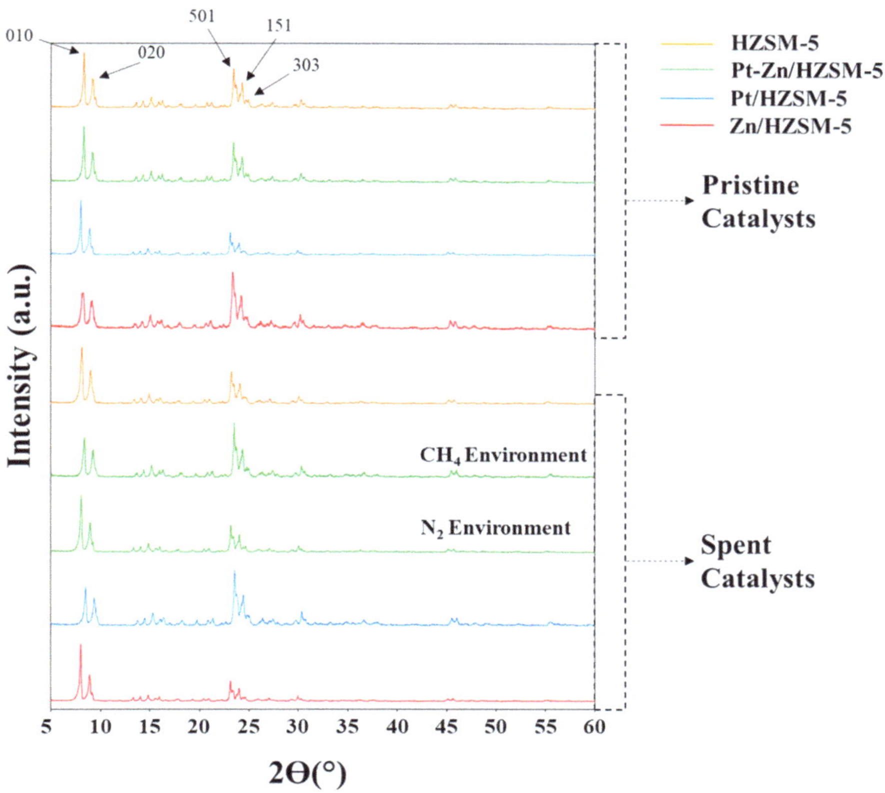

Fig. 4.10 XRD patterns for varying HZSM-5 (280:1) catalysts [33]

and 303 crystal planes, suggesting a difference in the number of atoms in the aforementioned planes [34]. It is therefore possible that Pt migration from the 010 crystal plane towards the 501, 151, and 303 crystal planes is occurring or that the migration of Pt during the reaction is shifting other atoms in the framework. The direct opposite effect is seen for the Zn/HZSM-5 catalyst where 501, 151, and 303 peaks are greatly reduced post reaction and the 010 plane intensity is increased. When both Pt and Zn are employed (green diffractograms), a nitrogen environment seems to slightly reduce 501, 151, and 303 crystal planes, whereas a methane environment increases these planes with a slight reduction in the 010 plane. Given that the presence of nitrogen as opposed to methane results in the formation of more gaseous products, it is possible that the presence of metals on the active 501, 151, and 303 planes inhibits the formation of these undesired gaseous products. Perhaps methane activation through metal interaction, in particular CH_4-Pt interactions as seen on Pt clusters by Xiao et al. [35] and which indeed see a reduced activation energy, is promoted over the aforementioned planes and increases the stability of Pt atoms at these positions with migration from the 010 plane, a result of a lack of methane activation at this plane.

DRIFT spectra for adsorbed pyridine on HZSM-5, Zn/HZSM-5, Pt/HZSM-5, and Pt-Zn/HZSM-5 are displayed in Fig. 4.11 and enable us to establish the nature

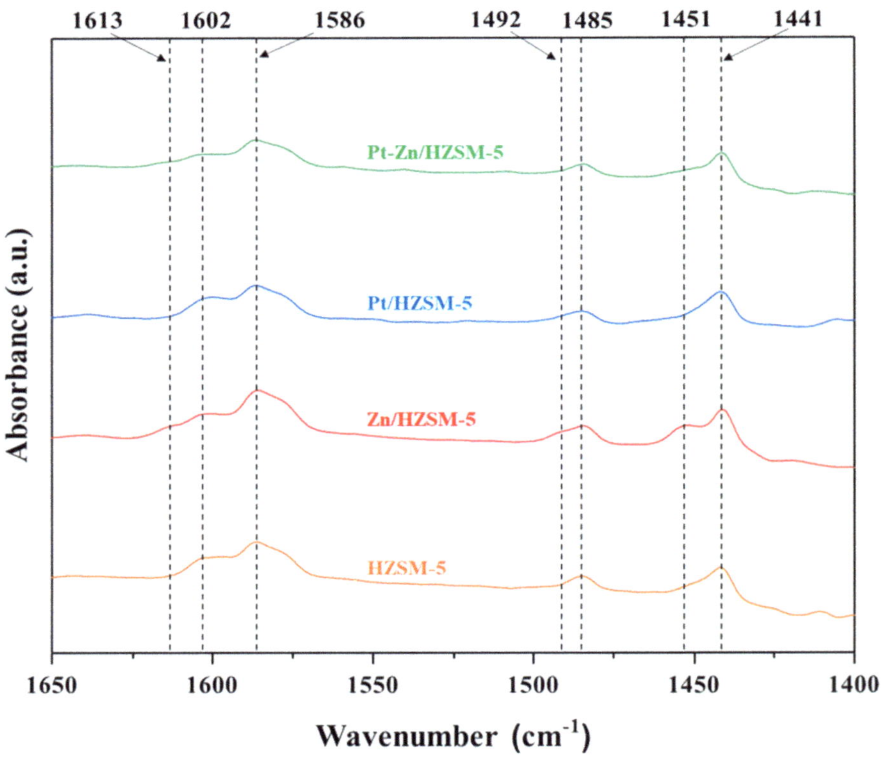

Fig. 4.11 Pyridine DRIFT spectra of varying HZSM-5 (280:1) catalysts collected at RT [33]

of the acid sites present on the catalyst surface. Three forms of acid sites are investigated here, including Brönsted, Lewis, and physiosorbed H-bonded sites, and so the region investigated lies between 1440 and 1620 cm^{-1}. Peaks observed at 1441 and 1451 cm^{-1} demonstrate Lewis acid sites [36], i.e., electron pair acceptors arising from molecularly coordinated pyridine (C-C stretch). The peaks at 1485 and 1492 cm^{-1} show us the contribution of both Brønsted and Lewis acid sites [37]. Brønsted acid sites, nominally proton donors, can usually be observed at approximately 1550 cm^{-1} [38]; however, given the relatively high SiO_2/Al_2O_3 ratio employed for these catalysts (280:1), it is possible that typical Brønsted sites are difficult to detect as a result of such a small amount of Al used to replace Si in the framework. We are encouraged to recall that this is a desired attribute of our catalysts given the need for less catalytic cracking. That being said, bands at 1586 and 1602 cm^{-1} are indicative of H-bonded pyridine, a sign of Brønsted acidity [39]. The small shoulder peak noted at 1613 cm^{-1} is that of pyridine adsorption on Lewis acid sites [40]. Intensities see little to no change given that all experiments were conducted on the same support and at RT with the exception of Pt loading resulting in a reduced peak intensity compared to Zn, which could be a result of Pt's high dispersion and uniform distribution, leading to the occupation of more acid sites. However, more notable changes come from the comparison of Zn-containing catalysts where all peaks are enhanced along with shoulder peaks in some cases (1451, 1492, and 1613 cm^{-1}), all of which are only present when Zn is employed. Since all peaks can be attributed to Lewis acid sites, it is suitable to conclude that Zn presence has increased the Lewis acidity compared to Brønsted acidity in all cases. This is understandable given the higher loading of Zn and that metal replacement of Si sites would result in such a phenomenon.

To further acidic inspection of the catalysts, and to even quantify the acidity of the catalysts used, NH_3-TPD can be employed (Fig. 4.12). Two distinct peaks are witnessed when no metal is loaded and when sole Pt is loaded onto HZSM-5, with peaks at approximately 150 and 350 °C showing desorption from weak and strong acidic sites, respectively. When comparing spectra (a) and (b), it is clear that upon Pt loading, the contribution made by weaker acid sites is reduced significantly and strong acid sites conversely increased. In previous work [21] as well as later on in this piece, it has been concluded that an increased performance through selectivity and yield is closely associated with metal loading, resulting in strong acid sites associating themselves with metal species to form weaker acidic sites. This helps explain why sole Pt loaded onto HZSM-5 results in one of the worst increases in performance when compared to Zn/HZSM-5 and Pt-Zn/HZSM-5. As seen in previous work [41], and upon inspection of Fig. 4.12c and d, strong acid sites have been dissociated and resulted in the formation of medium acid sites when Zn is loaded. This phenomenon is enhanced when sole Zn is employed as opposed to Pt-Zn, an expected occurrence given Pt's tendency to promote dispersion and thus result in a reduced effect on acidity. The addition of Pt to Zn has increased the contribution of strong acid sites in comparison to sole Zn loading which suggests that framework oxygens of the support help provide electron density to Pt. The reader is also referred to Table 4.1 where quantified values of acidity can be inspected in the form of 10^{-5} mol NH_3/g cat [42].

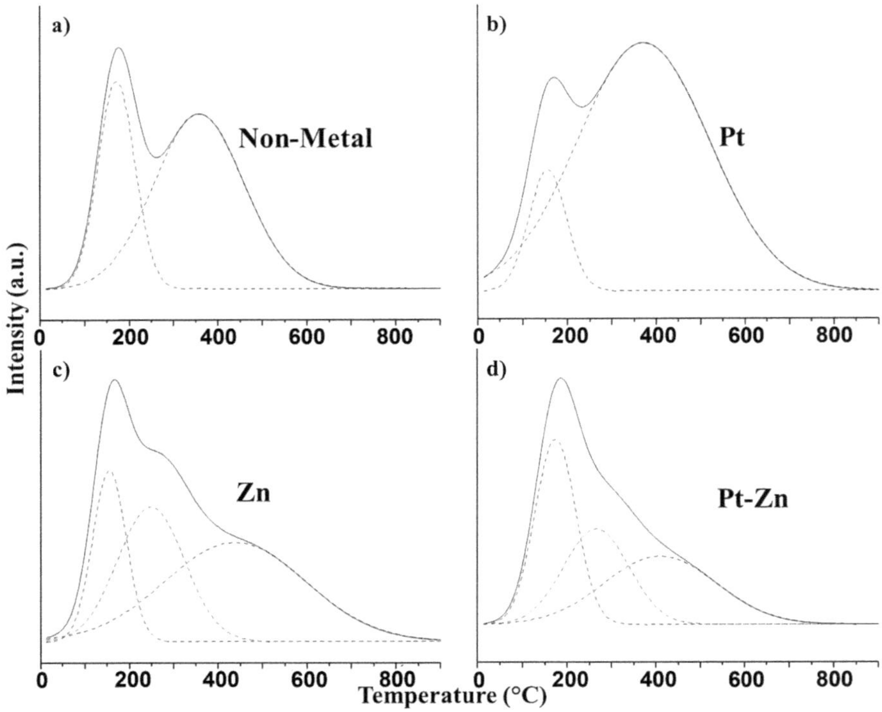

Fig. 4.12 NH_3-TPD spectra of (**a**) HZSM-5, (**b**) Pt/HZSM-5, (**c**) Zn/HZSM-5, and (**d**) Pt-Zn/HZSM-5 [33]

Table 4.1 The acidity of varying catalysts via NH_3-TPD ($\times 10^{-5}$ mol NH_3/g cat) and coke yield (µg of coke formed·h^{-1}) calculated via TGA-DSC [33]

Catalyst	Weak acid sites	Medium acid sites	Strong acid sites	Total	µg coke/(g cat h)
HZSM-5 (280:1)	3.5	0	7.1	10.6	5.9
Zn/HZSM-5 (280:1)	18.0	11.6	7.4	37.1	1.0
Pt/HZSM-5 (280:1)	4.6	0	32.6	37.2	8.3
Pt-Zn/HZSM-5 (280:1)	11.1	9.1	11.0	31.3	1.7
Pt-Zn	–	–	–	–	2.4

It is also important to probe the deactivation tendency of catalysts to assess their ability to perform for long durations in a reactor system given that less time spent regenerating catalysts is desired. Given the lack of sulfur-containing components in this system, the main focus of catalyst deactivation is that of carbon deposits, coke. Thus, TGA-DSC analysis has been conducted so as to gauge the level of deactivation imposed on the catalysts. Table 4.1 shows the rate of coke formation on the catalysts as obtained from the liquid feedstock. Not surprisingly, all catalysts under both CH_4 and N_2 environments exhibited very low coke yields because of little to no formation of polyaromatics (known precursors to coke). Another contributor to such impressively low deactivation is the low acidity of the support used, minimizing cracking levels. Pt is notorious for its promotion of coke formation, given its less selective tendencies, and so it is not surprising to see that it is responsible for higher levels of coke formation than all other catalysts, including nonmetal. Sole Zn loading shows the lowest coke yield which has been seen by other groups [43]. Interestingly, although the Pt-Zn catalyst results in higher levels of coking than sole Zn, the performance is clearly not inhibited, suggesting that the level of coke is not significant enough to block active sites on the Pt-Zn catalyst.

LSR feedstocks can contain large quantities of undesirable normal paraffins whose octane numbers are not sufficiently high to provide useful characteristics to combustion fuels. With respect to the model feedstock shown in this case study, it has been shown that it is feasible to upgrade such paraffins to more desirable i-paraffins as well as BTX. Of which, the latter not only contributes to boosting fuel octane numbers but also can be used as industrial chemical feedstocks of a higher value than their upgraded feeds. A catalyst in the form of Pt-Zn/HZSM-5 with low acidity has been used to isomerize and aromatize this LSR into the aforementioned products with superior performance with respect to total liquid yield (63%), BTX selectivity (94%), and BTX yield (22%). No polyaromatics or olefins are produced during the reaction, meaning that any other products formed (of which there are few) will not be detrimental to the product mixture during transportation or utilization. Not only is this experimental data promising for the future of LSR reforming, but it has also been conducted under a CH_4 environment which has shown to reduce coking, enhance liquid yield and increase BTX selectivity, likely a result of methane incorporation into the products, when compared to N_2. Such results mean great promise for the future of light oil reformation with potential use of CH_4, the main component of natural gas as opposed to expensive and non-abundant H_2.

4.3 Co-Feeding with Heavy Naphtha

Naphtha fractions obtained from petroleum refinement contain an abundant mixture of hydrocarbons in the form of paraffins, naphthenes, and aromatics in varying distributions [44], although usually in respective decreasing order of percent by weight. Unless converted, these fractions can be both detrimental to the environment and economically unviable for the industry, making them otherwise undesirable.

Naphtha reforming requires an element of hydrocracking to reduce the number of larger carbon number components and is also one of the reactions that requires hydrogen. One of the challenges associated with methane activation is the need for high-operating temperatures. But this inhibits selectivity towards desired products. Therein lies the problem and a strong desire within the scientific community to activate methane at lower temperatures so that selectivity is maintained along with affordable economics. This has indeed been achieved through the use of bifunctional catalysts at low temperatures and the presence of co-reactants where an increase in liquid product is seen under a methane environment when compared to nitrogen [37].

Another important aspect of naphtha reforming is the ability to produce useful aromatic products such as benzene (currently the second largest volume petrochemical), toluene, and xylenes (henceforth called BTX) [45–48]—not to mention hydrogen as a by-product which is seeing increased interest due to its application in many hydro-consuming processes such as fuel cells and even its potential as a future energy carrier. Thus, the prospect of producing these highly valued products used in polymer and other petrochemical syntheses from otherwise undesirable feeds has been investigated thoroughly since its conception by Vladimir Haensel in the 1950s [49]. In fact, between 1949 and 2013, there were over 600 publications dedicated to the process of naphtha reforming with 50% of these being dedicated to catalyst formulation and performance analysis [4].

Thanks to the works of many researchers [50–52], the key reactions that take place during the formation of aromatics from paraffins and naphthene have been elucidated as dehydrocyclization, isomerization, dehydrogenation, hydrodealkylation, and hydrocracking. With regard to catalyst formulations, H-type ZSM-5 (HZSM-5) has proved to be a promising catalyst for cracking hydrocarbons within the desired naphtha range through two mechanistic pathways. Protolytic cracking is the first and proceeds through the protonation of alkanes to provide carbenium ions which then collapse to give the final products [53]. This pathway occurs because of the Brønsted acid function of zeolites which is attained through aluminum substitution with silicon and a resulting proton for electronic neutrality as well as their Lewis acid function through extra-framework aluminum [54]. The second is through hydride transfer between a reactant and an adsorbed carbenium ion in which a smaller carbenium ion is then obtained through β-scission [54]. Ultimately, the HZSM-5 acid functions provide sites for isomerization as well as cracking, but further catalyst alterations are needed to provide the necessary tools for desirable reactions. Thus, a metal function can provide additional pathways that lead to the desired products. For example, noble metals are very successful at enabling hydrogenation and dehydrogenation reactions which are vital processes in the aromatization processes of naphtha feeds. However, coke formation is extensive when these metals are employed, and so this issue needs to be addressed [55]. Zn and Ga metal addition to HZSM-5 has shown some of the best results for the aromatization of paraffins [55–60] as well as reduced coking, with evidence of reduced undesirable cracked products and more aromatics. In addition to this, a promoter can allow for more flexibility for attaining desired reaction pathways and as a result, higher selectivity towards desirable products. As seen in the previous case study, bimetallic synergy can have a positive effect on performance. Catalyst modification can be taken a step further by the investigation of the effect of specific functions; for example, Ding et al. [61] investigated the effect of external acid sites by salinization of Mo/HZSM-5 using 3-aminopropyl-triethoxylsilane (APTES). This synthetic method allowed for effective blocking of external acid sites with SiO_2, formed after calcination and removal of bulky organosilicon molecules while leaving the acid sites within the internal pores free to contribute to catalytic reactions.

The following case study shows a systematic approach to identifying the effects of various catalyst properties on the ability of methane to co-aromatize with heavier naphtha feeds.

4.3.1 Experimental Studies of Heavy Paraffin Cross-Coupling with Methane

HZSM-5 is well established within the field of catalysis as a successful support for heterogeneous applications involved with hydrocarbon isomerization reactions [62], not to mention its synthetic flexibility with regard to different SiO_2:Al_2O_3 ratios within the framework, leading to well-controlled acidity [62]. Investigating a range of different acidities is a suitable way to gauge which support would provide the best results in terms of BTX yield and selectivity. Figure 4.13 shows the effect of HZSM-5 acidity on metrics such as BTX yield.

Liquid yield shows a negative trend with increased acidity, an understandable observation given that the increased aluminum content provides more Brønsted sites for proton donation and thus enables the formation of more gaseous products as a result of isomerization and cracking. Protolytic paraffin cracking can aid with an explanation for this in the form of a more acidic catalyst-protonating paraffins to

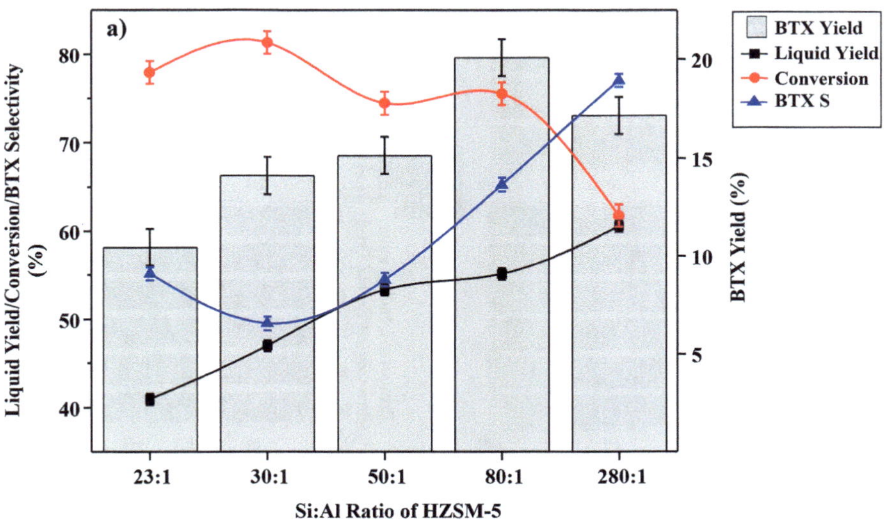

Fig. 4.13 A comparison of different Si:Al ratio HZSM-5 support catalysts and their effect on various quantifiable values [21]

form pentavalent carbonium ions, which in turn collapse to form smaller paraffins as well as olefins. It must be noted that this mechanism as proposed by Haag and Dessau, and now widely accepted, is only relevant when olefins are in small quantities [53], as in the case of the naphtha feedstock used here. At any rate, despite showing the greatest conversions, HZSM-5 (23:1 and 30:1) also demonstrates the poorest BTX selectivity and yields due to the formed gas products contributing to the calculations. Figure 4.14a demonstrates a small quantity of olefins in all cases but a complete reduction of naphthene content to zero when less acidic supports are employed, suggesting that aromatization of cyclic alkanes is inhibited upon increased support acidity under a methane environment. A general trend is almost evident with increasing ratios of BTX to other aromatics in the liquid fraction when the acidity is continually reduced, even as far as the HZSM-5 (280:1) catalyst. It would seem that the HZSM-5 (80:1) catalyst, however, is equipped with the desired number of acidic sites for obtaining BTX as a product with catalysts of lower and higher alumina ratios falling short when concerned with BTX yield. This could be because a reduced acidity does not provide enough acidic sites for protolytic cracking to occur to a significant degree, whereas higher acidities provide too much cracking, resulting in a kinetically favorable reaction towards gaseous products. As a result of this, a plateau can be seen with regard to superior performance from the HZSM-5 (80:1) catalyst under a methane environment, encouraging further catalyst modifications to be conducted on this support.

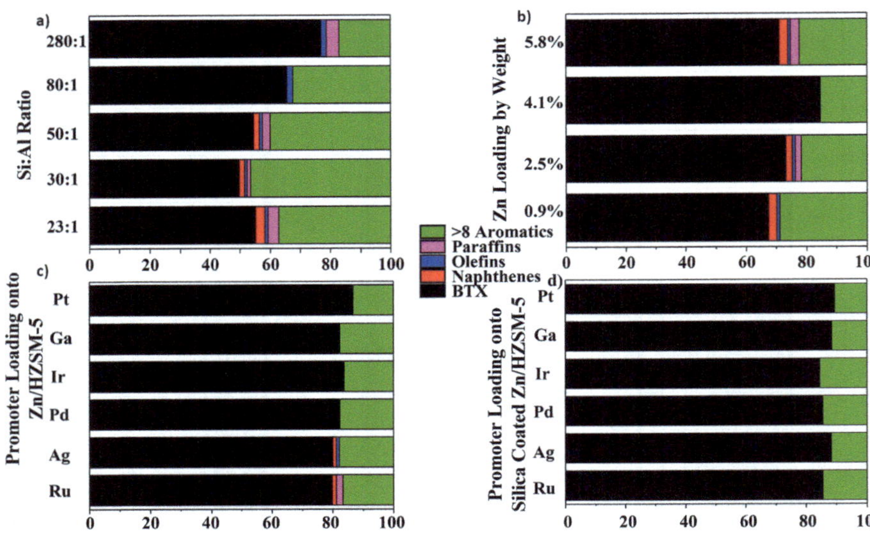

Fig. 4.14 (**a**) A comparison of different Si:Al ratio ZSM-5 supports and their respective effects on the liquid product distribution of upgraded naphtha [21]. (**b**) A comparison of different Zn loadings (calculated on a weight basis) onto ZSM-5 (80:1) and their respective effects on the liquid product distribution of upgraded naphtha. (**c**) A comparison of different promoter loadings on Zn/HZSM-5 (80:1) and their respective effects on the liquid product distribution of upgraded naphtha. (**d**) A comparison of different weight promoter loadings on Zn/HZSM-5 (80:1) with the external surface sites blocked and their respective effects on the liquid product distribution of upgraded naphtha

Metal addition to HZSM-5 has been extensively studied due to the increased stability, neutralized acidity, and improved selectivity towards BTX that can be observed if the metals chosen are suitable. Zn and Ga, in particular, have been of significant interest to those in the field of aromatization, with different pathways to those of conventional zeolites observed upon the study of a model compound. To observe which loading weight % is suitable for a heavy naphtha feedstock, Zn was loaded onto HZSM-5 (80:1) at nominal loadings of 1%, 3%, 5%, and 7% (actual loadings 0.9, 2.5, 4.1, and 5.84, respectively, were achieved through ICP-MS analysis), and the aforementioned reactions undertaken with graphical displays shown in the form of performance (Fig. 4.15). A marked increase in performance is instantly witnessed upon the loading of Zn to the catalyst in all cases, probably owing to zinc's attributes in the activation of methane and dehydrogenation of paraffinic components. Recalling that overall catalyst performance is gauged based on BTX selectivity and yield, it is clear to see that 4.1% Zn loading by weight onto HZSM-5 (80:1) yields the most promising results with an impressive improvement made even more noticeable when comparing BTX and non-BTX aromatic fractions. However, understandably, an increase of Zn loading beyond 4.1% by weight to 5.84% impedes the overall catalyst performance with results similar to that of unmodified zeolites observed earlier in this section. Clearly, Zn has a positive impact on performance, but when loading is increased too dramatically, too many acidic sites are blocked, and higher metal loadings even cause a deformation in the crystal structure of the support (shown later), ultimately altering the effectiveness initially achieved with the unique morphology of MFI zeolites. As a result of 4.1% Zn's superior performance when compared to other loadings under methane, 4.1%

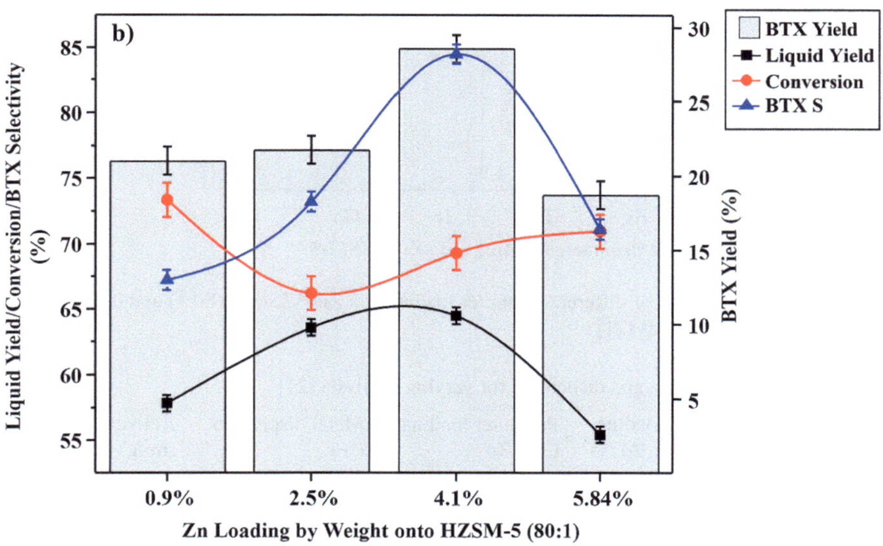

Fig. 4.15 A comparison of different Zn loadings (calculated by weight) on HZSM-5 (80:1) and their effect on various quantifiable values [21]

Zn is loaded in situ with other metal promoters to identify any positive synergetic effects between these metals in the next section.

Promoters and their contribution to aromatization reactions have also been studied. Zn as a base metal on HZSM-5 catalysts is shown below with various metal promoters. Figures 4.14d and 4.16 show quantifiable results and liquid product distribution, respectively. On addition of most metals (Ru, Ag, Pd, and Ir), the performance is actually reduced when compared to that of 4.1% Zn/HZSM-5 with only Ga and Pt showing an increase in BTX yields, no doubt a result of their well-established attributes as seen in literature [63]. However, when concerned with the poorer performing metal promoters, a likely result is due to increased loading of metals as mentioned earlier with the 5.84% by weight loading of Zn to the catalyst. This is likely what is being seen here and the performance of the catalyst is thus impeded. In the cases of Pt and Ga, however, these metals are efficient dispersion promoters [63], leading to increased dispersion of both Zn species and Ga and Pt themselves during catalyst synthesis. A higher metal dispersion is indeed witnessed in the case of Pt (Table 4.2) and is witnessed in literature [64].

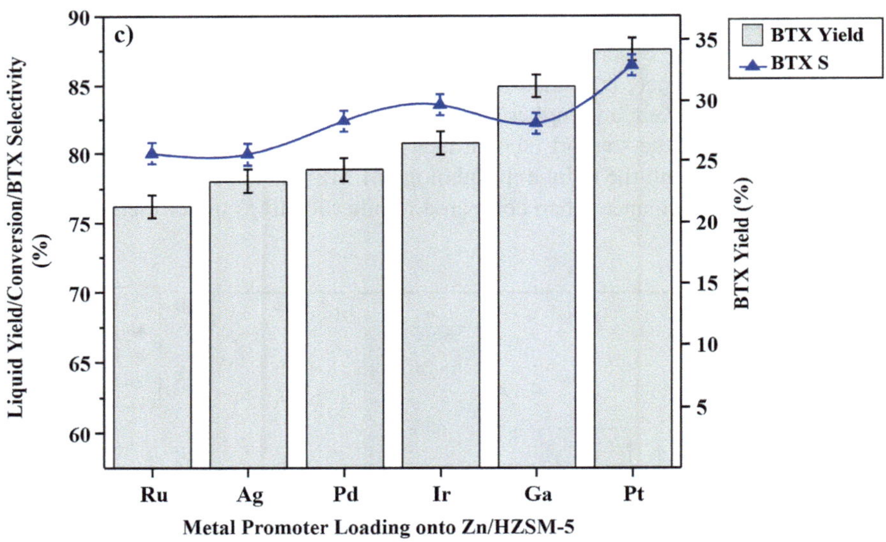

Fig. 4.16 A comparison of different promoter loadings on Zn/HZSM-5 (80:1) and their effect on various quantifiable values [21]

Table 4.2 Metal loadings and dispersion for varying catalysts [21]

Catalyst	Zn loading (Wt. %)	Promoter loading (Wt. %)	Metal dispersion (%)	Active metal site (m^2/g)
Ru-Zn/HZSM-5	3.46	0.09	3.8	1.37
Ag-Zn/HZSM-5	3.86	0.45	0.6	0.48
Pd-Zn/HZSM-5	3.06	0.51	6.5	2.50
Ir-Zn/HZSM-5	4.16	0.28	1.2	0.51
Pt-Zn/HZSM-5	3.90	0.57	7.2	1.97

External surface sites can be blocked using APTES, a bulky organosilicon, to conduct a preliminary investigation on the role that external sites have to play [61] when different metal promoters are employed. Interestingly, when looking at Fig. 4.17, the deactivation of the external sites has a negative effect when the Pt catalyst is employed and a positive effect for the others, albeit a small increase in some cases. Of particular note is the increase of performance on the Pd catalyst, where originally the performance was mediocre at best; upon blocking the external sites, the performance increases drastically. On the other hand, Pt and Ga catalysts, which gave promising results when free of blocking, seem to be negatively affected or only affected in a minor way, by the blocking of external sites. Perhaps Ga and Pt were not well distributed within the pores of the catalyst, preferring to locate on the external surface, and hence upon blocking of these external sites, these metal species are rendered ineffective. In fact, it has been suggested by Lou et al. [65] that Pd may prefer to diffuse into the internal pores of the catalyst, allowing the assumption that more effective reactions are permitted to occur if the external surface sites are blocked. At any rate, the performance is not increased dramatically in favor of BTX selectivity and BTX yield, and so the extra costs associated with catalyst synthesis may not be worthwhile for industrial applications. Thus, Pt-Zn/HZSM-5 (80:1) is investigated further using varying characteristic techniques as well as 4.1%Zn/HZSM-5 (80:1) and HZSM-5 (80:1) to establish the positive synergetic effect that

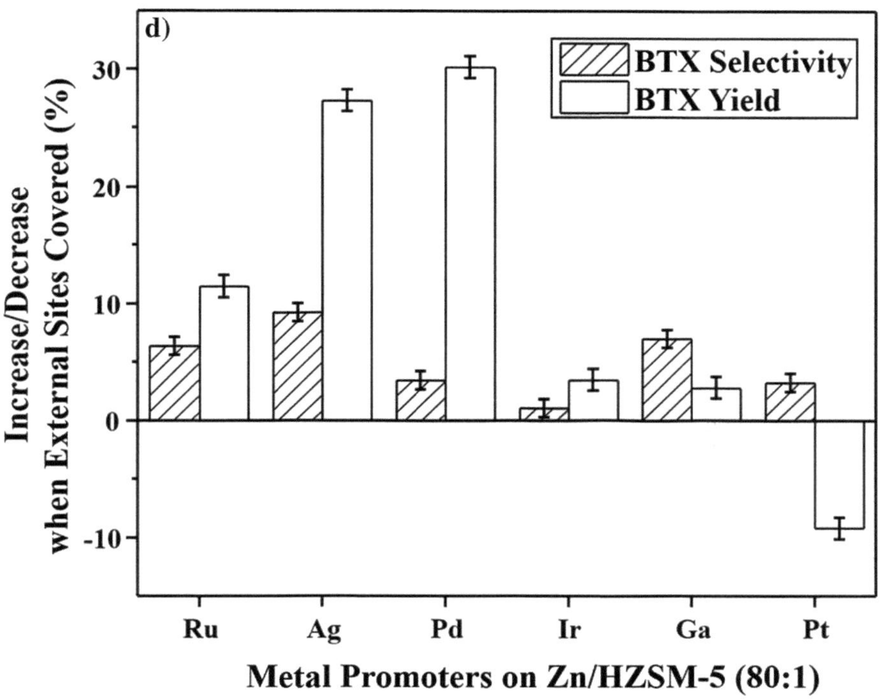

Fig. 4.17 The effect of external surface coverage on BTX selectivity and yield for promoter loaded Zn/HZSM-5 catalysts [21]

Pt and Zn have for these reactions. As previously mentioned, the enhanced dispersion of metallic functions as promoted by Pt is one reason for improved performance as well as the findings made by Mroczek et al. [66] where Pt-Zn/HZSM-5 resulted in the highest aromatization activity in their investigation on the aromatization of ethane. It was concluded that Pt and Zn can exhibit a stronger interaction between the corresponding metallic sites and show a higher activity towards aromatization. Despite this, the blocking of external sites deserves future investigation using model compounds on the reaction system used here to try and elucidate the reason for the interesting results observed.

Now that a base understanding is given for the aromatization of heavy naphtha feeds, we will discuss the role of a methane environment for the same reactions. This was done by using an inert nitrogen environment over the Pt-Zn/HZSM-5 (80:1) catalyst, and the results are shown in Figs. 4.18 and 4.19. When looking at the effect on liquid product distribution, fewer polyaromatics are formed under a methane environment, possibly a result of a reduced environment upon CH_4 activation and increased generation of H_2 which might stabilize the formed monoaromatics for preventing its further condensation to produce polyaromatics. As well as this, methane incorporation finds preference on the benzylic sites of benzene due to a slight increase in toluene and xylene components in the liquid product after reaction under methane compared to nitrogen and a slight reduction in benzene formation under methane, corroborated during other mechanistic studies [37]. As well as this, Fig. 4.19 demonstrates an increased liquid yield upon performing the reaction under a methane environment, a likely result of methane incorporation into the liquid products and an attractive reality for future industrial applications.

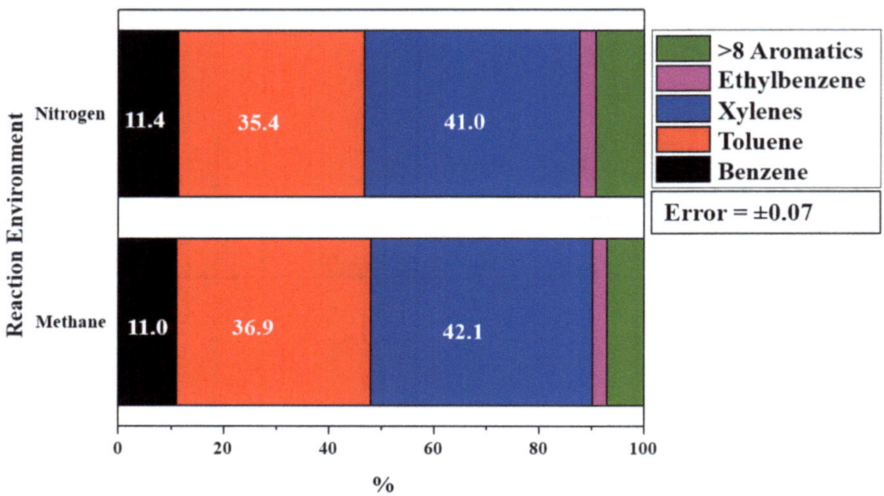

Fig. 4.18 A comparison of different environments and their effects on the distribution of components in the liquid product using a Pt-Zn/ZSM-5 catalyst [21]

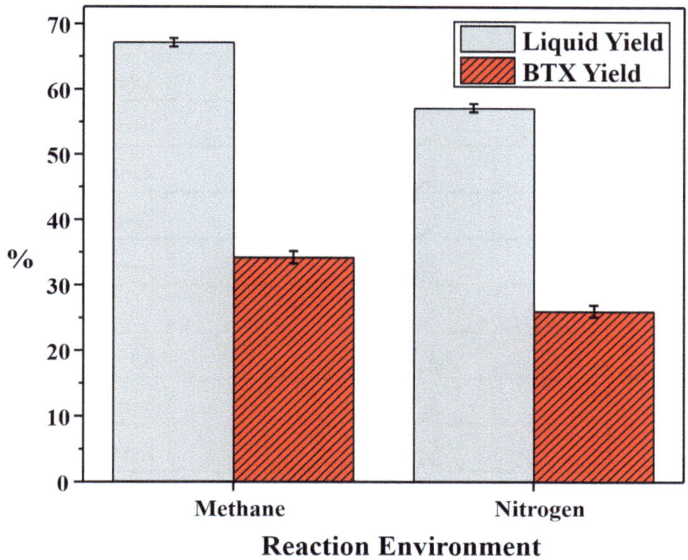

Fig. 4.19 A comparison of different environments and their effects on liquid and BTX yields using a Pt-Zn/ZSM-5 catalyst [21]

Methane clearly enhances the performance of the catalyst in all aspects and is even more successful using Pt-Zn/HZSM-5 than any other catalyst shown in this case study.

XRD patterns are displayed for all the aforementioned catalysts in Figs. 4.20 and 4.21, with XRD patterns for Pt-Zn/HZSM-5, 4.1%Zn/HZSM-5, and HZSM-5 and their spent counterparts shown in Fig. 4.22. Unfortunately, the metal loading is too low, or dispersion is too high to witness the characteristic diffraction peaks of Pt, Zn, and their oxidized forms; however, all catalysts (fresh and spent) contain the peaks associated with HZSM-5. Although loading of metals does reduce the crystallinity of the catalyst, the Pt promoter reduces the crystallinity to a lesser extent when compared to the catalyst where only Zn is loaded, suggesting that Pt can help to maintain crystallinity of these planes and thus, might increase performance. As well as this, when only Zn is loaded onto the catalyst, the peaks of the spent one are reduced significantly, suggesting that Zn migration during the reaction may be occurring and is affecting the structure of the catalyst. This is not the case when Pt is loaded as a promoter, where the crystal structure remains similar before and after the reaction. As well as this, Pt addition seems to provide more stability by comparing the diffraction patterns before and after reactions when compared to HZSM-5 and 4.1% Zn/HZSM-5 catalysts. The notable decrease in lower-angle peaks (approximately 7.5° and 9.5°) upon loading of 4.1% Zn and Pt-Zn is likely a result of significantly decreased micropores within the zeolite structure, a possible result of metal species blocking the inner pores.

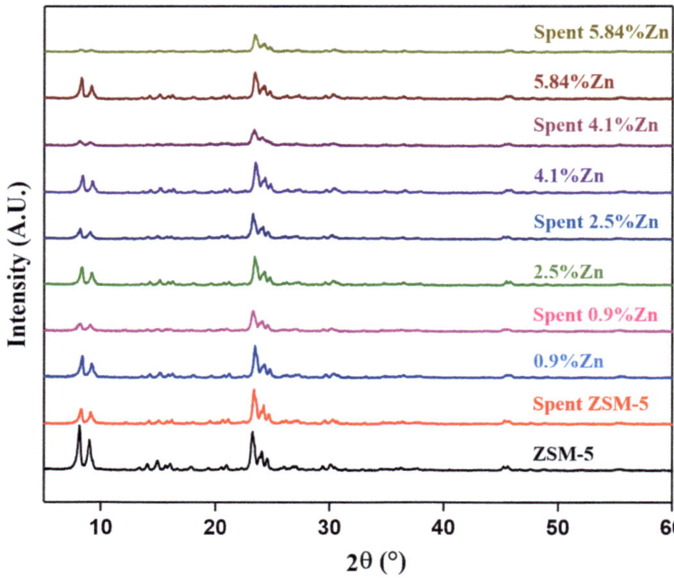

Fig. 4.20 XRD patterns of HZSM-5 and varying Zn metal loadings as well as their spent counterparts under a methane environment [21]

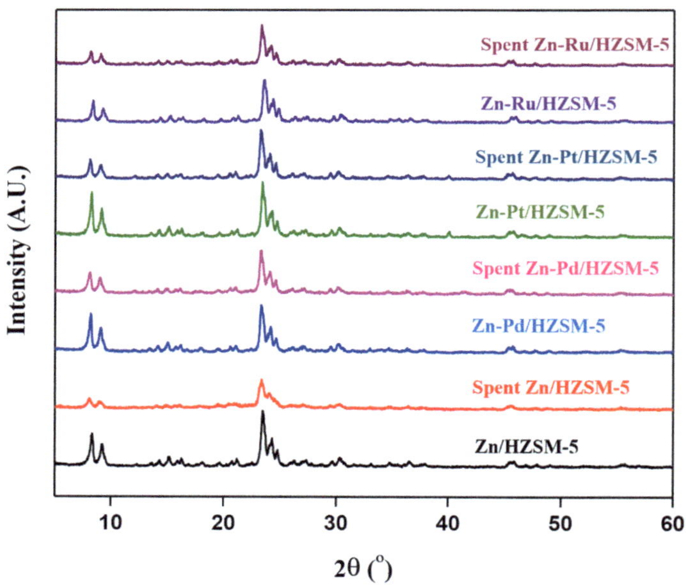

Fig. 4.21 XRD patterns of Zn/HZSM-5 and varying promoter loadings as well as their spent counterparts under a methane environment [21]

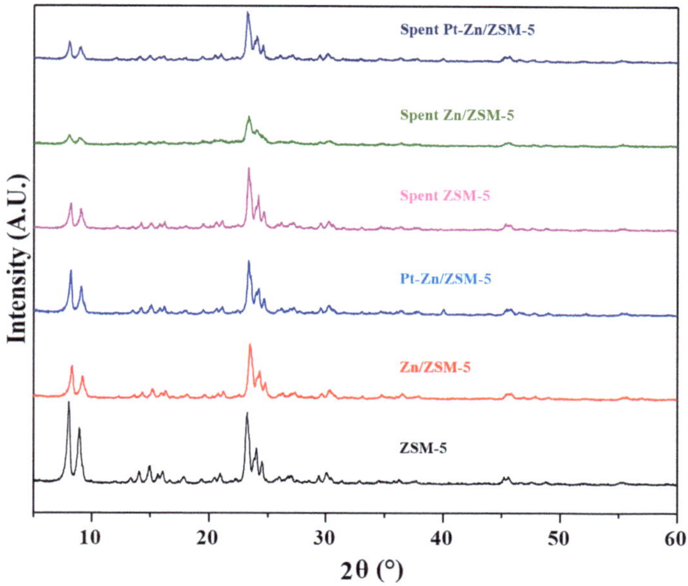

Fig. 4.22 XRD patterns of ZSM-5, Zn/ZSM-5, and Pt-Zn/ZSM-5 and their spent counterparts under a methane environment [21]

TEM is an effective microscopic technique for analyzing catalyst materials (Fig. 4.23) and can be used to witness the morphology of Pt-Zn/HZSM-5 after reactions under methane and nitrogen. In both cases, fringing is witnessed in the images, showing that the crystalline form of the catalyst stays intact after the reactions. Small particles can also be seen when referring to Fig. 4.23a and b, with average particle sizes of 1.7 nm for the catalyst under a methane environment and 2.9 nm for the catalyst under a nitrogen environment. The small particles are assumed to be that of Zn and/or Pt with smaller particle sizes attributed to less agglomeration under a methane environment when compared to nitrogen, a likely explanation for increased performance under methane. As well as this, a greater dispersion of particles is seen under methane compared to nitrogen, leading to the conclusion that despite platinum's ability to improve dispersion on the catalyst, methane also plays a role during the reaction that reduces metal agglomeration and enhances dispersion, leading to enhanced performance.

NH_3-TPD data is shown in Fig. 4.24 and Table 4.3 and allows the quantification of acidic sites for the three catalysts in question. Two distinct peaks are observed in the case of HZSM-5 at the low-temperature region (210 °C) and high-temperature region (420 °C) and are attributed to desorption from weak acidic sites and strong acidic sites, respectively [67]. Upon metal loading, however, the peak associated with strong acidic sites is depleted and in fact shifted towards lower temperatures (360 °C), whereas the peak associated with weak acid sites is intensified with the addition of a shoulder peak at approximate temperatures of 260 °C. This indicates that some of the strong acidic sites associate with the metal species to form weaker

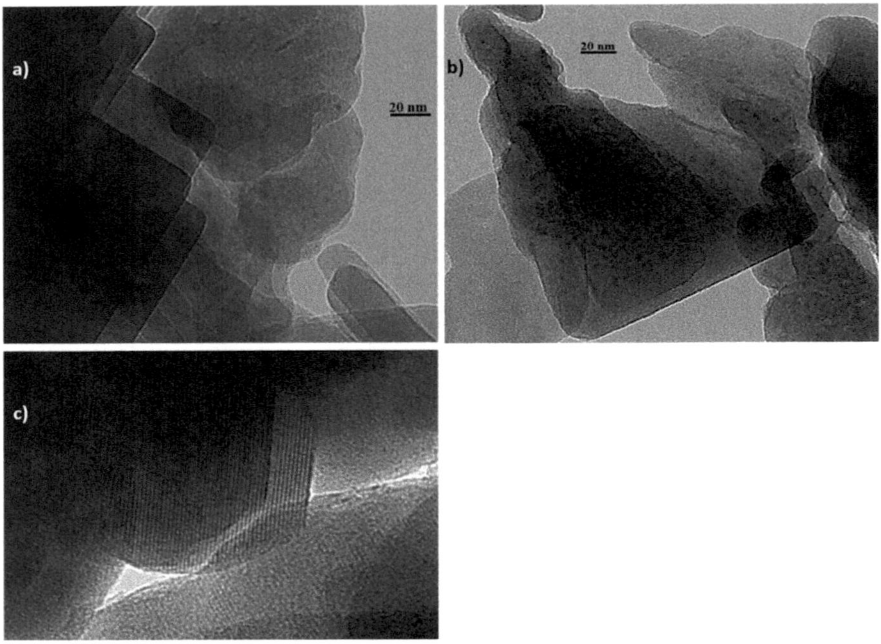

Fig. 4.23 TEM images of the spent Pt-Zn/HZSM-5 catalyst collected under. (**a**) CH_4 and (**b**) N_2 environments as well as (**c**) fringing of the Pt-Zn/HZSM-5 catalyst under CH_4 [21]

acidic sites, an effect that clearly enhances the performance of the catalysts compared to HZSM-5. Recalling observations made earlier in this case study, metal loadings in both cases (Zn and Pt-Zn) are shown to increase the catalyst performance in terms of BTX yields and are attributed to strong acidic sites which are replaced by weaker acidic sites in the form of Lewis sites provided by oxidized metal species. In addition to this, it has been suggested by Treesukol et al. [68] that Pt loaded on HZSM-5 would not be stable without the presence of Brønsted acid sites due to the redistribution of electron density in the Pt atom. This would help support the maintained crystallinity of Zn-Pt/HZSM-5 during reaction compared to Zn/HZSM-5, as the Brønsted acid sites help prevent agglomeration of Pt atoms but not in the case of Zn alone. Furthermore, it is suggested that framework oxygens of the zeolite framework provide electron density to the Pt atoms. This would in turn increase the acidity of strong acid sites within the catalyst as is witnessed in Table 4.3 and Fig. 4.24. However, the nature of ammonia does not allow us to discern the contribution of Brønsted and Lewis acid sites and so to support this data. Therefore, DRIFT spectra (Fig. 4.25) were also obtained using pyridine as the molecular probe. Peaks at 1445 cm^{-1} and 1614 cm^{-1} are attributed to Lewis acid sites [69] with the peak at 1597 cm^{-1} being assigned to H-bonded pyridine, and the peak at 1492 cm^{-1} is an indication of both Brønsted and Lewis acid sites [37]. The peak at 1542 cm^{-1} is only evident on the HZSM-5 catalyst and is a result of Brønsted acid sites [38]. This supports the NH_3-TPD data through an increase in Lewis acid sites when compared to Brønsted sites upon metal loading, perhaps indicating the interaction of Zn

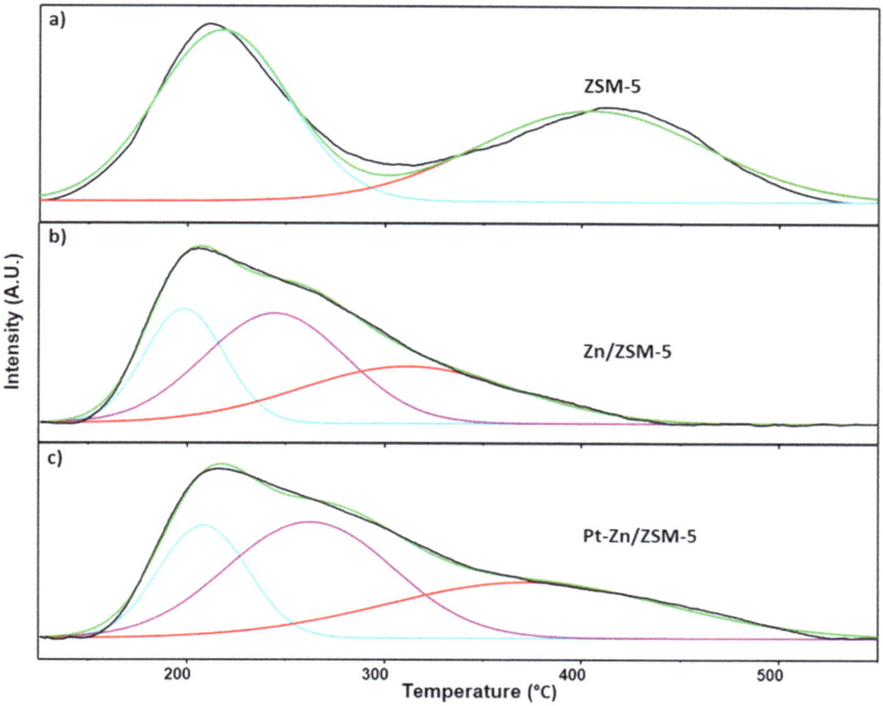

Fig. 4.24 NH_3-TPD profiles of (**a**) pure ZSM-5, (**b**) Zn/HZSM-5, and (**c**) Pt-Zn/ZSM-5 catalysts [21]

Table 4.3 The acidity of varying catalysts (μmol NH_3/g cat) [21]

Catalyst	Weak acidic sites	Medium acidic sites	Strong acidic sites	Total
HZSM-5 (80:1)	546	0	504	1050
Zn/HZSM-5 (80:1)	307	533	403	1242
Pt-Zn/HZSM-5 (80:1)	249	493	442	1185

species with Brønsted sites to form ZnO species and/or $[ZnOZn]^{2+}$ [70]. Lewis acid site depression in the case of Zn/HZSM-5 when compared to Pt-Zn/HZSM-5 is attributed to a poorer metal dispersion in the case of Zn/HZSM-5, leading to the less concentrated formation of Lewis acid sites.

Heavy naphtha upgrading to valuable chemical intermediates is an important process for industries to provide environmental and economic stability. The feasibility of this venture has been demonstrated during this case study in the form of a bimetallic heterogenous catalyst (Zn-Pt/HZSM-5) whose performance is exemplary for the formation of BTX from a complex naphtha feed under a methane environment. XRD, NH_3-TPD, CO chemisorption, and DRIFT analyses enable the characterization of catalytic materials to elucidate the reason for improved performances compared to conventional HZSM-5 and Zn/HZSM-5, by showing Pt's ability to promote metal dispersion and maintain catalyst crystallinity throughout reactions as well as

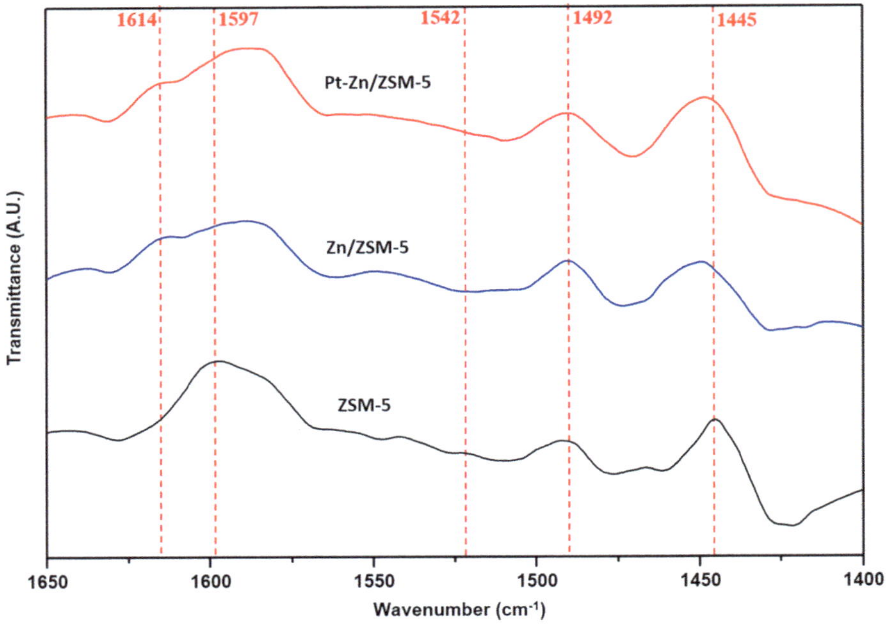

Fig. 4.25 DRIFT spectra of ZSM-5, Zn/ZSM-5, and Pt-Zn/ZSM-5 using pyridine as the molecular probe for adsorption, followed by desorption at 400 °C [21]

increase stability and provide the catalyst with the suitable acidic sites to enhance selectivity to BTX. As well as this, methane as an alternative source of hydrogen, an attractive prospect compared to current reformation techniques, is being realized in studies such as the outlined above. Experimental investigations have been discussed so far in the form of case studies, the next section will outline current mechanistic understandings for the cross-coupling of methane with paraffins, the dominant hydrocarbon type found in downstream refinery streams such as naphtha.

4.4 Mechanisms in Cross-Coupling of Methane with Hydrocarbons

The mechanism for methane activation is believed to be like that of the direct MDA reaction discussed in Chap. 3. However, the methyl radicals can now be incorporated into higher carbon number species due to the co-feeds which are being simultaneously activated at a higher rate due to their lower energies of activation compared to methane. Olefins, naphthenes, and paraffins will thus be discussed so that the reader may better understand their pathways towards isomers and aromatics. Based on experimental observations, it may be assumed that the methyl radicals resulting from methane activation are involved (without the exact incorporation mechanism being known).

4.4.1 *Mechanism of Olefin Aromatization*

Due to the reactive nature of the C=C bond found in olefins, the product distribution ranging during their upgrading is vast and often difficult to control. The works of Chen et al. over a Ni-Mo/HZSM-5 catalyst revealed a complex mechanism for olefin aromatization through olefin interconversion where the olefins undergo dimerization, cracking, and isomerization steps [71]. During olefin interconversion, a pair of light olefins (C_3–C_6) may dimerize to form single higher olefins which may then isomerize and/or crack. It should be noted that isomerization products of light olefins result from cracked higher carbon number olefin cracking and not from the direct isomerization of light olefins themselves. Olefins may also undergo aromatization through a series of steps: dehydrogenation, hydrogen transfer, and cyclization. In this aromatization mechanism, the olefinic species are first cyclized over acid sites and then undergo hydrogen transfer over the same sites. Dehydrocyclization is thought to occur over the metal sites as opposed to acid sites. Cyclic dienes and aromatics can then be formed by subsequent hydrogen transfer and dehydrogenation [72]. It is also likely that the olefins are saturated which is also a desirable outcome for fuel improvement. During this mechanism, the olefins undergo hydrogenation over metal sites and hydrogen transfer reactions over acidic sites. Figure 4.26 below summarizes these processes.

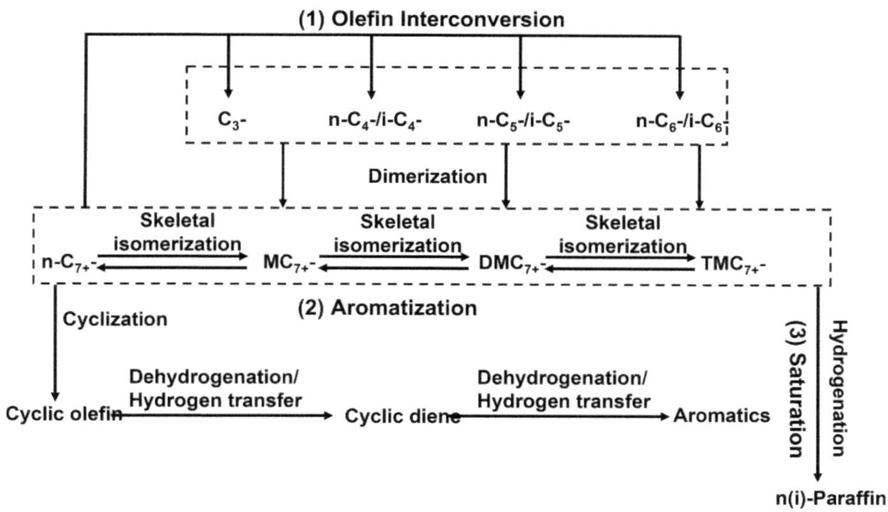

Fig. 4.26 Schematic representation showing the reaction pathways of olefins during the FCC naphtha hydroisomerization and hydroaromatization over a Ni-Mo/Al_2O_3-HZSM-5 bifunctional catalyst. Adapted with permission from (Fuel Processing Technology 2015, 130, 117–126). Copyright (2015) Elsevier

4.4.2 Mechanism of Naphthene Aromatization

Studies have been conducted to elucidate the role of both acid and metal catalytic active sites on the aromatization of naphthenes [73]. It was shown that dehydrogenation occurs mainly on the Ni active centers of a Ni/HZSM-5 catalyst and that a synergetic effect occurs between Ni and the Brønsted acid sites to yield the overall dehydrogenation and aromatization reactions. Hydrocracking and isomerization of methylcyclohexane (MCH) takes place in several steps: first is that the saturated cycloalkanes are transformed to their respective olefins via dehydrogenation over the metal sites, and then the olefin intermediates are protonated to carbenium ions on the Brønsted acid sites, which either isomerize or crack. Finally, these unsaturated products can be further hydrogenated to the saturated rings (cyclic paraffins) or the saturated lower molecular paraffins (principally propane).

4.4.3 Mechanism of Paraffin Aromatization

A detailed mechanism was proposed by Jarvis and Harrhy et al. for the aromatization of n-octane, one of the heavier and most abundant alkanes found in naphtha [74]. Although this study was not investigated under a methane environment, the mechanism for aromatization of paraffins still holds and provides valuable insight into the role of single-atom Pt sites. By combining their observations, it was concluded that the C_6 ring closure mechanism (C_1–C_6 direct ring closure over a single site) for paraffins occurs through the following mechanism over Pt/ZSM-5 catalyst (Fig. 4.27): the terminal methyl of n-octane approaches Pt cluster to which the sp^3 hybridized carbon binds, steric hinderance of the inner pore forces the opposite terminal methyl group towards the same Pt cluster. However, the formation of a C_8 metallocycle is unfavorable compared to the most stable C_6 metallocycle, assisted

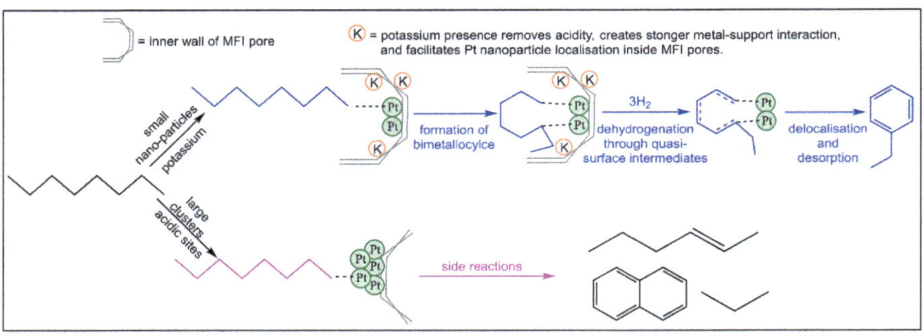

Fig. 4.27 Proposed mechanism for n-octane reaction over Pt nanoparticles inside the pores of KMFI and large cluster on the external surface of MFI. Bimetallocycle mechanism for the reaction of n-octane over Pt/KMFI (KX) (blue). Pt cluster mechanism for the reaction of n-octane over Pt/MFI (magenta) [74]

by puckering. 2,7 ring closure can also occur, to a lesser extent as the C_2 in n-octane, although less likely to bond to the Pt cluster than terminal methyls, is still preferable over other carbons. The resulting C_6 bimetallocycle, formed over Pt nanoparticles, undergoes subsequent dehydrogenation to form a bimetallo-ethylcyclohexene followed by bimetallo-ethylcyclohexadiene and finally a bimetallo-ethylcyclohexatriene. Note that all these steps occur on the same site, and no desorption and readsorption is thought to occur due to the absence of cyclopentenes in the product [74], which would form if the ethylcyclohexene were to desorb and readsorb as shown in the previous subsection. Furthermore, the absence of free intermediates in the product stream of low temperature reactions strengthens this hypothesis. Finally, the species becomes unbound from the Pt cluster to give the final product, ethylbenzene or o-xylene, if 2,7 ring closure has occurred. It is reiterated that there are thought to be no gas-phase intermediates during this process and that the aromatization of n-alkanes likely occurs over the same Pt nanoparticles for the duration of the reaction.

4.5 Highlights and Summary on the Cross-Coupling of Methane

A series of naphtha feedstocks have been evaluated over years with extensive studies conducted on conventional naphtha and light straight run naphtha. Two case studies have been provided here for the aromatization of light and heavy naphtha feeds. It should be noted that there are many possible bimetallic catalysts that can be suitable for the cross-coupling of methane with paraffinic feeds; however, these case studies focused on Pt-Zn intermetallic nanoparticles on ZSM-5. It was found that the electronic changes occurring on these metals as well as a unique level of acidity provided to the support upon the loading of these metals can be attributed to their success. These catalysts and feedstocks have been subjected to batch reactor and continuous fixed bed flow reactor systems to provide an idea of their applicability on an industrial scale. Methane has been compared to nitrogen-controlled experiments and methane provides an improved selectivity through potential incorporation into liquid products, reduced metal particle agglomeration, and reduced coking on the catalyst materials.

When heavy naphtha (C_6–C_{12}) was employed as the co-feed with methane, the following conclusions were made:

- Pt promotes metal dispersion, assessed by pulsed CO chemisorption.
- Catalyst crystallinity can be maintained throughout the reaction, obtained by XRD.
- Pt-Zn presence may provide the necessary ratio and strength of acidic sites to promote the reaction.
- Methane presence increased liquid yield and BTX yield compared to nitrogen environment, possibly a result of methane incorporation into the liquid products.

Light naphtha in the form of LSR (C_5–C_7) was also employed as a co-feed with methane over Pt-Zn/ZSM-5 with the following conclusions:

- Reduced acidity is more effective for the upgrading of LSR compared to heavier naphtha.
- No formation of polyaromatics or olefins is witnessed, meaning that any unwanted products are not detrimental to the processes used.
- Methane environment reduces coking, enhances liquid yield, and increases BTX selectivity compared to nitrogen environment.
- Pt-Zn presence provides desirable distribution and strength of acid sites once again.

Although much is known about the mechanisms for different hydrocarbons to aromatics, the incorporation of methane and subsequent mechanisms is lacking. Given the greatly reduced thermodynamic constraints of the cross-coupling methane technology, the process is far more promising than the direct MDA reaction mentioned in Chap. 3. However, more work is needed as well as collaboration between commercial outfits and Research and Development in order to scale up technologies such as those shown here. Further innovation on a larger scale using technologies such as those shown here may provide a real solution for the utilization and activation of methane in natural gas.

References

1. Q. Li et al., Investigation on the light alkanes aromatization over Zn and Ga modified HZSM-5 catalysts in the presence of methane. *Fuel* **219**, 331–339 (2018). https://doi.org/10.1016/j.fuel.2018.01.104
2. W. Nabgan, M. Rashidzadeh, B. Nabgan, The catalytic naphtha reforming process: Hydrodesulfurization, catalysts and zeoforming. Environ. Chem. Lett. **16**(2), 507–522 (2018). https://doi.org/10.1007/s10311-018-0707-x
3. Haensel, United States Patent office. (1951)
4. M.R. Rahimpour, M. Jafari, D. Iranshahi, Progress in catalytic naphtha reforming process: A review. Appl. Energy **109**, 79–93 (2013) USPTO. https://doi.org/10.1016/j.apenergy.2013.03.080
5. R. J. Argauer. Crystalline Zeolite ZSM-5 and Method of Preparing the Same. 1972
6. F. Yaripour, Z. Shariatinia, S. Sahebdelfar, A. Irandoukht, Conventional hydrothermal synthesis of nanostructured H-ZSM-5 catalysts using various templates for light olefins production from methanol. J. Nat. Gas Sci. Eng. **22**, 260–269 (2015). https://doi.org/10.1016/j.jngse.2014.12.001
7. L. Meng et al., Direct synthesis of hierarchical ZSM-5 zeolite using cetyltrimethylammonium as structure directing agent for methanol-to-hydrocarbons conversion. Cat. Sci. Technol. **7**(19), 4520–4533 (2017). https://doi.org/10.1039/c7cy01603d
8. A. Hart, G. Leeke, M. Greaves, J. Wood, Down-hole heavy crude oil upgrading by CAPRI: Effect of hydrogen and methane gases upon upgrading and coke formation. Fuel **119**, 226–235 (2014). https://doi.org/10.1016/j.fuel.2013.11.048
9. F. Feijoo et al., The future of natural gas infrastructure development in the United states. *Appl. Energy* **228**, 149–166 (2018). https://doi.org/10.1016/j.apenergy.2018.06.037

10. V.R. Choudhary, Low-temperature nonoxidative activation of methane over H-Galloaluminosilicate (MFI) zeolite. Science (80-.) **275**(5304), 1286–1288 (1997). https://doi.org/10.1126/science.275.5304.1286
11. L. Wang, L. Tao, M. Xie, G. Xu, J. Huang, Y. Xu, Dehydrogenation and aromatization of methane under non-oxidizing conditions. Catal. Lett. **21**(1–2), 35–41 (1993). https://doi.org/10.1007/BF00767368
12. Y. Xu, S. Liu, X. Guo, L. Wang, M. Xie, Methane activation without using oxidants over Mo/HZSM-5 zeolite catalysts. Catal. Lett. **30**(1–4), 135–149 (1994). https://doi.org/10.1007/BF00813680
13. V.R. Choudhary, A.K. Kinage, T.V. Choudhary, Low-temperature nonoxidative activation of methane over H- galloaluminosilicate (MFI) zeolite. Science (80-.) **275**(5304), 1286–1288 (1997). https://doi.org/10.1126/science.275.5304.1286
14. A. Chica, A. Corma, Hydroisomerization of pentane, hexane, and heptane for improving the octane number of gasoline. J. Catal. **187**(1), 167–176 (1999). https://doi.org/10.1006/jcat.1999.2601
15. P. Ghosh, K.J. Hickey, S.B. Jaffe, Development of a detailed gasoline composition-based octane model. Ind. Eng. Chem. Res. **45**(1), 337–345 (2006). https://doi.org/10.1021/ie050811h
16. Y. Fan, X. Bao, G. Shi, W. Wei, J. Xu, Olefin reduction of FCC gasoline via hydroisomerization aromatization over modified HMOR/HZSM-5/Hβ composite carriers. Appl. Catal. A Gen. **275**(1–2), 61–71 (2004). https://doi.org/10.1016/j.apcata.2004.07.046
17. Y. Ono, A survey of the mechanism in catalytic isomerization of alkanes. *Catal. Today* **81**(1), 3–16 (2003). https://doi.org/10.1016/S0920-5861(03)00097-X
18. X.R. Chen, Y.Q. Du, C.L. Chen, N.P. Xu, C.Y. Mou, Highly active and stable n-pentane isomerization catalysts without noble metal containing: Al- or Ga-promoted tungstated zirconia. Catal. Lett. **111**(3–4), 187–193 (2006). https://doi.org/10.1007/s10562-006-0146-3
19. M.A. Arribas, A. Martínez, Simultaneous isomerization of n-heptane and saturation of benzene over Pt/Beta catalysts - the influence of zeolite crystal size on product selectivity and sulfur resistance. Catal. Today **65**(2–4), 117–122 (2001). https://doi.org/10.1016/S0920-5861(00)00576-9
20. Z.P. Cano et al., Batteries and fuel cells for emerging electric vehicle markets. *Nature Energy* **3**(4), 279–289 (2018). https://doi.org/10.1038/s41560-018-0108-1
21. J. Jarvis, A. Wong, P. He, Q. Li, H. Song, Catalytic aromatization of naphtha under methane environment: Effect of surface acidity and metal modification of HZSM-5. Fuel **223**, 211–221 (2018). https://doi.org/10.1016/j.fuel.2018.03.045
22. X. Lin, Y. Fan, G. Shi, H. Liu, X. Bao, Coking and deactivation behavior of HZSM-5 zeolite-based FCC gasoline hydro-upgrading catalyst. Energy Fuels **21**(5), 2517–2524 (2007). https://doi.org/10.1021/ef0700634
23. A. Haryanto, S. Fernando, N. Murali, S. Adhikari, Current status of hydrogen production techniques by steam reforming of ethanol: A review. Energy Fuels **19**(5), 2098–2106 (2005). https://doi.org/10.1021/ef0500538
24. P. Tang, Q. Zhu, Z. Wu, D. Ma, Methane activation: The past and future. Energy Environ. Sci. **7**(8), 2580–2591 (2014). https://doi.org/10.1039/C4EE00604F
25. V. Abdelsayed, M.W. Smith, D. Shekhawat, Investigation of the stability of Zn-based HZSM-5 catalysts for methane dehydroaromatization. Appl. Catal. A Gen. **505**, 365–374 (2015). https://doi.org/10.1016/j.apcata.2015.08.017
26. J. Liu et al., The crucial role of skeleton structure and carbon number on short-chain alkane activation over Zn/HZSM-5 catalyst: an experimental and computational study. *Catal. Lett.* **148**(7), 2069–2081 (2018). https://doi.org/10.1007/s10562-018-2394-4
27. Y. Zhao, T.Y. Gao, Y.J. Wang, Y.J. Zhou, G.Q. Huang, Zinc supported on alkaline activated HZSM-5 for aromatization reaction. React. Kinet. Mech. Catal. **125**(2), 1085–1098 (2018). https://doi.org/10.1007/s11144-018-1426-9

28. A. Roshanaei, S.M. Alavi, Using two-zone fluidized bed reactor in propane aromatization over Zn/HZSM-5 catalyst. Fuel Process. Technol. **176**, 197–204 (2018). https://doi.org/10.1016/j.fuproc.2018.03.020
29. A. Samanta, X. Bai, B. Robinson, H. Chen, J. Hu, Conversion of light alkane to value-added chemicals over ZSM-5/metal promoted catalysts. Ind. Eng. Chem. Res. **56**(39), 11006–11012 (2017). https://doi.org/10.1021/acs.iecr.7b02095
30. N. Parsafard, M.H. Peyrovi, M. Jarayedi, Catalytic study and kinetic modeling of the n-heptane isomerization over Pt/Al-HMS/HZSM-5 hybrid catalysts. Energy Fuels **31**(6), 6389–6396 (2017). https://doi.org/10.1021/acs.energyfuels.7b00657
31. X. Bai, A. Samanta, B. Robinson, L. Li, J. Hu, Deactivation mechanism and regeneration study of Ga-Pt promoted HZSM-5 catalyst in ethane dehydroaromatization. Ind. Eng. Chem. Res. **57**(13), 4505–4513 (2018). https://doi.org/10.1021/acs.iecr.7b05094
32. T.E. Tshabalala, M.S. Scurrell, Aromatization of n-hexane over Ga, Mo and Zn modified H-ZSM-5 zeolite catalysts. Catal. Commun. **72**, 49–52 (2015). https://doi.org/10.1016/j.catcom.2015.06.022
33. J. Jarvis, P. He, A. Wang, H. Song, Pt-Zn/HZSM-5 as a highly selective catalyst for the co-aromatization of methane and light straight run naphtha. Fuel **236**, 1301–1310 (2019). https://doi.org/10.1016/j.fuel.2018.09.109
34. W. Ding, Y. Cui, J. Li, Y. Yang, W. Fang, Promoting effect of dual modification of H-ZSM-5 catalyst by alkali treating and Mg doping on catalytic performances for alkylation of benzene with ethanol to ethylbenzene. RSC Adv. **4**(91), 50123–50129 (2014). https://doi.org/10.1039/C4RA07918C
35. L. Xiao, L. Wang, Methane activation on Pt and Pt4: A density functional theory study. J. Phys. Chem. B **111**(7), 1657–1663 (2007). https://doi.org/10.1021/jp065288e
36. M.B. Sayed, R.A. Kydd, R.P. Cooney, A Fourier-transform infrared spectral study of H-ZSM-5 surface sites and reactivity sequences in methanol conversion. J. Catal. **88**(1), 137–149 (1984). https://doi.org/10.1016/0021-9517(84)90059-9
37. P. He, R. Gatip, M. Yung, H. Zeng, H. Song, Co-aromatization of olefin and methane over Ag-Ga/ZSM-5 catalyst at low temperature. Appl. Catal. B Environ. **211**, 275–288 (2017). https://doi.org/10.1016/j.apcatb.2017.04.052
38. C.A. Emeis, Determination of integrated molar extinction coefficients for infrared absorption bands of pyridine adsorbed on solid acid catalysts. J. Catal. **141**(2), 347–354 (1993). https://doi.org/10.1006/jcat.1993.1145
39. B.S. Liu et al., Characteristic and mechanism of methane dehydroaromatization over Zn-based/HZSM-5 catalysts under conditions of atmospheric pressure and supersonic jet expansion. J. Phys. Chem. C **115**(34), 16954–16962 (2011). https://doi.org/10.1021/jp2027065
40. B.-T. Lønstad Bleken et al., Probing the surface of nanosheet H-ZSM-5 with FTIR spectroscopy. *Phys. Chem. Chem. Phys.* **15**(32), 13363 (2013). https://doi.org/10.1039/c3cp51280k
41. Q. Li et al., Catalytic co-aromatization of methane and heptane as an alkane model compound over Zn-Ga/ZSM-5: A mechanistic study. *Appl. Catal. B Environ.* **236** (2018). https://doi.org/10.1016/j.apcatb.2018.05.006
42. L. Rodríguez-González, F. Hermes, M. Bertmer, E. Rodríguez-Castellón, A. Jiménez-López, U. Simon, The acid properties of H-ZSM-5 as studied by NH3-TPD and27Al-MAS-NMR spectroscopy. Appl. Catal. A Gen. **328**(2), 174–182 (2007). https://doi.org/10.1016/j.apcata.2007.06.003
43. M. Rasouli, N. Yaghobi, Synthesis and characterization of platinum impregnated Zn-ZSM5 nanocatalysts for xylene isomerization reactions. *Catal. Lett.* (2018). https://doi.org/10.1007/s10562-018-2439-8
44. S.K. Saxena, N. Viswanadham, M.O. Garg, Cracking and isomerization functionalities of bi-metallic zeolites for naphtha value upgradation. Fuel **107**, 432–438 (2013). https://doi.org/10.1016/j.fuel.2013.01.076
45. J. Yao, R. le van Mao, L. Dufresne, Conversion of n-butane into BTX aromatics on new hybrid catalysts. Appl. Catal. **65**(1), 175–188 (1990). https://doi.org/10.1016/S0166-9834(00)81596-1

46. J. Song et al., The effect of Fe on Pt particle states in Pt/KL catalysts. Appl. Catal. A Gen. **492**, 31–37 (2015). https://doi.org/10.1016/j.apcata.2014.12.017
47. R. Ahuja et al., Catalytic dehydroaromatization of n-alkanes by pincer-ligated iridium complexes. Nat. Chem. **3**(2), 167–171 (2011). https://doi.org/10.1038/nchem.946
48. D. Seddon, Paraffin oligomerisation to aromatics. Catal. Today **6**(3), 351–372 (1990). https://doi.org/10.1016/0920-5861(90)85009-D
49. M. Sterba, V. Haensel, Catalytic reforming. Ind. Eng. Chem. **15**, 1–17 (1976). https://doi.org/10.1021/i360057a002
50. J. Fling, I. Wang, Dehydrocyclization of C6C8 n-paraffins to aromatics over TiO2ZrO2 catalysts. J. Catal. **130**(2), 577–587 (1991). https://doi.org/10.1016/0021-9517(91)90137-S
51. B.S. Kwak, W.M.H. Sachtler, W.O. Haag, Catalytic conversion of propane to aromatics: Effects of adding ga and/or Pt to HZSM-5. J. Catal. **149**(2), 465–473 (1994). https://doi.org/10.1006/jcat.1994.1313
52. F.M. Dautzenberg, J.C. Platteeuw, Isomerization and dehydrocyclization of hexanes over monofunctional supported platinum catalysts. J. Catal. **19**(1), 41–48 (1970). https://doi.org/10.1016/0021-9517(70)90294-0
53. S. Kotrel, H. Knözinger, B.C. Gates, The Haag-Dessau mechanism of protolytic cracking of alkanes. Microporous Mesoporous Mater. **35–36**, 11–20 (2000). https://doi.org/10.1016/S1387-1811(99)00204-8
54. J.L. Motz, H. Heinichen, W.F. Holderich, Influence of extra-framework alumina in H-[Al]ZSM-5 zeolite on the direct hydroxylation of benzene to phenol. Stud. Surf. Sci. Catal. **105**, 1053–1060 (Jan. 1997). https://doi.org/10.1016/S0167-2991(97)80739-2
55. V.R. Choudhary, C. Sivadinarayana, A.K. Kinage, P. Devadas, M. Guisnet, H-Gallosilicate (MFI) propane aromatization catalyst influence of calcination temperature on acidity, activity and deactivation due to coking. Appl. Catal. A Gen. **136**(2), 125–142 (1996). https://doi.org/10.1016/0926-860X(95)00277-4
56. P. Mériaudeau, C. Naccache, Gallium based MFI zeolites for the aromatization of propane. Catal. Today **31**(3–4), 265–273 (1996). https://doi.org/10.1016/S0920-5861(96)00017-X
57. D.B. Lukyanov, T. Vazhnova, Aromatization activity of gallium containing MFI and TON zeolite catalysts in n-butane conversion: Effects of gallium and reaction conditions. Appl. Catal. A Gen. **316**(1), 61–67 (2007). https://doi.org/10.1016/j.apcata.2006.09.019
58. N. Viswanadham, A.R. Pradhan, N. Ray, S.C. Vishnoi, U. Shanker, T.S.R. Prasada Rao, Reaction pathways for the aromatization of paraffins in the presence of H-ZSM-5 and Zn/H-ZSM-5. Appl. Catal. A Gen. **137**(2), 225–233 (1996). https://doi.org/10.1016/0926-860X(95)00287-1
59. V.R. Choudhary, S.A. Mulla, S. Banerjee, Aromatization of n-heptane over H-AlMFI, Ga/H-AlMFI, H-GaMFI and H-GaAlMFI zeolite catalysts: influence of zeolitic acidity and non-framework gallium. Microporous Mesoporous Mater. **57**, 317–322 (2003). https://doi.org/10.1016/S1387-1811(02)00606-6
60. X. Su, W. Zan, X. Bai, G. Wang, W. Wu, Synthesis of microscale and nanoscale ZSM-5 zeolites: Effect of particle size and acidity of Zn modified ZSM-5 zeolites on aromatization performance. Cat. Sci. Technol. **7**(9), 1943–1952 (2017). https://doi.org/10.1039/c7cy00435d
61. W. Ding, G.D. Meitzner, E. Iglesia, The effects of Silanation of external acid sites on the structure and catalytic behavior of Mo/H–ZSM5. J. Catal. **206**(1), 14–22 (2002). https://doi.org/10.1006/jcat.2001.3457
62. Y. Ono, Transformation of lower alkanes into aromatic hydrocarbons over ZSM-5 zeolites. Catal. Rev. **34**(3), 179–226 (1992). https://doi.org/10.1080/01614949208020306
63. M.N. Mikhailov, I.V. Mishin, L.M. Kustov, A.L. Lapidus, Structure and reactivity of Pt/GaZSM-5 aromatization catalyst. Microporous Mesoporous Mater. **104**(1–3), 145–150 (2007). https://doi.org/10.1016/j.micromeso.2007.01.023
64. D. Schanke, "Study of Pt-promoted cobalt CO hydrogenation catalysts," Fuel Energy Abstr., vol. 37, no. 3. p. 176, 1996, doi: https://doi.org/10.1016/0140-6701(96)88441-4
65. Y. Lou, P. He, L. Zhao, H. Song, Refinery oil upgrading under methane environment over PdOx/H-ZSM-5: Highly selective olefin cyclization. Fuel **183**, 396–404 (2016). https://doi.org/10.1016/j.fuel.2016.06.074

66. U. Mroczek, W. Reschetilowski, K. Pietzsch, K.H. Steinberg, Aromatization of ethane on Pt and Pd supported Zn- and Mn-ZSM-5 zeolites. React. Kinet. Catal. Lett. **43**(2), 539–544 (1991)
67. F. Lónyi, J. Valyon, On the interpretation of the NH3-TPD patterns of H-ZSM-5 and H-mordenite. Microporous Mesoporous Mater. **47**(2–3), 293–301 (2001). https://doi.org/10.1016/S1387-1811(01)00389-4
68. P. Treesukol, K. Srisuk, J. Limtrakul, T.N. Truong, Nature of the metal-support interaction in bifunctional catalytic Pt/H-ZSM-5 zeolite. J. Phys. Chem. B **109**(24), 11940–11945 (2005). https://doi.org/10.1021/jp0511348
69. J. Fermoso et al., Lamellar and pillared ZSM-5 zeolites modified with MgO and ZnO for catalytic fast-pyrolysis of eucalyptus woodchips. Catal. Today **277**, 171–181 (2016). https://doi.org/10.1016/j.cattod.2015.12.009
70. H. Hernando et al., Biomass catalytic fast pyrolysis over hierarchical ZSM-5 and Beta zeolites modified with mg and Zn oxides. Biomass Convers. Biorefinery **7**(3), 289–304 (2017). https://doi.org/10.1007/s13399-017-0266-6
71. Z. Chen, J. Xu, Y. Fan, G. Shi, X. Bao, Reaction mechanism and kinetic modeling of hydroisomerization and hydroaromatization of fluid catalytic cracking naphtha. Fuel Process. Technol. **130**(C), 117–126 (2015). https://doi.org/10.1016/j.fuproc.2014.09.037
72. Y. Fan, J. Yin, G. Shi, H. Liu, X. Bao, Mechanistic pathways for olefin hydroisomerization and aromatization in fluid catalytic cracking gasoline hydro-upgrading. Energy Fuels **23**(6), 3016–3023 (2009). https://doi.org/10.1021/ef900030h
73. Y. Song et al., Aromatization and isomerization of methylcyclohexane over Ni catalysts supported on different supports. Green Energy Environ. **4**(1), 75–82 (2019). https://doi.org/10.1016/j.gee.2018.05.003
74. J.S. Jarvis et al., Inhibiting the dealkylation of basic Arenes during n-alkane direct aromatization reactions and understanding the C6Ring closure mechanism. ACS Catal. **10**(15), 8428–8443 (2020). https://doi.org/10.1021/acscatal.0c02361

Chapter 5
Catalytic Upgrading of Heavy Oil Resources Under Methane

5.1 Introduction

Because of the rapid technology advancement and economic growth, the processing of conventional oil resources cannot meet the global demand in the last few decades. A lot of governments and companies are investing in the extraction of unconventional oil resources, which are abundant but cannot be exploited using traditional methods. Unconventional petroleum resources can be classified into three categories based on their density and viscosity, which are heavy oil (900–1000 kg/m^3, 10^2–10^5 mPa·s), extra heavy oil (>1000 kg/m^3, <10^5 mPa·s), and bitumen (>1000 kg/m^3, >10^5 mPa·s) [1].

Canada's natural bitumen reserves are estimated to be 174 billion barrels. In 2009, the daily bitumen production from oil sands was 1.49 million barrels. Less than 20% of the reserves can be extracted via surface mining. Alternative extraction processes like steam-assisted gravity drainage (SAGD) are required to access the rest 80% of bitumen that has a deposit depth greater than 75 m. A normal SAGD operation produces bitumen at 5000–200,000 barrels per day, while 100,000 barrels per day of diluent will be consumed during operation [2].

One more adverse feature of the bitumen produced from SAGD operations is that it has a heavy density (1.0073 g/cm^{-3} and API gravity of 8.9) and a high dynamic viscosity (2×10^5–2×10^6 cp) under atmospheric pressure [2]. Pipeline transportation needs a density lower than 0.94 g/cm^3 and a dynamic viscosity lower than 330 cP (tested at 7.5–17 °C) [2, 3]. To meet the specification for pipeline transportation in the United States, the extracted bitumen from SAGD processes must be diluted by a solvent such as naphtha, gas condensate, or synthetic crude oil (SCO). However, before entering the refinery, the diluent needs to be removed from the diluted bitumen and recycled via pipeline or reused to produce fuels. Therefore, the elimination of diluent could be beneficial from the financial and operating perspectives.

H. Song et al., *Methane Activation and Utilization in the Petrochemical and Biofuel Industries*, https://doi.org/10.1007/978-3-030-88424-6_5

Hydrotreating is a traditional process used to enhance the oil quality, which could reduce the viscosity, density, and sulfur content. Although hydrogen is widely used in bitumen upgrading process, the operation cost of hydrotreatment processes is usually high due to the consumption of high-pressure hydrogen (>8 MPa) [4], which is mainly produced from the natural gas via steam reforming. A novel idea is to replace hydrogen with another hydrogen donor. As the main component of natural gas, methane is a cheaper and more abundant candidate when compared with hydrogen. Therefore, it would be more profitable if a novel technology can be developed to enhance the properties of the SAGD-extracted bitumen by selecting methane as the atmosphere for upgrading processes.

5.2 Nonoxidative Methane Activation in Heavy Oil Upgrading

As mentioned in previous chapters, methane is much more difficult to be activated due to its stable molecular structure compared to hydrogen. To make the aforementioned heavy oil upgrading with methane occurring at the comparative or even less severe conditions (e.g., lower reaction temperature and reduced operation pressure) used for hydrotreatment process, a special catalyst should be carefully designed. The created catalyst should be able to initiate the methane activation first to make it participate in the upgrading reaction. The activation of methane is initiated by the cracking of the first C-H bond in its tetrahedral structure, which requires a high activation energy (104 kcal/mol). Nonetheless, when the first C-H bond is cracked, the generated free radical is very active, and the rest of the C-H bonds will break automatically in a sequential reaction. Therefore, to realize the heavy oil upgrading with methane, the most important part is to design a catalyst that can effectively decrease the energy barrier of methane activation for less extreme reaction conditions.

In the past few decades, oxidative (OCM) and nonoxidative coupling (NOCM) of methane have been vastly studied for methane utilization. For oxidative approach, the reaction is under an oxidative environment so that the methane is converted for ethylene production [5–7]. Although hundreds of catalysts have been synthesized for OCM, the low product yields (<25%) still cannot meet the requirement of industrial application. Different from OCM, lots of studies have been conducted for the conversion of methane under a nonoxidative environment [8–10]. However, there is limited industrial application due to its low conversion and high reaction temperature (700 °C) [11, 12]. Thanks to the studies by Choudhary et al., the enhanced activation of methane at mild conditions (i.e., 400–600 °C) is observed when higher alkanes and/or alkenes are co-fed with methane [13]. Based on this discovery, similar results are reported using various catalysts when methane was co-fed with a series of hydrocarbons, like ethane, propane, pentane, hexane, light gasoline, liquefied petroleum gas, and even oxygenated hydrocarbons such as methanol [14–19].

These results provide the possibility to partially upgrade bitumen with methane as the hydrogen donor. The conventional oil resources like heavy oil or bitumen are composed of various hydrocarbon compounds, which could facilitate the methane activation. Except for methane activation, the instability and incompatibility of crude oil are another big concerns, which caused by the formation of degradable products. These unstable species could result in the formation of precipitate, which could separate oil product into different phases. This issue needs to be addressed because a stable and homogeneous single-phase oil product without deposit formation is critical for the pipeline transportation and long-period storage.

Among the catalysts applied for methane utilization, most of them are ZSM-5 zeolites, because of its promising aromatization performance [20, 21]. ZSM-5 is also widely used for other processes in petrochemical industry, such as fluid catalytic cracking. Adjustable acidity is one advantage of zeolite catalyst. Within the framework of zeolite, the proton species in the Si(OH)Al group could work as the Bronsted acidic site in the catalytic processes, which is regarded as the active sites for the reactions like cracking, isomerization, and alkylation [22–25]. The zeolite acidity is greatly affected by the concentration of Bronsted sites, which is determined by the silica-to-alumina ratio of the framework. Therefore, the reaction performance could be optimized by tuning the zeolite acidity. For heavy oil upgrading with methane, ZSM-5 could be a promising support because of its good performance of nonoxidative coupling of methane, excellent mechanical and thermal stability, and adjustable acidity. Zeolite Y is also selected as the candidate due to their larger pore sizes, which may benefit the conversion of large molecules in heavy oil [26, 27]. The sulfated ZrO_2 is a support with high surface acidity that has been repeatedly proven in other studies [28–30], which may enhance the cracking of heavy oil or bitumen.

In general, the target of partial upgrading of heavy oil and bitumen is to achieve oil product with improved quality, in terms of lower viscosity, density, acidity, asphaltene content, and average molecular weight. Meanwhile, a higher content of gasoline and diesel fractions and a controllable coke formation are also expected. Over optimized catalyst, co-fed methane can be nonoxidatively activated to provide hydrogen for heavy oil upgrading under low-temperature and low-pressure reaction conditions. This idea has been explored in several studies, which will be introduced in this chapter.

5.2.1 Heavy Oil Upgrading Under Methane

It is very attractive to convert viscous heavy oil, like bitumen, extracted from Canadian oil sand to pipeline transportable oil product. As mentioned, the conventional hydrotreatment processes consumed expensive hydrogen under a pressure of 15–20 MPa. It will be economically friendly if the heavy oil upgrading can be realized with a much cheaper H-donor such as methane. It has been proved that the catalytic heavy oil upgrading under methane is achievable at low temperature (380 °C) and low pressure (5 MPa) [31]. In our previous studies, a Canadian bitumen provided by Syncrude Canada Ltd. was firstly selected as the feedstock. Catalysts used for methane activation under mild condition are usually acidic catalyst such as zeolites with metal loading (e.g., Ag, Zn). It is reported that there is a clear synergistic effect between methane participation and bitumen partial upgrading when proper catalyst is charged.

As shown in Tables 5.1, 5.2, and 5.3, with methane participation, the lowest viscosity, density, acidity, asphaltene content, and molecular weight are achieved over Ag-Zn/ZSM-5, as well as the highest gasoline and diesel contents and highest liquid yield. Compared to a series of control experiments, methanotreating process shows promising performance of carbon chain cracking reaction. Activated by Ag-Zn/ZSM-5 catalyst, the viscosity of bitumen is greatly decreased (from 848,080 cP to 413.7 cP) under methane atmosphere, which is only slightly higher than H_2 run (280.2 cP) and very close to the required value in the pipeline transportation regulation of Canada and the United States. When methane is present, the liquid yield is always higher than its N_2 counterpart, especially when Ag-Zn/ZSM-5 catalyst is loaded, indicating the incorporation of methane into the collected oil product. As shown in Table 5.1, all experiments show controllable coke generation (<1 wt.%), which will not cause severe clogging for pipeline transportation. Therefore, these results prove that methane can be effectively activated at mild condition over zeolite catalyst to upgrade heavy oil for enhanced mobility.

Table 5.2 includes gasoline and diesel fractions of oil product that are collected from all experiments listed in Table 5.1 apart from the H_2 one. It is exciting that the gasoline content and the total light oil fraction (gasoline and diesel) of the oil produced from the bitumen upgrading with methane over Ag-Zn/ZSM-5 are higher than that of all other control runs. A high content of light oil would lighten the burden of downstream refining and accordingly bring a decreased cost. Besides, when Ag-Zn/ZSM-5 is loaded, more than twofold increase in the gasoline fraction from catalytic upgrading with methane gives us another proof of the synergistic effect between bitumen upgrading and methane participation. With the optimized catalyst, activated methane can participate in the cracking of heavy oil molecules to produce light fraction.

Table 5.1 Performance of bitumen upgrading under various environments [31]

Entry	Atmosphere	Coke yield (wt.%)	Liquid yield (wt.%)	Viscosity (cP at 25 °C)
Bitumen	–	–	–	848,080
–	N_2	0.60	94.5	1718.3
–	CH_4	0.55	96.5	1617.1
HZSM-5	CH_4	0.80	93.8	1374.4
Ag-Zn/ZSM-5	N_2	0.86	92.9	1276.0
Ag-Zn/ZSM-5	CH_4	0.75	97.8	413.7
Ag-Zn/ZSM-5	H_2	0.62	93.1	280.2

Table 5.2 Gasoline and diesel fractions of oil samples collected before and after bitumen upgrading under various environments [31]

Entry	Atmosphere	Gasoline (wt.%)	Diesel (wt.%)	Total gasoline and diesel (wt.%)
Bitumen	–	0.19	11.96	12.15
–	N_2	6.88	20.81	27.69
–	CH_4	6.95	23.74	30.69
HZSM-5	CH_4	6.47	25.26	31.73
Ag-Zn/ZSM-5	N_2	6.28	26.57	32.85
Ag-Zn/ZSM-5	CH_4	13.38	23.39	36.77

Table 5.3 Properties of the oil samples collected before and after bitumen upgrading under various environments [31]

	Liquid product properties				
Entry	Density (g/cm^3)	TAN (mg KOH/g)	Water content (wt.%)	Average molecular weight (g/mol)	Asphaltene content (wt.%)
Bitumen	1.0275	2.59	0.159	700	22.04
N_2	0.9957	0.51	0.147	527	16.81
CH_4	0.9871	0.24	0.162	541	16.12
CH_4 + ZSM-5	0.9762	0.26	0.171	524	14.84
N_2 + Ag-Zn/ ZSM-5	0.9755	0.39	0.158	458	14.43
CH_4 + Ag-Zn/ ZSM-5	0.9668	0.03	0.185	330	12.32

As shown in Table 5.3, density is a critical property that directly related to the oil quality. The lowest density is observed over the oil sample that produced from the methanotreatment with Ag-Zn/ZSM-5. The significant reduction of oil density well agrees with the viscosity decrease and the enhancement of the light oil fraction show in Table 5.2. In addition, acidity is another property that will affect the oil quality. The fuel with high acidity could be corrosive to the wall of the pipeline and storage tank, which may cause the leakage during transportation and storage. As shown in Table 5.3, the total acid number (TAN) is analyzed for oil product collected from all control experiments. The oil obtained from the methane run with Ag-Zn/ZSM-5 catalyst exhibits the lowest acidity (0.0256 mg KOH/g sample), which is only 1% of the acidity of the feedstock. Compared with the methane run, the oil product collected from the N_2 counterpart has a much higher TAN, demonstrating the synergistic effect between bitumen upgrading and methane participation with an optimized catalyst.

As shown in Table 5.3, the increased water content is observed when low TAN is achieved with methanotreatment over optimized Ag-Zn/ZSM-5 catalyst. As exemplified in Fig. Fig. 5.1, water may be generated from hydrodeoxygenation reactions with hydrogen produced from the activated methane. The promising performance of Ag-Zn/ZSM-5 catalyst for partially heavy oil upgrading is further proved when comparing the average molecular weights of oil products generated from all control

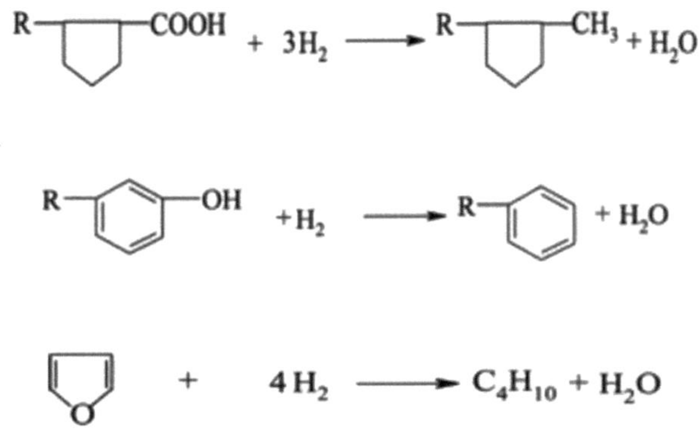

Fig. 5.1 Hydrodeoxygenation reactions taking place in the presence of hydrogen formed from activated methane [31] (Reproduced from Ref. [31] with permission from the Royal Society of Chemistry)

experiments. As shown in Table 5.3, the oil sample collected from methanotreatment over Ag-Zn/ZSM-5 catalyst has the lowest average molecular weight, which is even lower than that of support-only run, suggesting that Ag and Zn loadings could assist the cracking of heavy oil to generate lighter molecules.

As mentioned previously, asphaltene is the main contributor to the high viscosity of bitumen, which is also the most difficult species to be upgraded in raw bitumen. Therefore, to assess the ability of catalytic methanotreating on asphaltene reduction, the asphaltene content of raw bitumen and collected oil products are reported in Table 5.3. Compared with the raw bitumen, the oil product from catalytic upgrading with methane over Ag-Zn/ZSM-5 has a much lower asphaltene content. The asphaltene content reduction further evidenced the beneficial effect of the methane participation and proper catalyst presence in heavy oil upgrading.

The stability and compatibility of fuel are critical during long-distance transportation and long-period storage. Fuel stability is the potential of a fuel to change during storage and transportation. The fuel with bad stability usually has a high asphaltene content, which could cause sludge formation and pipeline blockage. Fuel compatibility is the potential of a fuel to produce precipitate when mixed with other fuels. Spot test is an important tool to study the stability and compatibility of the oil. As shown in Fig. 5.2, spot tests over the oil product collected from the experiments listed in Table 5.1 are conducted. To gain a better understanding of the spot test results shown in Fig. 5.2, text and graphical references (Table 5.4 and Fig. 5.3) described in ASTM D4740 are provided. From Fig. 5.2, it can be easily found that the oil produced from bitumen upgrading under methane atmosphere over Ag-Zn/ZSM-5 shows the best stability and compatibility in all control experiments, which is very close to the product from the H_2 experiment. These results suggest that catalytic upgrading of heavy oil with methane can produce stable and compatible oil suitable for pipeline transportation and storage.

Additionally, elemental analyses are performed for all collected oil products, to determine the performance of saturation, desulfurization, and denitrification reactions. As shown in Table 5.5, the highest H/C ratio is achieved over the oil sample collected from the catalytic methanotreatment run, indicating that methane can be effectively activated over Ag-Zn/ZSM5 to saturate heavy oil molecules. In addition,

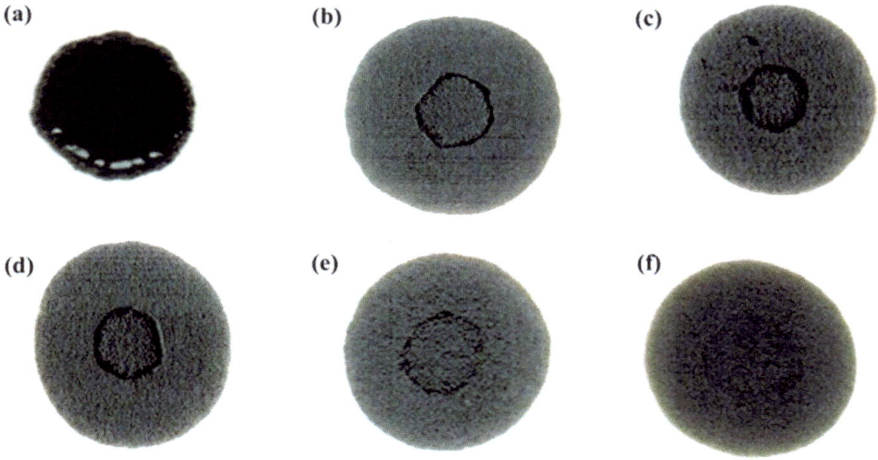

Fig. 5.2 Stability test of (**a**) bitumen and oil products collected under (**b**) N2, (**c**) CH4, (d) CH4 with ZSM-5, (**e**) CH4 with catalyst Ag-Zn/ZSM-5, and (**f**) H2 with catalyst Ag-Zn/ZSM-5 [31] (Reproduced from Ref. [31] with permission from the Royal Society of Chemistry)

Table 5.4 Reference spot description for oil sample compatibility and stability measurement [31]

Reference spot no.	Characterizing features
1.	Homogeneous spot (no inner ring)
2.	Faint or poorly defined inner ring
3.	Well-defined inner ring, only slightly darker than the background
4.	Well-defined inner ring, thicker than the ring in reference spot no. 3 and somewhat
5.	Very dark solid or nearly solid area in the center. The central area is much darker

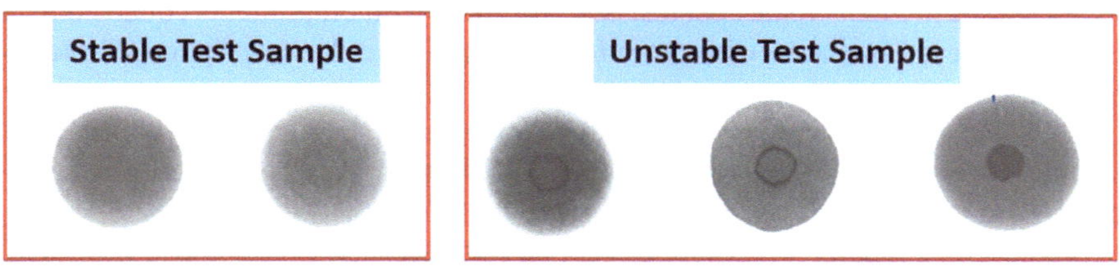

Fig. 5.3 Reference spot definition and visualization for oil sample compatibility and stability measurement [31] (Reproduced from Ref. [31] with permission from the Royal Society of Chemistry)

a slight reduction of sulfur and nitrogen contents are observed when compared to raw bitumen. These results indicate a promising hydrogenation performance in catalytic bitumen upgrading with methane, while the performance of desulfurization and denitrification reactions is limited.

Table 5.5 Elemental analysis of oil samples collected before and after bitumen upgrading [31]

Oil sample	Carbon (wt.%)	Hydrogen (wt.%)	H/C atomic ratio	Nitrogen (wt.%)	Sulfur (wt.%)
Bitumen	81.09	10.29	1.52	1.85	6.48
N_2	81.96	10.44	1.53	1.74	5.58
CH_4	81.88	10.28	1.51	1.78	5.81
CH_4 + ZSM-5	81.93	10.19	1.49	1.84	5.74
N_2 + Ag-Zn/ZSM5	81.66	10.48	1.54	1.73	5.89
CH_4 + Ag-Zn/ZSM5	81.34	11.18	1.65	1.72	5.58

5.2.2 *Effect of Support and Surface Acidity*

To investigate the effect of support on the upgrading reaction, Peng et al. prepared a series of zeolite catalyst, which includes HZSM-5, zeolite Y, and zeolite A [32]. The widely used sulfated ZrO_2 with high surface acidity is also prepared for comparison. The catalytic heavy oil upgrading performance of these catalysts are tested at 380 °C and 50 bar of methane.

By employing HZSM-5 support with various silica-to-alumina ratios, the effect of the HZSM-5 acidity on the catalytic bitumen upgrading performance is investigated. As shown in Table 5.6, it is observed that when HZSM-5 with a silica-to-alumina ratio of 30 is applied for catalytic bitumen under methane, the viscosity of the collected oil product is effectively reduced from 8.5×10^5 to 963 mPa·s, suggesting that the cracking of carbon chain and cyclic structure is achieved under this reaction condition. Except for viscosity, coke formation is another important factor that determines the quality of bitumen upgrading process. As mentioned in the previous chapter, if the content of coke particles is higher than 1 wt.%, oil may create sediment and cause the clogging of transportation pipeline. Over HZSM-5 (30:1), the viscosity is effectively reduced, while an acceptable amount of coke is generated (0.73 wt.%).

As shown in Fig. 5.4, the spot test is performed for all oil products collected from HZSM-5 runs. Compared with the reference figure and the description (Table 5.4 and Fig. 5.3), the inner ring is observed in Fig. 5.4b–f, indicating that the stability of collected oil sample is not satisfied. Despite the instability of all collected oil sample, HZSM-5 (30) shows the best catalytic bitumen cracking performance compared with other HZSM-5 support.

The surface acidity of HZSM-5 supports is investigated by collecting FT-IR spectra of the supports upon the pyridine adsorption. As shown in Fig. 5.5, the peaks around 1448 cm^{-1} are assigned to Lewis acid sites [33, 34], while the signal near 1540 cm^{-1} is attributed to Brönsted acid sites upon pyridine adsorption [35]. The signal around 1491 cm^{-1} is assigned to the 19a mode of pyridinium H-bonded with pyridine and the zeolite framework. The peaks around 1600 cm^{-1} are originated from the H-bonded pyridine. The signals near 1643 cm^{-1} are caused by the 8a and 8b modes of pyridinium hydrogen-bonded with pyridine and the zeolite framework

Table 5.6 Viscosity and coke yield of the oil product collected over HZSM-5 with variable SiO_2/Al_2O_3 molar ratios under methane environment [32]

	Bitumen	No catalyst	HZSM-5 (23:1)	HZSM-5 (30:1)	HZSM-5 (50:1)	HZSM-5 (80:1)	HZSM-5 (280:1)
Viscosity (mPa·s, 25 °C)	$8.5 \pm 0.2 \times 10^5$	$1.62 \pm 0.08 \times 10^3$	$1.52 \pm 0.08 \times 10^3$	$9.6 \pm 0.5 \times 10^2$	$1.40 \pm 0.07 \times 10^3$	$1.20 \pm 0.06 \times 10^3$	$1.46 \pm 0.07 \times 10^3$
Coke yield (wt.%)	–	0.55	0.69	0.73	0.84	1.28	0.89

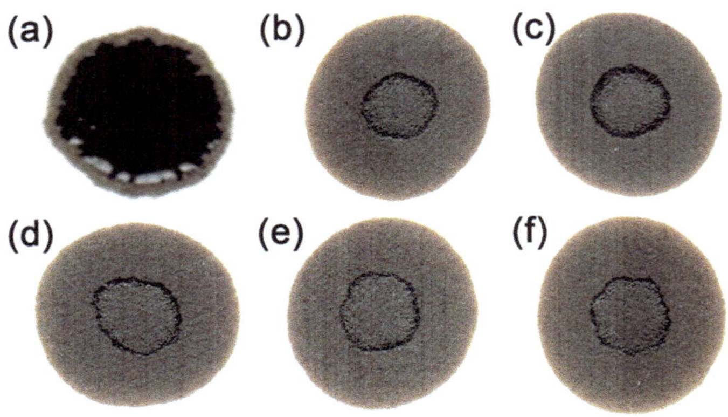

Fig. 5.4 Stability test of (**a**) bitumen and oil products obtained using HZSM-5 with SiO_2/Al_2O_3 molar ratios of (**b**) 23:1, (**c**) 30:1, (**d**) 50:1, (**e**) 80:1, and (**f**) 280:1 as the catalyst [32] (Reproduced from Ref. [32] with permission from Elsevier)

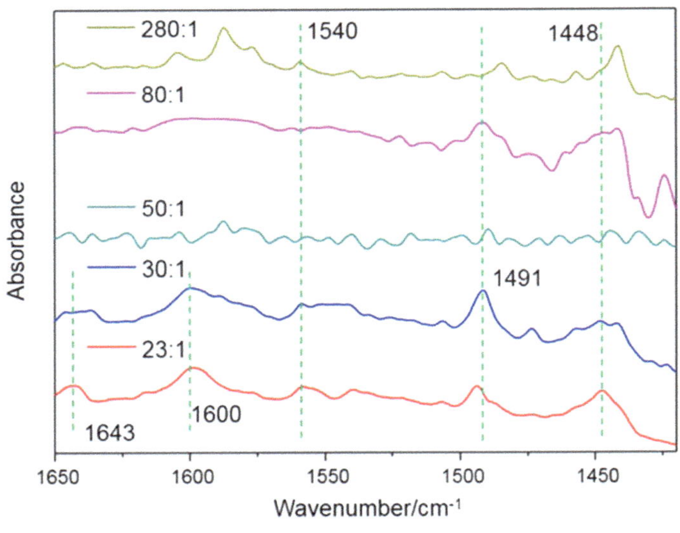

Fig. 5.5 DRIFT spectra collected over HZSM-5 with various Si/Al ratios upon pyridine adsorption [32] (Reproduced from Ref. [32] with permission from Elsevier)

[36, 37]. Among the HZSM-5 support with various silica-to-alumina ratio, HZSM-5 with a ratio of 23 and 30 shows the highest quantity of Brönsted acid sites, which are closely related to the cracking of hydrocarbons. These sites also facilitate the alkylation and isomerization reactions. Among all HZSM-5 supports, HZSM-5 (30:1) shows the strongest peak around 1491 cm^{-1}, which is attributed to the interaction between pyridine and -OH groups. In general, the high concentration of -OH site and Brönsted acid site for interaction with reactants could explain the promising catalytic performance observed on HZSM-5 (30:1).

The effect of the silica-to-alumina ratio in the zeolite framework on catalytic bitumen upgrading is further studied by loading Ag on HZSM-5 support. The catalytic upgrading performance of these Ag/ZSM-5 catalysts are tested. As shown in Table 5.7, the lowest viscosity of 715 mPa·s is observed when 1%Ag/HZSM-5 (30:1) is applied. As shown in Fig. 5.6, the spot test results also confirm that the oil product collected from Ag/HZSM-5 (30:1) has the best stability and compatibility among all collected oil samples. Compared with Fig. 5.4 and Table 5.6, it is not difficult to find that silver loading not only enhances the catalytic performance of bitumen upgrading reaction but also boosts the stability and compatibility of oil product. Based on these results, it is confirmed that Ag/HZSM-5 (30:1) is the best catalyst with the optimized silica-to-alumina ratio.

Besides ZSM-5, zeolite Y is another type of zeolite which is widely applied in petroleum industry. Compared with ZSM-5, the larger pore size of zeolite Y might provide an advantage for converting larger molecules. Zeolite Y wish larger surface area and pore volume may tolerate higher amount of coke and has better diffusion efficiency for oil molecules. Therefore, zeolite Y could be a promising candidate for catalytic bitumen upgrading.

To test the effect of pore size and surface acidity, the catalytic performance of several zeolite Y (HY) supports with different silica-to-alumina ratios is tested at the same condition mentioned previously. The viscosity of oil product and the coke yield of each run are reported in Table 5.8. The lowest viscosity is observed over HY(30:1). However, when compared with HZSM-5 (30:1), HY(30:1) experiment ends up with a higher coke yield and a slightly more viscous oil product. As shown in Fig. 5.7, all spot test images show a clear ring, indicating that the oil product collected over HY supports has poor stability and compatibility.

The sulfated ZrO_2 (S-ZrO_2) is a catalyst support with high surface acidity, which has been reported in other studies [28–30]. Therefore, S-ZrO_2 is included in this study because of its high surface acidity that might enhance bitumen partial cracking. The performance of catalytic bitumen upgrading is tested over ZrO_2 and 1%Ag/ZrO_2. As shown in Table 5.9, the addition of S-ZrO_2 shows limited effect to the viscosity reduction. However, S-ZrO_2 could effectively control coke generation. In addition, silver loading over S-ZrO_2 support doesn't provide obvious enhancement in terms of viscosity reduction and coke generation. Except HY, zeolite A is another type of widely used zeolite with excellent thermal stability under high temperature up to 700 °C, which is also chosen as the support for catalytic bitumen upgrading. Unfortunately, as shown in Table 5.9, the viscosity of collected oil product has no

Table 5.7 Viscosity and coke yield of the oil product collected over Ag/HZSM-5 with variable SiO_2/Al_2O_3 molar ratios under methane environment [32]

	Ag/HZSM-5 (23:1)	Ag/HZSM-5 (30:1)	Ag/HZSM-5 (50:1)	Ag/HZSM-5 (80:1)	Ag/HZSM-5 (280:1)
Viscosity (mPa·s, 25 °C)	$1.27 \pm 0.06 \times 10^3$	$7.2 \pm 0.4 \times 10^2$	$1.32 \pm 0.07 \times 10^3$	$1.18 \pm 0.09 \times 10^3$	$1.58 \pm 0.07 \times 10^3$
Coke yield (%)	0.97 ± 0.05	1.72 ± 0.08	0.70 ± 0.04	0.75 ± 0.04	0.92 ± 0.05

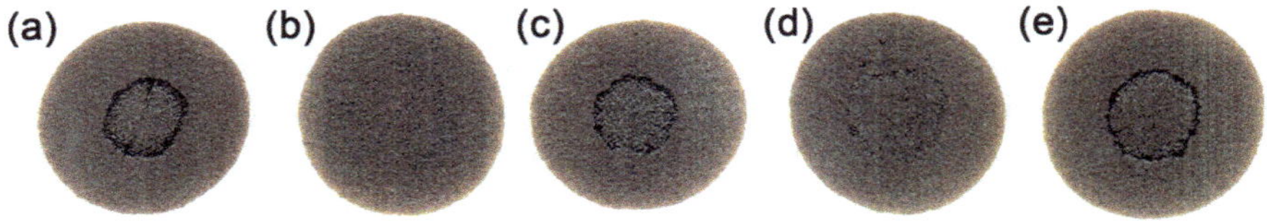

Fig. 5.6 Stability test of oil products obtained using 1%Ag/HZSM-5 with SiO_2/Al_2O_3 molar ratios of (**a**) 23:1, (**b**) 30:1, (**c**) 50:1, (d) 80:1, and (**e**) 280:1 as the catalyst [32] (Reproduced from Ref. [32] with permission from Elsevier)

Table 5.8 Viscosity and coke yield of the oil product collected over HY with variable SiO_2/Al_2O_3 molar ratios under methane environment [32]

	HY (5.2:1)	HY (30:1)	HY (60:1)	HY (80:1)
Viscosity (mPa·s, 25 °C)	$1.39 \pm 0.07 \times 10^3$	$1.24 \pm 0.06 \times 10^3$	$1.38 \pm 0.07 \times 10^3$	$1.54 \pm 0.08 \times 10^3$
Coke yield (%)	0.79 ± 0.04	0.95 ± 0.05	1.14 ± 0.06	1.16 ± 0.06

obvious reduction compared with the no catalyst run, indicating that zeolite A is not suitable for bitumen partial upgrading reaction.

The DRIFT pattern of HZSM-5, HY and HA are also collected by conducting the adsorption of pyridine to study these zeolite supports in the difference of surface acidity. The DRIFT results are shown in Fig. 5.8. On the spectra of HZSM-5 and HY, the peaks assigned to Lewis acid sites are located at 1448 and 1439 cm^{-1}, respectively. On the HZSM-5 DRIFT pattern, the peaks at 1491 and 1600 cm^{-1} are because of the ring-stretching mode of the H-bonded pyridine [36]. On the DRIFT spectra collected over HY, these two peaks shift to 1482 and 1581 cm^{-1}, indicating that HZSM-5 and HY show different interactions between zeolite framework and pyridine molecule. This phenomenon explains that the interaction between support and bitumen molecule is also different over various supports, which could further affect the catalytic performance of upgrading reaction and the quality of oil product [38, 39]. Based on these results, the promising catalytic performance of HZSM-5 catalysts could be contributed by the high surface acidity and proper interaction between the reactants and zeolite structure.

In conclusion, Peng et al. evaluate the effect of catalyst support on the catalytic performance of bitumen partial upgrading. Among all tested supports, HZSM-5 with a silica-to-alumina ratio of 30 achieve the lowest coke yield, while the collected oil product has the lowest viscosity and best stability. DRIFT results further clarify that the high surface acidity of HZSM-5 (30:1) and the proper interaction between the reactants and zeolite structure could be the reason of its good catalytic performance. For future study, this research triggers an urgent need to further investigate the involved catalytic reaction mechanism to assist catalyst design in a logical way for further enhanced upgrading performance.

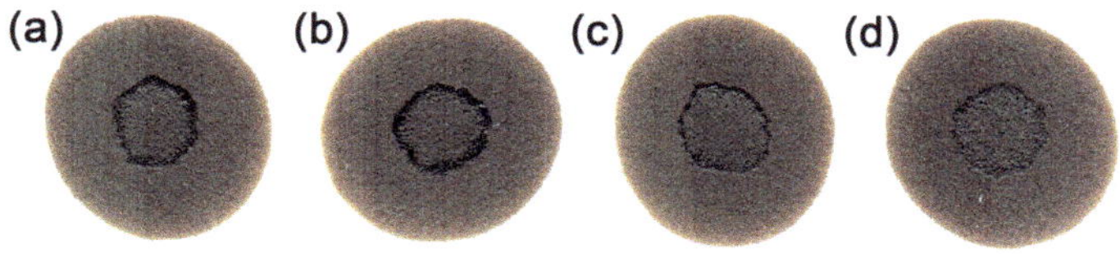

Fig. 5.7 Stability test of oil products obtained using HY with SiO_2/Al_2O_3 molar ratios of (**a**) 5.2:1, (**b**) 30:1, (**c**) 60:1, and (**d**) 80:1 as the catalyst under methane environment [32] (Reproduced from Ref. [32] with permission from Elsevier)

Table 5.9 Viscosity and coke yield of the oil product collected over S-ZrO_2, 1%Ag/S-ZrO_2, and zeolite A [32]

	No catalyst	S-ZrO_2	1%Ag/S-ZrO_2	Zeolite A
Viscosity (mPa·s, 25 °C)	$1.62 \pm 0.08 \times 10^3$	$1.50 \pm 0.07 \times 10^3$	$1.45 \pm 0.07 \times 10^3$	$1.56 \pm 0.08 \times 10^3$
Coke yield (%)	0.55	0.04	0.28	0.73

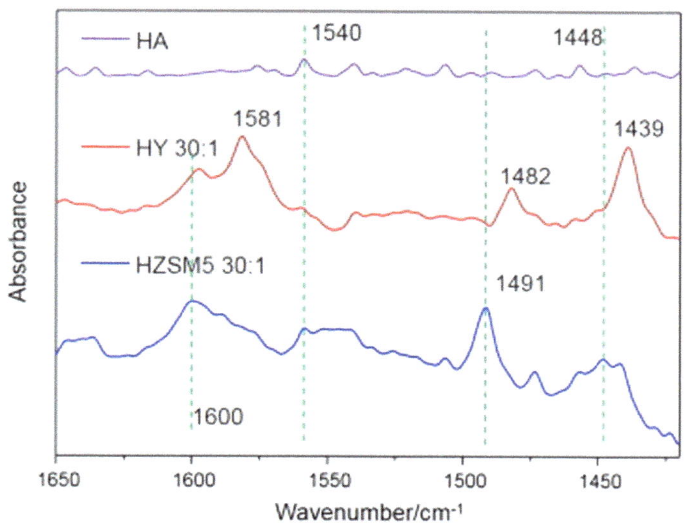

Fig. 5.8 DRIFT spectra collected over HZSM-5, HY, and HA upon pyridine adsorption [32] (Reproduced from Ref. [32] with permission from Elsevier)

5.3 Heavy Oil Upgrading with Methane: Model Compound Studies

Because of the complexity of heavy oil system, model compound-based reactions are critical to study the reaction mechanism of catalytic heavy oil upgrading with methane. N-butylbenzene has both a n-butyl group and an aromatic ring, which is widely used as a model compound of asphaltene to study the upgrading mechanism in heavy oil research [40, 41]. Comparing the normal paraffins or paraffins with branches, n-butylbenzene would produce the product with less components, accounting for the stable structure of phenyl group.

As shown in Tables 5.10 and 5.11 and Fig. 5.10, n-butylbenzene conversions are conducted over various catalysts, and 5.0 MPa of methane or nitrogen is co-fed with n-butylbenzene. The final pressure of reactor is recorded in Table 5.10, as well as the yield of gas and liquid products. Without catalyst loading, the conversions of n-butylbenzene under methane and nitrogen almost have no difference in final pressure and the yield of gas and liquid product, indicating limited effect of methane without catalyst loading. When bare support HZSM-5 is packed, the over-cracking of model compound is observed, confirmed by the most elevated final pressure and highest gas yield. These results indicate that acidic catalyst could boost the cracking of n-butylbenzene for gas product generation. The lowest gas yield is observed with the conversions over Ag-Zn/ZSM-5, which is the previously mentioned optimized catalyst for the methanotreatment of heavy oil. However, the final pressure of

Table 5.10 Performance of N-butylbenzene upgrading under various environments at 5.0 MPa and 380 °C [31]

Catalyst or support	Atmosphere	Final pressure (380 °C, MPa)	Gas yield (wt.%)	Liquid yield (wt.%)
–	N_2	7.39	2.4	97.6
–	CH_4	7.30	2.4	97.6
HZSM-5	N_2	9.30	5.6	94.4
HZSM-5	CH_4	9.38	5.5	94.5
Ag-Zn/ZSM-5	N_2	7.95	2.3	97.7
Ag-Zn/ZSM-5	CH_4	8.20	2.0	98.0

Table 5.11 Composition of liquid products and conversions of N-butylbenzene and methane [31]

	Liquid product properties (wt.%)					
Compound	N2	CH_4	ZSM-5, N2	ZSM-5, CH_4	Ag-Zn/ZSM-5, N2	Ag-Zn/ZSM-5, CH_4
Benzene	0.58	0.61	29.50	28.98	37.86	45.88
Methylbenzene	30.34	31.33	1.70	1.28	2.33	2.28
Ethylbenzene	1.36	1.32	2.32	4.10	0	0.82
Styrene	19.59	19.78	0	0	0.71	0.12
Isopropylbenzene	0	0	8.94	9.79	1.74	1.35
N-propylbenzene	0	0	0	0	5.97	5.51
Pentylbenzene	3.05	2.95	1.76	4.00	3.39	2.16
Heptylbenzene	0	0	9.00	7.37	15.56	13.22
Octylbenzene	0	0	3.66	1.74	20.74	16.26
Nonylbenzene	0	0	0	2.75	3.81	2.59
Butylbenzene*	0.89	0.81	88.71	86.43	21.94	25.04
Methane*	–	0	–	0	–	10.84

Ag-Zn/ZSM-5 runs is neither the lowest nor the highest. For these results, the possible explanation could be that for the reactions over Ag-Zn/ZSM-5, light hydrocarbons like benzene are formed from activated methane.

The composition of liquid products are further analyzed by GC-MS. As summarized in Table 5.11, it is noteworthy that benzene is one of the products in all of the model reactions for n-butylbenzene. From the first two columns, it seems that the reaction atmosphere (N_2 or CH_4) has an insignificant effect for the conversions without the catalyst or bare support participation. However, when Ag-Zn/ZSM-5 is added, the participation of methane results in greatly higher selectivity of benzene in the formed liquid product, compared to its N_2 counterpart. It is well agreed with the high methane conversion (10.84%) when Ag-Zn is added. These results are further evidenced that except for the benzene produced from n-butylbenzene, more benzene can be formed from the nonoxidative coupling of methane when proper catalyst is involved. Aromatization of methane for benzene production has been vastly studied by other researchers [11, 13, 42, 43].

More information are provided in Table 5.11. For the conversion with the participation of Ag-Zn/ZSM-5 catalyst, the ratio of n-propylbenzene to styrene and isopropylbenzene to styrene are 8.4 and 2.5 for nitrogen case, while the ratios of n-propylbenzene to styrene and propylbenzene to styrene are 45.9 and 11.3 for methane one. Therefore, the higher ratio of propylbenzene to styrene when butylbenzene is converted under methane atmosphere demonstrates that the activated methane could assist the hydrogenation reaction for unsaturated molecules such as styrene. The hydrogen generated from methane may come from the methane aromatization reaction.

In addition to the previous n-butylbenzene study, styrene is also selected as the model compound of the olefin fraction in raw heavy oil. Thus, to study the hydrogenation of unsaturated compound, the catalytic upgrading of styrene is conducted under methane over Ag-Zn/ZSM-5. As shown in Table 5.12, the composition of collected liquid products is carefully analyzed by GC-MS. The result shows promising hydrogenation performance. Unsaturated styrene is effectively converted to saturated toluene, ethylbenzene, and isopropylbenzene.

As shown in Table 5.11 and Fig. 5.9, styrene can be easily produced from the catalytic cracking of n-butylbenzene. Without proper H-donor like hydrogen and methane, styrene can also be formed from the conventional thermal cracking.

Table 5.12 Composition of liquid products from styrene upgrading with methane over Ag-Zn/ZSM-5 [31]

Compound	Liquid product distribution (wt.%)
Benzene	0.73
Toluene	25.53
Ethylbenzene	42.53
Isopropylbenzene	12.75
N-propylbenzene	3.25
Butylbenzene	1.10

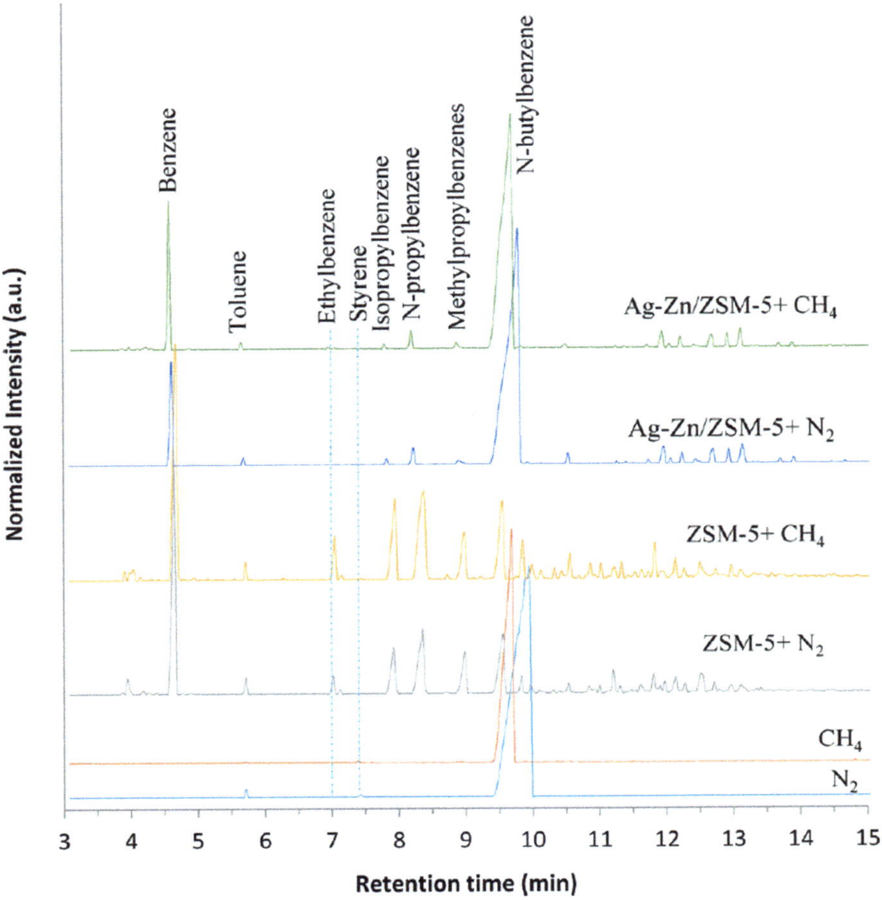

Fig. 5.9 TIC chromatographs of the liquid products collected from N-butylbenzene upgrading under different environments and catalysts charged [31] (Reproduced from Ref. [31] with permission from the Royal Society of Chemistry)

Because of the presence of unsaturated bond, styrene can be polymerized to form heavy compound with high molecular weight and even solid coke in transporting pipeline. Therefore, the hydrogenationof the unsaturated hydrocarbon, one of the main products formed during the methanotreating that previously mentioned in this chapter, could greatly enhance the stability of oil product for pipeline transportation and storage.

To obtain a deeper understanding of the mechanism of styrene upgrading reaction, Fourier transform infrared spectroscopy (FTIR) operated at diffuse reflectance mode is applied to study the evolution of adsorbed styrene on catalyst surface when styrene upgrading is happening. As shown in Fig. 5.10, the diffuse reflectance infrared Fourier transform spectroscopy (DRIFT) spectra is collected from

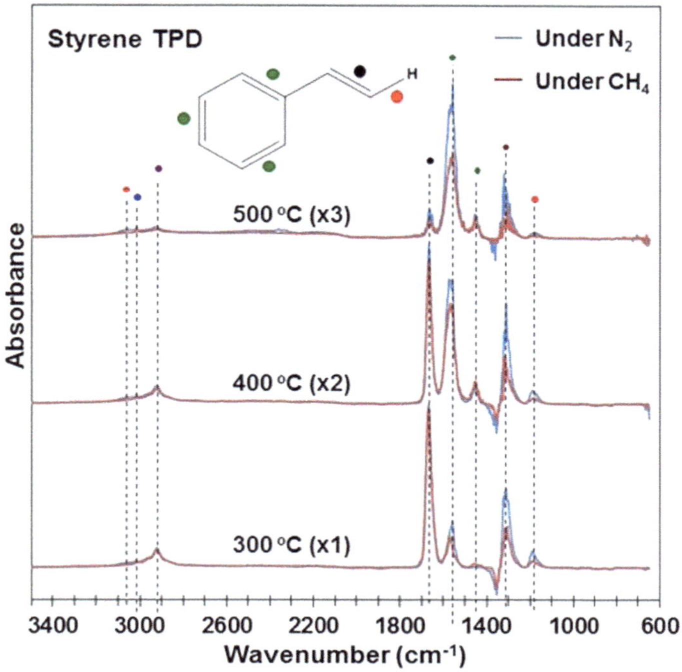

Fig. 5.10 DRIFT spectra collected at different temperatures during styrene saturation under various gas environments over Ag-Zn/ZSM-5 [31] (Reproduced from Ref. [31] with permission from the Royal Society of Chemistry)

temperature-programmed desorption (TPD) experiment of styrene when Ag-Zn/ZSM-5 is loaded into an in situ environmental chamber. At different operation temperatures (300 °C, 400 °C, and 500 °C), the spectra collected under methane environment are in red, while those for nitrogen counterpart are in blue. As is presented, several IR-active vibration peaks are detected in the range of 1000–3000 cm^{-1}. These peaks are classified into the vibration of different bonds and labelled by dots in various colors. Red dots represents the vibration of terminal C-H bond (C-H stretching at 1080 cm^{-1} and C-H bending at 1185 cm^{-1}) located in the branch of aromatics [44]. For better understanding, all bonds are also labelled in the molecular structure of styrene, as shown in Fig. 5.10. The peak tagged with the blue dot is assigned for the C-H stretching at 3015 cm^{-1} because of the presence of methane in the atmosphere [45]. The peaks that in the range of 2800–3000 cm^{-1} are regarded as the C-H stretching from CH_x group in the molecules of adsorbed styrene and methane, which are labelled with purple dots. Represented by black dot, the C=C bond vibration in the styrene unsaturated branch is observed around 1695 cm^{-1}, while the vibration of C=C double bond in the phenyl ring of aromatics is labelled with dots in green at 1448 and 1560 cm^{-1} [44]. The peaks labelled with brown dot are assigned to the C-O bond bending from the styrene that adsorbed on the catalyst surface. For C-O bond vibration, the oxygen could be provided by the catalyst matrix.

Further investigations are performed over the peak intensity difference of the side chain C=C bond (at 1695 cm^{-1}) and phenyl C=C bond (at 1448 and 1560 cm^{-1}) at different temperatures between methane and nitrogen experiments. It is noteworthy that the peak intensity of C=C vibration from the vinyl group (labelled with the black dot) and aromatics (labelled with the green dot) are weaker in the methane experiment, demonstrating that the presence of methane reduces the amount of unsaturated compound. More significant difference is observed at 400 and 500 °C compared with 300 °C, which may account for the boosted methane activation at higher reaction temperature. These phenomena further indicate that methane presence can benefit the conversion of unsaturated compound such as styrene when suitable catalyst is charged.

To further confirm the methane incorporation into the catalytic heavy oil upgrading, benzene is also selected as the model compound. Due to the stable structure of benzene, the cracking of benzene is very difficult to conduct at the reaction temperature (380 °C) that used for catalytic heavy oil upgrading. If we could observe aromatic molecules with branch in product such as toluene, it can be undoubtedly confirmed that methane is incorporated into the produced oil product. Although not all species of produced compounds are listed in Table 5.13, a series of aromatic with branches are detected in the collected liquid product from benzene model reaction over Ag-Zn/ZSM-5, indicating the methane incorporation of methane into the final liquid product over proper catalyst.

Based on all results from mentioned model reactions, the activated methane could effectively saturate olefin compound such as styrene. The methyl group generated from activated methane could form benzene via the nonoxidative coupling or incorporate into the liquid product to produce hydrocarbons with branches. It has been proved that hydrocarbon with long carbon chain can be cracked under methane over Ag-Zn/ZSM-5. In general, these model compound studies further evidenced that methane is incorporated into the liquid product of heavy oil upgrading.

To further investigate the physical and chemical properties of Ag-Zn/ZSM-5 catalyst, a series of characterization including X-ray diffraction (XRD), transmission electron microscopy (TEM), and X-ray photoelectron spectroscopy (XPS) are performed over fresh catalyst or spent catalyst from model reactions, and the processed results are shown in Figs. 5.11, 5.12, 5.13, and 5.14.

Table 5.13 Composition of liquid products from benzene upgrading with methane over Ag-Zn/ZSM-5 [31]

Compound	Liquid product distribution (wt.%)
Toluene	9.31
Ethylbenzene	7.81
Isopropylbenzene	4.20
N-propylbenzene	9.83
2-methyl-1-propylbenzene	1.96
1-methyl-4-propylbenzene	3.12
Butylbenzene	6.18

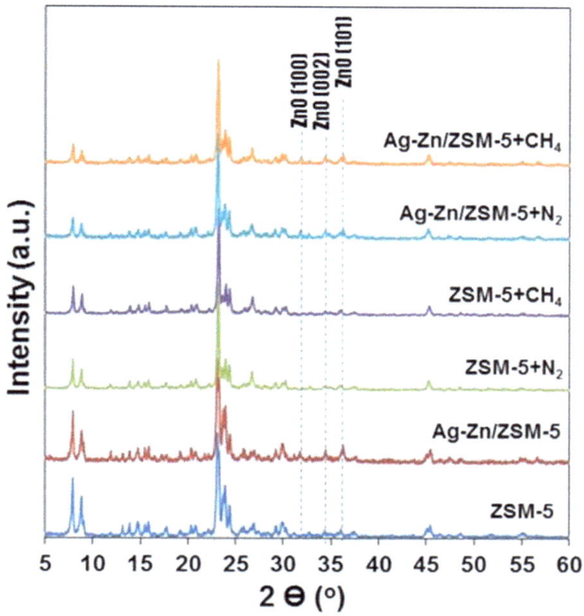

Fig. 5.11 XRD patterns of ZSM-5 and Ag-Zn/ZSM-5 before and after N-butylbenzene upgrading [31] (Reproduced from Ref. [31] with permission from the Royal Society of Chemistry)

As shown in Fig. 5.11, the XRD patterns are collected from fresh ZSM-5 and Ag-Zn/ZSM-5, as well as the spent ones from n-butylbenzene upgrading under methane or nitrogen environment. Comparing the XRD pattern of fresh ZSM-5 and Ag-Zn/ZSM-5, the metal loading does not have obvious effect on the XRD pattern of ZSM-5, implying that the zeolite structure hasn't changed after metal loading. In the XRD pattern of Ag-Zn/ZSM-5, there are three peaks assigned for ZnO diffraction. However, it is noteworthy that all these three peaks become weaker after reaction. The possible explanation is that, during the catalytic upgrading of n-butylbenzene under methane or nitrogen, ZnO is partially reduced to metallic Zn because of the hydrogen that generated from activated methane or n-butylbenzene. Another explanation is that loaded zinc species has a better dispersion over catalyst surface after the catalytic upgrading reaction. Besides, the diffraction peaks of silver species are not observed over fresh or spent Ag-Zn/ZSM-5 catalyst, which may because of its low loading (1%) and/or a good dispersion.

Using TEM equipped with energy-dispersive X-ray (EDX) mapping, the morphology of zeolite and the particle sizes of loaded metal species are studied over fresh Ag-Zn/ZSM-5. As shown in Fig. 5.12, a TEM image is collected for Ag-Zn/ZSM5 (Fig. 5.12a). Another two EDX mapping are performed at Ag-focused area (Fig. 5.12b) and Zn-concentrated spot (Fig. 5.12c). Putting all these results together, it can be noticed that there are two different species of loaded particles on ZSM-5 support. As confirmed by EDX mapping, Ag_2O has potential to form large particles (10–20 nm), while ZnO has a better dispersion over zeolite surface with smaller particle size (<10 nm). Comparing the TEM image of fresh and spent catalysts, the

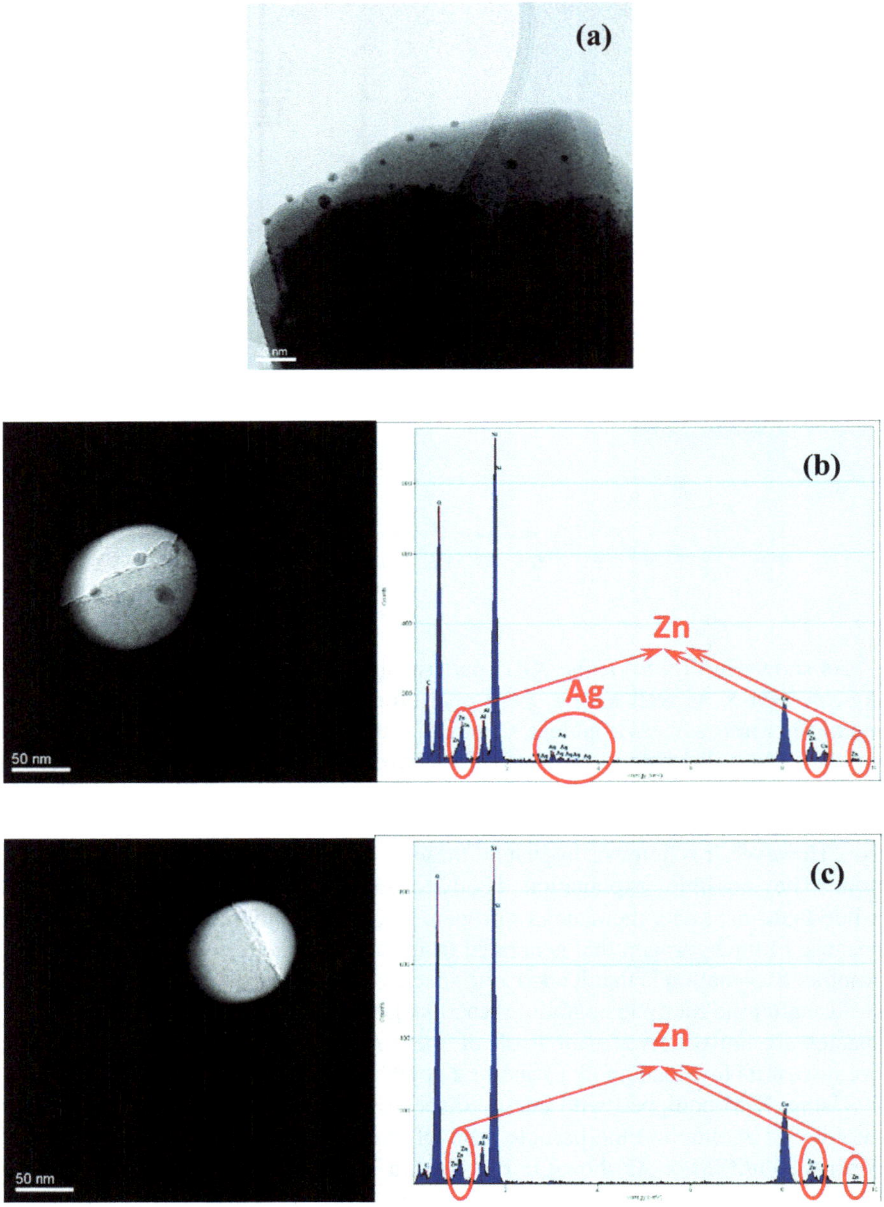

Fig. 5.12 (**a**) TEM image of Ag-Zn/ZSM-5 and its corresponding EDX spectra at (**b**) Ag2O-focused area and (**c**) ZnO-concentrated spot [31] (Reproduced from Ref. [31] with permission from the Royal Society of Chemistry)

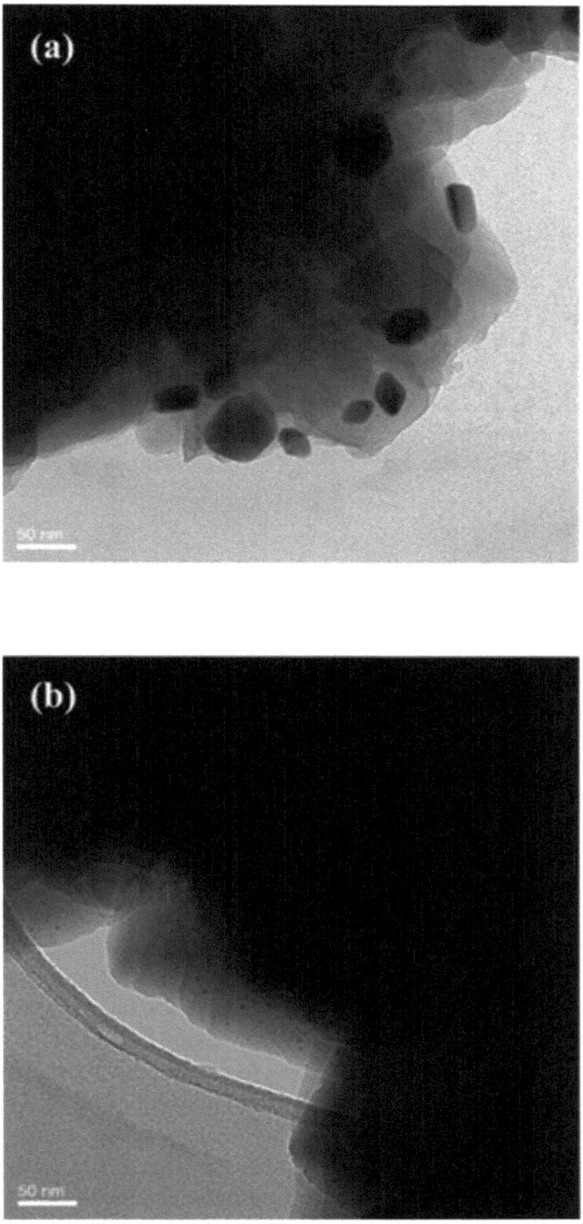

Fig. 5.13 TEM images of spent Ag-Zn/ZSM-5 collected after catalytic N-butylbenzene cracking under the environment of (**a**) N2 and (**b**) CH_4 [31] (Reproduced from Ref. [31] with permission from the Royal Society of Chemistry)

zeolite framework has no obvious change after catalytic upgrading reaction at high-reaction temperature and pressure. The TEM images well agreed with the XRD results, which further confirms the excellent stability of Ag-Zn/ZSM5 during upgrading reactions. However, over spent catalyst after catalytic upgrading under

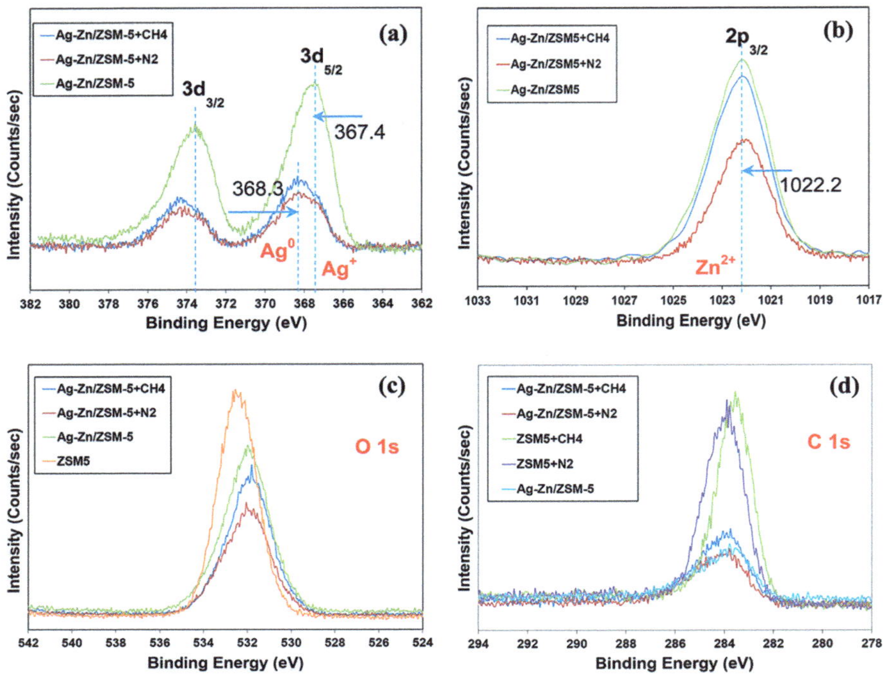

Fig. 5.14 XPS spectra of HZSM-5 and Ag-Zn/ZSM-5 before and after N-butylbenzene under different environments at (**a**) Ag 3d, (**b**) Zn 2p, (**c**) O 1 s, and (**d**) C 1 s regions. [31] (Reproduced from Ref. [31] with permission from the Royal Society of Chemistry)

N_2 atmosphere (Fig. 5.13a), the loaded metal species has larger average particle size than its CH_4 counterpart (Fig. 5.13b). This phenomenon indicates that activated methane is beneficial for the dispersion of metal species, which could further enhance the performance of catalytic upgrading reaction.

To gain the information of the valence of each element contained on the surface of Ag-Zn/ZSM-5 catalyst, XPS analysis is conducted over fresh and spent Ag-Zn/ZSM-5 by performing the Ag 3d, Zn 2p, O 1s and C 1s scans. As shown in Fig. 5.14a, the reduced Ag^+ peak intensity is observed over spent catalysts, which can be explained by the reduced surface concentration of Ag caused from the diffusion of silver from external surface to the inner pores. Additionally, the Ag^+ peaks on spent catalysts which shift to the high binding energy, because of the partial reduction of Ag_2O to metallic silver by hydrogen generated from activated hydrocarbons. After taking a closer look at Fig. 5.14a, the Ag^0 peak of spent catalyst collected from methane reaction has a slightly higher binding energy than its nitrogen counterpart, indicating that activated methane could generate extra hydrogen for a more thorough reduction. As shown in Fig. 5.14b, the peak intensity of Zn^{2+} is also reduced after upgrading reaction, due to the diffusion of zinc species to the inner pores of the zeolite.

The similar results are found in Fig. 5.14c; the surface concentration of oxygen is reduced after metal loading and upgrading reaction under nitrogen. However, the O surface concentration has no obvious change after methane reaction, which accounts for its better catalytic performance. Moreover, the bare ZSM-5 support has significant lower binding energy of O due to the introduction of other metal oxide species. As shown in Fig. 5.14d, the signal of carbon species is also detected over the fresh catalyst and spent catalyst sample collected from upgrading reaction under methane or nitrogen. Over fresh catalyst sample, the C signal may be because of the physically adsorbed CO_2 or other C-containing species. For spent bare ZSM-5 samples, there is a strong carbon signal due to the severe coke formation. However, the fresh and spent catalysts have no obvious difference when metal species are loaded, demonstrating that the introduction of metal loading could suppress the coke formation.

In general, based on the results collected from characterization and previously mentioned mechanism studies, it can be confirmed that silver and zinc are present in the Ag-Zn/ZSM-5 catalyst in the form of Ag^+ and Zn^{2+} cations, which play a critical role in the catalytic heavy oil upgrading under methane. A hypothesis of reaction mechanism is provided in the Fig. 5.15. With the assistance of the loaded metal species, methane molecule is activated by interacting with the catalyst. Once a C-H

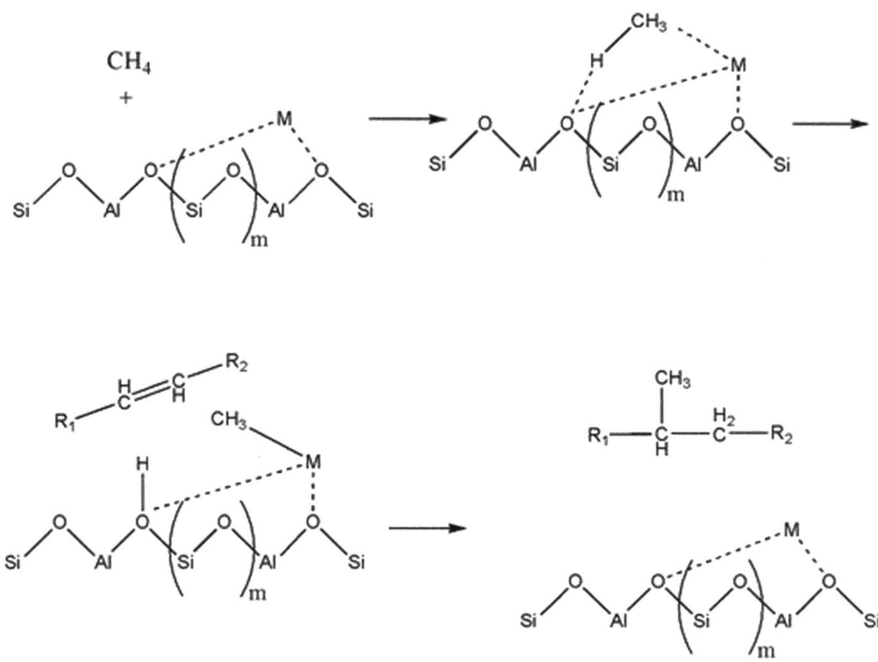

Fig. 5.15 The hypothetic reaction mechanism of methane activation and addition to the broken pieces formed during hydrocarbon cracking over Ag-Zn/ZSM-5 (M = Zn^{2+} or Ag^+) [31] (Reproduced from Ref. [31] with permission from the Royal Society of Chemistry)

bond is broken, the H atom bonds with the O atom of the zeolite framework and the formed methyl group interact with the metal cation. The generated proton and methyl group are highly active, which will participate the cracking of long hydrocarbon molecules or the olefin saturation reactions.

5.4 Conclusions

Because of the scarcity of conventional oil resources and increased global energy consumption, unconventional oil resources with abundant reserves such as heavy oil and bitumen are drawing more attention from governments and industries in the last decades. However, due to the high density and viscosity of unconventional oil, it is critical to conduct catalytic upgrading reaction to produce transportable oil product with enhanced quality.

For heavy oil upgrading process, conventional hydrotreatment involves high pressure of expensive hydrogen. To solve this issue, it is confirmed that catalytic upgrading of heavy oil can be conducted under methane atmosphere. Several related studies are described in this chapter. Zeolite-based catalysts are vastly applied for heavy oil methanotreatment because of its flexible surface acidity, strong mechanical and thermal stability, and abundant porous structure. Among all zeolite support introduced in this chapter, HZSM-5 with a silica-to-alumina ratio of 30 is the most desired one because of its excellent performance in heavy oil cracking. Over HZSM-5 (30), the quality of collected oil product is greatly enhanced under methane environment in terms of viscosity, density, total acid number, average molecular weight, stability, and compatibility. It is also confirmed that the metal loading (e.g., Zn or Ag) could further boost the catalytic performance.

In addition, methane participation in the upgrading reaction has been proved by conducting model compound studies. Over desired zeolite catalyst, it is proved that methane is a promising hydrogen donor for heavy oil cracking, which can be activated at moderate temperature and pressure. However, due to the complex composition of bitumen, there are limited mechanism studies for methanotreating processes. It is urgent to further investigate the involved catalytic reaction mechanism to assist catalyst design in a logical way for further enhanced upgrading performance in the future.

References

1. M.R. Gray, *Upgrading Oilsands Bitumen and Heavy Oil* (University of Alberta, Edmonton, AB, 2015)
2. G. Belyk, D. Burgart, B. Jablonski, J. Heida, T. Kaiser, R. Bernar, B. Whitelaw, *Heavy Oil 101. Participant Handbook* (Canadian Heavy Oil Associations, 2013)
3. M. Mech, A comprehensive guide to the alberta oil sands. Understanding the environmental and human impacts, export implications, and political, economic, and industry implications, and political, economic, and industry influences. Green Party of Canada 2011

4. E. Furimsky, *Catalysts for Upgrading Heavy Petroleum Feeds* (Elsevier, 2007)
5. G.J. Hutchings, M.S. Scurrell, J.R. Woodhouse, Oxidative coupling of methane using oxide catalysts. Chem. Soc. Rev. **18**, 251–283 (1989)
6. J.S. Lee, S.T. Oyama, Oxidative coupling of methane to higher hydrocarbons. Catal. Rev. Sci. Eng. **30**(2), 249–280 (1988)
7. J.H. Lunsford, The catalytic oxidative coupling of methane. Angew. Chem. Int. Ed. Engl. **34**(9), 970–980 (1995)
8. Y.G. Kolyagin, I.I. Ivanova, V.V. Ordomsky, A. Gedeon, Y.A. Pirogov, Methane activation over Zn-modified MFI zeolite: NMR evidence for Zn– methyl surface species formation. J. Phys. Chem. C **112**(50), 20065–20069 (2008)
9. Y. Xu, X. Bao, L. Lin, Direct conversion of methane under nonoxidative conditions. J. Catal. **216**(1–2), 386–395 (2003)
10. Y. Xu, L. Lin, Recent advances in methane dehydro-aromatization over transition metal ion-modified zeolite catalysts under non-oxidative conditions. Appl. Catal. A Gen. **188**(1–2), 53–67 (1999)
11. X. Guo, G. Fang, G. Li, H. Ma, H. Fan, L. Yu, C. Ma, X. Wu, D. Deng, M. Wei, Direct, non-oxidative conversion of methane to ethylene, aromatics, and hydrogen. Science **344**(6184), 616–619 (2014)
12. D. Wang, J.H. Lunsford, M.P. Rosynek, Characterization of a Mo/ZSM-5 catalyst for the conversion of methane to benzene. J. Catal. **169**(1), 347–358 (1997)
13. V.R. Choudhary, A.K. Kinage, T.V. Choudhary, Low-temperature nonoxidative activation of methane over H-galloaluminosilicate (MFI) zeolite. Science **275**(5304), 1286–1288 (1997)
14. O.A. Anunziata, G.G. Mercado, L.B. Pierella, Improvement of methane activation using n-hexane as co-reactant over Zn/HZSM-11 zeolite. Catal. Commun. **5**(8), 401–405 (2004)
15. O.A. Anunziata, G.V.G. Mercado, L.B. Pierella, Catalytic activation of methane using n-pentane as co-reactant over Zn/H-ZSM-11 zeolite. Catal. Lett. **87**(3), 167–171 (2003)
16. V.R. Choudhary, K.C. Mondal, S.A.R. Mulla, Simultaneous conversion of methane and methanol into gasoline over bifunctional Ga-, Zn-, In-, and/or Mo-modified ZSM-5 zeolites. Angew. Chem. **117**(28), 4455–4459 (2005)
17. A.A. Gabrienko, S.S. Arzumanov, I.B. Moroz, A.V. Toktarev, W. Wang, A.G. Stepanov, Methane activation and transformation on Ag/H-ZSM-5 zeolite studied with solid-state NMR. J. Phys. Chem. C **117**(15), 7690–7702 (2013)
18. M.V. Luzgin, V.A. Rogov, S.S. Arzumanov, A.V. Toktarev, A.G. Stepanov, V.N. Parmon, Understanding methane aromatization on a Zn-modified high-silica zeolite. Angew. Chem. **120**(24), 4635–4638 (2008)
19. M.V. Luzgin, V.A. Rogov, S.S. Arzumanov, A.V. Toktarev, A.G. Stepanov, V.N. Parmon, Methane aromatization on Zn-modified zeolite in the presence of a co-reactant higher alkane: How does it occur? Catal. Today **144**(3–4), 265–272 (2009)
20. S. Al-Khattaf, S.A. Ali, A.M. Aitani, N. Žilková, D. Kubička, J. Čejka, Recent advances in reactions of alkylbenzenes over novel zeolites: The effects of zeolite structure and morphology. Catal. Rev. **56**(4), 333–402 (2014)
21. H. Kitagawa, Y. Sendoda, Y. Ono, Transformation of propane into aromatic hydrocarbons over ZSM-5 zeolites. J. Catal. **101**(1), 12–18 (1986)
22. S.M. Csicsery, Shape-selective catalysis in zeolites. Zeolites **4**(3), 202–213 (1984)
23. W.E. Farneth, R.J. Gorte, Methods for characterizing zeolite acidity. Chem. Rev. **95**(3), 615–635 (1995)
24. Y.V. Kissin, Chemical mechanisms of catalytic cracking over solid acidic catalysts: Alkanes and alkenes. Catal. Rev. **43**(1–2), 85–146 (2001)
25. B. Xu, C. Sievers, S.B. Hong, R. Prins, J.A. van Bokhoven, Catalytic activity of Brønsted acid sites in zeolites: Intrinsic activity, rate-limiting step, and influence of the local structure of the acid sites. J. Catal. **244**(2), 163–168 (2006)
26. A. Primo, H. Garcia, Zeolites as catalysts in oil refining. Chem. Soc. Rev. **43**(22), 7548–7561 (2014)
27. W. Vermeiren, J.P. Gilson, Impact of zeolites on the petroleum and petrochemical industry. Top. Catal. **52**(9), 1131–1161 (2009)

28. J.B. Joo, A. Vu, Q. Zhang, M. Dahl, M. Gu, F. Zaera, Y. Yin, A sulfated ZrO2 hollow nanostructure as an acid catalyst in the dehydration of fructose to 5-hydroxymethylfurfural. ChemSusChem **6**(10), 2001–2008 (2013)
29. N.A. Khan, D.K. Mishra, I. Ahmed, J.W. Yoon, J.-S. Hwang, S.H. Jhung, Liquid-phase dehydration of sorbitol to isosorbide using sulfated zirconia as a solid acid catalyst. Appl. Catal. A Gen. **452**, 34–38 (2013)
30. C. Liu, S. Lee, D. Su, Z. Zhang, L. Pfefferle, G.L. Haller, Synthesis and characterization of nanocomposites with strong interfacial interaction: Sulfated ZrO2 nanoparticles supported on multiwalled carbon nanotubes. J. Phys. Chem. C **116**(41), 21742–21752 (2012)
31. A. Guo, C. Wu, P. He, Y. Luan, L. Zhao, W. Shan, W. Cheng, H. Song, Low-temperature and low-pressure non-oxidative activation of methane for upgrading heavy oil. Cat. Sci. Technol. **6**(4), 1201–1213 (2016)
32. P. He, Y. Luan, L. Zhao, W. Cheng, C. Wu, S. Chen, H. Song, Catalytic bitumen partial upgrading over Ag-Ga/ZSM-5 under methane environment. Fuel Process. Technol. **156**, 290–297 (2017)
33. S.M.T. Almutairi, B. Mezari, E.A. Pidko, P.C.M.M. Magusin, E.J.M. Hensen, Influence of steaming on the acidity and the methanol conversion reaction of HZSM-5 zeolite. J. Catal. **307**, 194–203 (2013)
34. D. Yi, H. Huang, X. Meng, L. Shi, Adsorption–desorption behavior and mechanism of dimethyl disulfide in liquid hydrocarbon streams on modified Y zeolites. Appl. Catal. B Environ. **148**, 377–386 (2014)
35. W. Xu, S.J. Miller, P.K. Agrawal, C.W. Jones, Zeolite topology effects in the alkylation of phenol with propylene. Appl. Catal. A Gen. **459**, 114–120 (2013)
36. R. Buzzoni, S. Bordiga, G. Ricchiardi, C. Lamberti, A. Zecchina, G. Bellussi, Interaction of pyridine with acidic (H-ZSM5, H-β, H-MORD zeolites) and superacidic (H-Nafion membrane) systems: An IR investigation. Langmuir **12**(4), 930–940 (1996)
37. V.P. Glazunov, S.E. Odinokov, Infrared spectra of pyridinium salts in solution—I. the region of middle frequencies. Spectrochim. Acta A: Mol. Spectrosc. **38**(4), 399–408 (1982)
38. T.F. Degnan, The implications of the fundamentals of shape selectivity for the development of catalysts for the petroleum and petrochemical industries. J. Catal. **216**(1), 32–46 (2003)
39. J. Jae, G.A. Tompsett, A.J. Foster, K.D. Hammond, S.M. Auerbach, R.F. Lobo, G.W. Huber, Investigation into the shape selectivity of zeolite catalysts for biomass conversion. J. Catal. **279**(2), 257–268 (2011)
40. P.E. Savage, M.T. Klein, Asphaltene reaction pathways. 2. Pyrolysis of n-pentadecylbenzene. Ind. Eng. Chem. Res. **26**(3), 488–494 (1987)
41. P.E. Savage, D.J. Korotney, Pyrolysis kinetics for long-chain n-alkylbenzenes: Experimental and mechanistic modeling results. Ind. Eng. Chem. Res. **29**(3), 499–502 (1990)
42. T. Baba, Y. Abe, Metal cation–acidic proton bifunctional catalyst for methane activation: Conversion of 13CH4 in the presence of ethylene over metal cations-loaded H-ZSM-5. Appl. Catal. A Gen. **250**(2), 265–270 (2003)
43. T.V. Choudhary, E. Aksoylu, D. Wayne Goodman, Nonoxidative activation of methane. Catal. Rev. **45**(1), 151–203 (2003)
44. B. Bachiller-Baeza, J.A. Anderson, FTIR and reaction studies of styrene and toluene over silica–zirconia-supported heteropoly acid catalysts. J. Catal. **212**(2), 231–239 (2002)
45. T. Nobukawa, M. Yoshida, S. Kameoka, S.-I. Ito, K. Tomishige, K. Kunimori, In-situ observation of reaction intermediate in the selective catalytic reduction of N2O with CH4 over Fe ion-exchanged BEA zeolite catalyst for the elucidation of its reaction mechanism using FTIR. J. Phys. Chem. B **108**(13), 4071–4079 (2004)

Chapter 6
Nonthermal Plasma (NTP)-Assisted Catalytic Conversion of Methane and Other Hydrocarbons

6.1 Introduction

As the US Energy Information Administration (EIA) published in September 2019, the average price of natural gas is $2.22 per million British thermal units (MMBtu), while gasoline is $21.3 per MMBtu ($2.56 per gallon and 0.12 MMBtu per gallon). The huge price differential between natural gas and gasoline demonstrates the great economic potential of a cost-effective conversion of low-cost hydrocarbon. In this chapter, the low-cost hydrocarbon resource is defined as the hydrocarbon with small carbon number, vast reserve, and low cost, including light paraffins, light olefins, simple aromatics, natural gas, liquefied petroleum gas, and light naphtha. These resources are available in Canada with vast reserve. Therefore, converting low-cost hydrocarbon into high-value transportation fuel or petrochemicals is considered as a very attractive strategy.

However, light hydrocarbons like methane, ethane, and propane are quite stable at room or even high temperature. The bond dissociation energy (BDE) is widely used to measure the energy that is required to break the bond and generate free radicals. The BDE of CH_3-H, $CH_3CH_2CH_2$-H, $(CH_3)_2CH$-H, and CH_3CH_2-CH_3 are 439.3 kJ/mol, 422.2 kJ/mol, 410.5 kJ/mol, and 370.3 kJ/mol, respectively [1]. To utilize these stable chemicals, the conventional methods are thermal or thermal catalytic processes, including reforming [2–5], thermal cracking [6], C-C coupling [7, 8], and the partial oxidation of hydrocarbon [9]. Due to the high-operation temperature (normally higher than 700 °C), high energy consumption and fast catalyst deactivation are inevitable.

With the addition of catalyst, conventional thermo-catalytic approaches focus on converting methane to syngas [10], ethane and ethylene [8], C_1–C_2 oxygenates [11], and aromatics [12] in an oxidative or nonoxidative way. Unfortunately, these processes are thermodynamically unfavorable at low temperature. High temperature (>400 °C) is employed to facilitate reactions, leading to severe coke formation (up

H. Song et al., *Methane Activation and Utilization in the Petrochemical and Biofuel Industries*, https://doi.org/10.1007/978-3-030-88424-6_6

to 30% [13]) and a relatively low yield (<25%) of desired products [14]. Therefore, it is a challenge to utilize light hydrocarbons at mild condition with high conversion and excellent selectivity of desired product.

Plasma-assisted catalysis is regarded as a possible solution to solve the previously mentioned problems. Described as the fourth state of matter, plasma is comprised of electrons and ions in its simplest form. Over 99% of matters is plasma in the universe. Depending on the gas-phase temperature, the atmospheric plasma could be classified into three types, which is high-temperature plasma, low-temperature thermal plasma, and low-temperature nonthermal plasma (hereinafter called "nonthermal plasma" or "NTP") [15]. In high-temperature plasma, the thermal equilibrium is almost reached for all species (charged and neutral), while the gas temperature is higher than 10^7 K. Low-temperature thermal plasma also reached the thermal equilibrium with a much lower temperature (~10^4 K).

To the contrary, as known as nonequilibrium plasma, the nonthermal plasma is in thermal nonequilibrium state. The gas temperature is relatively low (~300 K) when compared with the electron temperature (10,000–100,000 K). A series of active species exist in the nonthermal plasma, while the gas-phase temperature can maintain around the room temperature (~300 K). Therefore, some thermodynamic unfavorable reactions can be realized in nonthermal plasma at low temperature, which is normally impossible for thermal process. Cohn et al. tried to produce hydrogen from diesel reforming with both thermal and nonthermal plasma [16]. The thermal one consumed a significantly higher amount of energy than the nonthermal method with similar hydrogen yield. The reason is that for thermal plasma, most energy are used to heating the particle instead of generating plasma. Therefore, nonthermal plasma is a more attractive method.

Plasma also has different configurations, depending on the method used for plasma generation, which includes DC torch, dielectric barrier discharge (DBD), pulse corona discharge (PCD), and more. These methods result in the differences in the properties like gas temperature, the appearance of plasma, electron temperature, discharging gas species, electron density, and the degree of ionization. In all of these, the DBD configuration is one of the most effective nonthermal plasma sources, which attract more and more interest from industrial application because of its scalability to the large system [17].

The plasma-assisted catalysis refers to the combination of a gaseous plasma with a catalytic material. This type of mixture may have synergistic effect and could enhance the performance in some way when compared with plasma or catalysis individually. The first reported plasma-assisted catalysis study was probably conducted in 1954, where a coating of platinum on the discharge tube walls could enhance the overall yield of hydrazine from the discharge of ammonia [18]. More importantly, placing the catalyst in the plasma zone could generate a synergistic effect, which benefits the formation of the desired products. The combination of a catalyst with plasma species could be the key to improve the selectivity. A numerical simulation study shows that the activated methane is a long-lived excited species. And, because these species are highly reactive on transition metal catalyst, a synergistic effect between plasma and Ni-based catalyst is achieved [19].

Plasma-assisted catalysis is a promising technique that could be scaled up and potentially commercialized for methane utilization [20]. Particularly, due to the nonequilibrium characteristics, nonthermal plasma (NTP) could promote thermodynamically unfavorable chemical reactions to occur at relatively low temperatures, even at near room temperature [21]. Methane is regarded as the most widely studied molecule in plasma-assisted catalysis. Over the last decades, a lot of studies about plasma-assisted methane utilization were conducted, ranging from the methane reforming for syngas production [20, 22–29], decomposition of hydrocarbons for hydrogen production [30–33], and partial oxidation of methane [34, 35]. Nevertheless, these processes still end up with the syngas, H_2 and C_2 hydrocarbons. A second process like Fischer-Tropsch process is required to synthesize liquid chemicals. Although it is theoretically possible, this two-step process is relatively complex. The high capital cost is inevitable, which makes reforming process less attractive. The other processes still end up with H_2 and C_2 hydrocarbons like ethylene and acetylene. Therefore, it is still a challenge to achieve the direct liquefication of light alkanes. There also are other issues like high energy consumption and severe coke formation.

Carbon deposition is another critical issue, especially for the reforming of pure methane. A large amount of carbon will be generated for the complete pyrolysis of methane. In the case that catalyst is incorporated, coke will block the hole and function site of the catalyst and make the catalyst deactivated. On the other hand, the deposited carbon could be very conductive, which could influence the plasma discharge [36].

All in all, it is extremely challenging for the direct liquefication of light hydrocarbon in a more environmentally friendly and economically profitable way. Here, we will introduce several researches about NTP-assisted directly hydrocarbon liquefication. The overall objective of these studies is to introduce a series of effective catalyst to achieve upgrading and direct liquefication of light hydrocarbon at mild condition with a higher energy efficiency and lower coke formation rate.

6.2 NTP-Assisted Catalytic Liquefaction of Natural Gas

Natural gas is an abundant, cheap, and underutilized resource with increasingly proved reserves as well as an agricultural by-product. Currently, natural gas is mainly used as a source of energy for heating, cooking, and power generation. Increasing worldwide energy demands with the discovery of huge shale deposits has stimulated research interests in the conversion of natural gas to high-value transportation fuel or chemicals. However, the chemical inertness of methane (the principal component of natural gas, >75%) [37] renders a challenge for the chemical transformation of natural gas. The nontraditional methods including plasma-assisted transformation and photocatalytic conversion are developed for methane utilization at mild conditions. Plasma is a promising technique that could be scaled up and potentially commercialized for methane utilization [20].

Meng et al. present a novel nonthermal plasma-assisted photocatalytic conversion of simulated natural gas (methane and propane in this case) at near ambient conditions [38]. A series of catalysts were prepared and shaped into pellets with a diameter of 2 mm and 5–10 mm long. Nonthermal plasma catalytic conversion of methane and propane produces various C_6–C_9 alkane compounds as the liquid products (Fig. 6.1) and H_2, C_2H_6, and C_3H_8 as the major gas products with a small amount of C_2H_4, C_2H_2, and C_4–C_5 hydrocarbons and coke. Careful GC-MS analysis shows that the liquid products contain C_6–C_9 alkanes with a predominantly branched structure, which could be potentially used as high-value aviation fuel for aircrafts [39].

As shown in Figs. 6.2 and 6.3, similar conversions of methane or propane are observed over different supports (SiO_2, γ-Al_2O_3, HZSM-5, UZSM-5) and various metal (Ag, Ga, Pd, Pt)-modified UZSM-5 catalysts. This is understandable because methane and propane are exclusively activated by plasma at such low reaction temperature (~100 °C as shown in Fig. 6.4). Hydrothermally synthesized UZSM-5 support has similar crystal structure, surface area, and pore volume as those of commercial HZSM-5, which is confirmed by XRD (Fig. 6.5) and physisorption analysis (Table 6.1). It is worth noting that methane conversion (1.3–4.1%) is much lower than propane conversion (26.9–36.4%), which is attributed to the higher bond dissociation energy of C-H in methane (439.3 kJ/mol), compared to that of C-H (422.2 kJ/mol, 410.5 kJ/mol) or C-C (370.3 kJ/mol) in propane [1]. However, the product distribution varies heavily depending on the charged support as shown in Fig. 6.2 and Table 6.2. There is no detectable liquid product over SiO_2 or γ-Al_2O_3, while more than 30 C% yield of liquid products is obtained over bare zeolite such

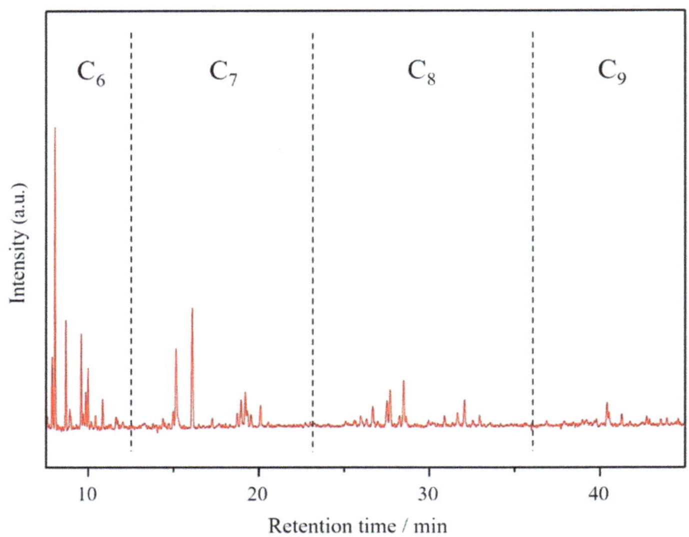

Fig. 6.1 GC-MS analysis of liquid product collected from the conversion of methane and propane over Ga/UZSM-5 with NTP (Reprinted with permission from (Ref. [38]. Copyright (2020) American Chemical Society)

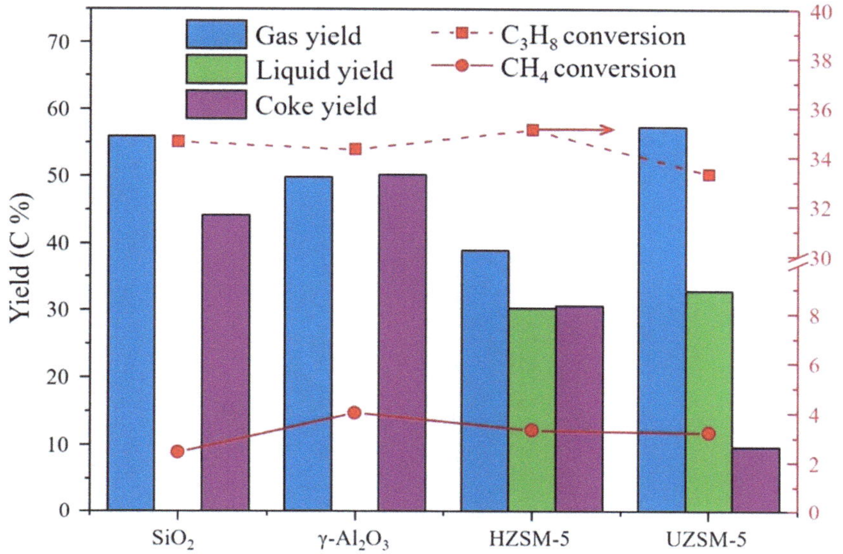

Fig. 6.2 The conversion and yields of products (gas, liquid, and coke) over different supports (Reprinted with permission from (Ref. [38]. Copyright (2020) American Chemical Society)

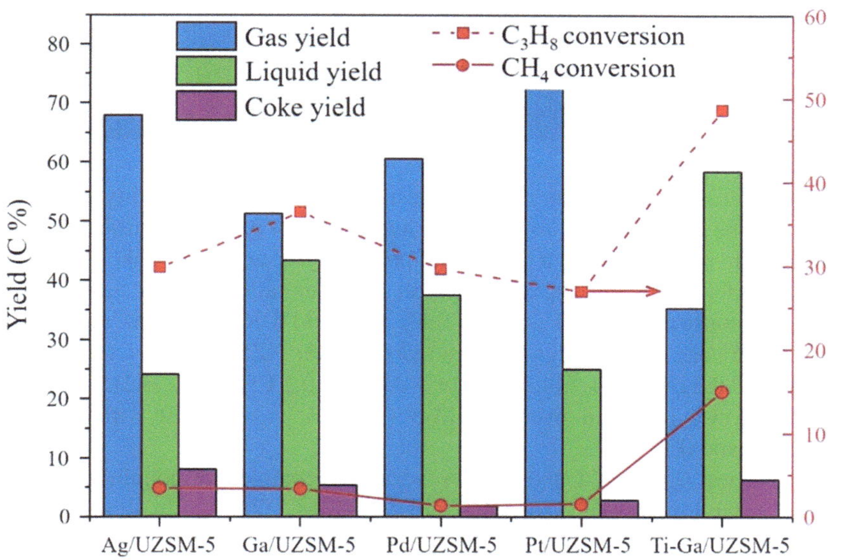

Fig. 6.3 The conversion and yields of products (gas, liquid, and coke) over various metal-modified UZSM-5 catalysts (Reprinted with permission from (Ref. [38]. Copyright (2020) American Chemical Society)

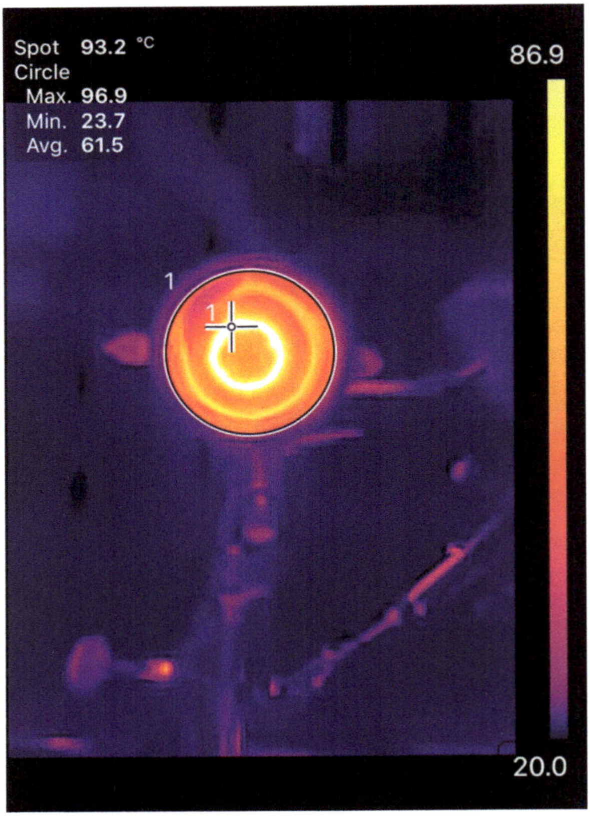

Fig. 6.4 The measurement of reactor temperature distribution by an infrared camera after 40-minute reaction (Reprinted with permission from (Ref. [38]. Copyright (2020) American Chemical Society)

as HZSM-5 and UZSM-5. This result suggests that the support's acidity and morphology may be critical for the formation of hydrocarbons with high carbon number (C_6–C_9). The hypothesis is that a high liquid product yield and a low coke formation were achieved when using the catalyst with a higher ratio of weak acidic sites, which is demonstrated by NH_3-TPD (Table 6.3) and DRIFT results (Fig. 6.6). Moreover, the coke yield is significantly reduced over HZSM-5 (30.7 C%), compare to that of the over SiO_2 (44.1 C%) or γ-Al_2O_3 (50.2 C%, Table 6.2). Coke formation is further inhibited over UZSM-5 (9.7 C% coke yield) with uniform cylindrical shape (Fig. 6.7).

Metal species, Ag, Ga, Pd, and Pt, are reported to promote C-C coupling reaction and coke reduction [40–42], thereby, introduced on UZSM-5 in this study to improve the yield of liquid products. As shown in Fig. 6.3 and Table 6.4, coke formation can be further suppressed by modification with metal species. However, compared to UZSM-5 (Table 6.2), the addition of Ag or Pt decreases the yield of C_6–C_9 liquid products (by >8.0 C%), instead more C_2–C_4 gas products (increased by >10.5 C%) are formed. Pd loading leads to a slight increase on both liquid and gas yield by lowering the coke deposition (1.9 C%). Ga addition reduces the yield of

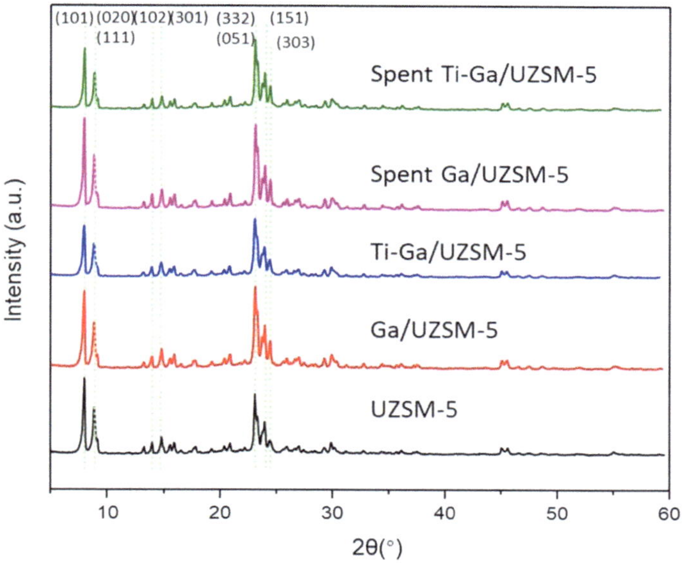

Fig. 6.5 XRD pattern of fresh and spent catalysts (Reprinted with permission from (Ref. [38]. Copyright (2020) American Chemical Society)

Table 6.1 Physisorption results of various catalysts

	BET surface area (m^2/g)	Pore volume (cm^3/g)
UZSM-5 powder	362 ± 10	0.231
UZSM-5 pellet	321 ± 5	0.258
Ga/UZSM-5	320 ± 6	0.258
Ti-Ga/UZSM-5	311 ± 5	0.242
Ga/UZSM-5 spent	326 ± 8	0.268
Ti-Ga/UZSM-5 spent	307 ± 7	0.248

both coke and gas products, thus, promotes the formation of liquid products (increased by 10.5 C%). Ga/UZSM-5 is thus selected as the preferred catalyst for further investigations.

The roles of plasma and a catalyst played in the co-conversion of methane and propane are verified by conducting equivalent reactions with/without plasma or a catalyst. As shown in Table 6.5, it is expected that there is no reaction without plasma as evidenced by no conversion of methane or propane in the catalyst-only mode. When plasma is applied, approximately 0.125 mmol methane and/or propane molecules per kilojoules are converted regardless of the catalyst under stable conditions. This supports the previous hypothesis that methane and propane are exclusively activated by plasma. For the test without a catalyst, SiO_2 is placed in the

Table 6.2 The conversion, the yields of products (gas, liquid, and coke), and energy efficiency (EE) over different supports

		SiO_2	γ-Al_2O_3	HZSM-5	UZSM-5
Conversion (%)	CH_4	2.4	4.0	3.3	3.2
	C_3H_8	34.6	34.3	35.1	33.3
Yield of H_2 (%)		25.1	26.8	29.5	27.3
Gas product yield (C%)	C_1	0.0	0.0	0.0	0.0
	C_2	24.4	22.5	20.2	27.9
	C_3	16.7	14.1	3.7	15.9
	C_4	11.0	8.7	8.2	9.6
	C_5	3.8	4.5	4.7	4.0
	Sum	55.9	49.8	36.8	57.4
Liquid product yield (C%)	C_6	0.0	0.0	9.5	13.6
	C_7	0.0	0.0	10.7	11.3
	C_8	0.0	0.0	4.7	6.6
	C_9	0.0	0.0	3.5	2.8
	$\geq C_{10}$	0.0	0.0	6.3	0.0
	Sum	0.0	0.0	32.5	32.9
Coke yield (C%)		44.1	50.2	30.7	9.7
EE (mmol/kJ)		0.108	0.104	0.107	0.101

Table 6.3 NH_3-TPD results of various catalysts

	Type of acidic sites (μmol NH_3/g)		
Catalysts	Weak (<200 °C)	Medium (200–400 °C)	Strong (>400 °C)
UZSM-5	38	0	0
Ga/UZSM-5	41	0	0
Ti-Ga/UZSM-5	44	0	0

reactor as the inert material to reduce the effect of mass or heat transfer on the reaction. No liquid products collected in the plasma-only mode implies that the catalyst with an appropriate acidity is essential for C-C coupling reaction to form hydrocarbons with longer chain or more branches. Without propane, methane activation by plasma generates methyl (CH_3•) or methylene (•CH_2•) radicals, which either combine to form C_2–C_3 species or further undergo C-H cleavage to form coke as evidenced by the observed high yield of C_2–C_3 hydrocarbons (69.0 C%, Table 6.5) with 21.2 C% of coke under pure CH_4 run. Without methane, propane activation produces a variety of C_1–C_3 radicals, such as CH_3•, CH_3-CH_2•, •CH_2-CH_2-CH_2•, and CH_3-CH_2-CH_2•. Radical-radical condensation reactions between these radicals generate C_2–C_9 hydrocarbons as evidenced by 67.1 C% of C_2–C_5 hydrocarbons as gas products and 28.9 C% of C_6–C_9 hydrocarbons as liquid products from pure C_3H_8 run. Such low coke formation (4.1 C%) may be because propane could generate more radicals available for radical combination; hence, less radicals suffer excessive C-H cleavage to form coke. Compared to sole CH_4 or C_3H_8 run, co-feeding C_3H_8 with CH_4 gives less C_2–C_5 products (51.3 C%), more C_6–C_9 products

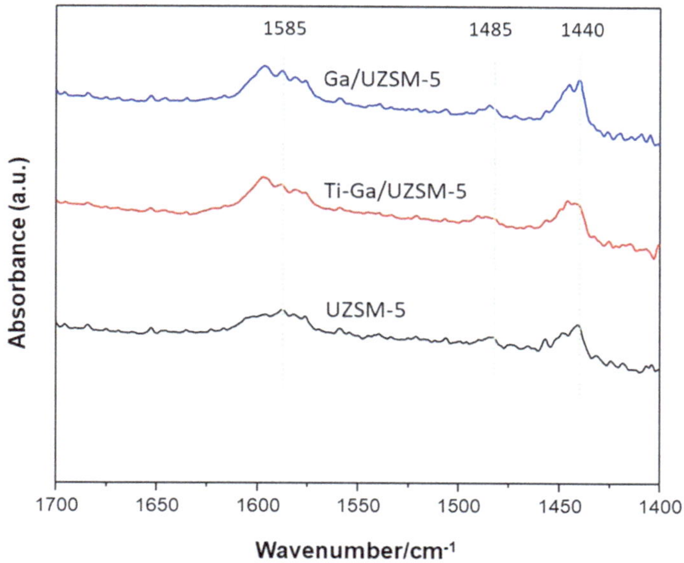

Fig. 6.6 DRIFT spectra of catalyst upon pyridine adsorption (Reprinted with permission from (Ref. [38]. Copyright (2020) American Chemical Society)

Fig. 6.7 SEM image of UZSM-5 showing the uniform cylindrical morphology (Reprinted with permission from (Ref. [38]. Copyright (2020) American Chemical Society)

(43.4 C%), and limited coke formation (5.3 C%). The presence of propane in methane offers more C_1–C_3 radicals to terminate the further cleavage of methyl radicals from methane; therefore, more branched hydrocarbons are formed as evidenced by a higher yield of C_7–C_8 hydrocarbons. This approach is also applied to convert natural gas into gasoline range hydrocarbons. Natural gas as recovered at the wellhead consists of mostly methane (73–85%), C_2–C_5 hydrocarbons (12–15%), and other

Table 6.4 The conversion, the yields of products (gas, liquid, and coke), and EE over UZSM-5 catalyst modified with various metal species

Catalyst		Ag/UZSM-5[a]	Ga/UZSM-5[a]	Pd/UZSM-5[a]	Pt/UZSM-5[a]	Ti-Ga/UZSM-5[b]
Conversion (%)	CH_4	3.4	3.3	1.3	1.5	15.0
	C_3H_8	29.8	36.4	29.6	26.9	48.6
Yield of H_2 (%)		34.3	23.6	7.3	10.6	28.0
Gas product yield (C%)	C_1	0.0	0.0	0.0	0.0	0.0
	C_2	35.1	25.9	45.7	55.3	21.7
	C_3	16.9	13.7	0.0	0.0	6.4
	C_4	11.6	8.4	10.9	12.5	5.2
	C_5	4.3	3.2	4.0	4.4	2.0
	Sum	67.9	51.3	60.6	72.2	35.3
Liquid product yield (C%)	C_6	10.0	15.9	14.6	11.3	15.1
	C_7	7.9	15.3	11.1	8.0	17.4
	C_8	4.4	10.2	6.8	5.5	19.7
	C_9	2.6	2.6	5.0	2.4	6.4
	$\geq C_{10}$	0.0	0.0	2.1	0.0	0.0
	Sum	24.1	43.4	37.5	25.0	58.4
Coke yield (C%)		8.0	5.3	1.9	2.8	6.3
EE (mmol/kJ)		0.101	0.115	0.093	0.087	0.134

impurities [37, 43]. To simplify, a simulated natural gas (the molar ratio of CH_4/C_3H_8 = 4) is prepared. The nonthermal plasma-assisted conversion of simulated natural gas over Ga/UZSM-5 (Table 6.6) gives 55.2 C% of gas products (C_2–C_5 hydrocarbons), 37.9 C% of liquid products (C_6–C_9 hydrocarbons), and 6.9 C% of coke. The yield of liquid products drops slightly due to more methane in the feed; however, the coke formation is still well controlled.

Another direction being taken to break the high thermodynamic barrier for methane activation is photocatalysis, which employs photoenergy to promote reactions at low temperatures. Such low temperature could minimize the energy consumption and reduce the side reactions such as coke formation [44]. As evidenced in Fig. 6.8, UV radiation can be self-generated by nonthermal dielectric barrier discharge plasma as evidenced by UV/Vis spectrum [45, 46].

Ga/UZSM-5 modified with TiO_2 (a widely used one as the photoactive material) is synthesized as the photocatalyst to further enhance the performance on the conversion of simulated natural gas. The metal species are homogeneously dispersed on the catalyst with good stability, confirmed by the ICP-OES, XPS, and STEM results (Figs. 6.9, 6.10, 6.11, 6.12, and 6.13). As shown in the last and the second-to-last columns of Tables 6.4 and 6.5, respectively, compared to Ga/UZSM-5, methane conversion and coke yield over Ti-Ga/UZSM-5 are quite similar; however, higher propane conversion (increased by 11.2%) and more C_6–C_9 hydrocarbons (the yield boosted from 37.9 C% to 58.4 C%) are obtained. A plausible explanation could be that the photocatalyst Ti-Ga/UZSM-5 notably enhances the energy utilization efficiency through the incorporation of the energy present in the coexisting UV/

Table 6.5 The effect of operating modes and catalysts on the conversion, the yields of products (gas, liquid, and coke), and energy efficiency (EE) in the reaction

Reaction Condition		$CH_4 + C_3H_8$	Pure CH_4	Pure C_3H_8	Catalyst only	Plasma only
Conversion (%)	CH_4	3.3	23.7	0.0	0.0	2.4
	C_3H_8	36.4	0.0	23.5	0.0	34.6
Yield of H_2 (%)		23.6	48.1	19.2	0.0	25.1
Gas product yield (C%)	C_1	0.0	0.0	10.3	0.0	0.0
	C_2	25.9	51.9	28.8	0.0	24.4
	C_3	13.7	17.1	16.7	0.0	16.7
	C_4	8.4	4.8	7.4	0.0	11.0
	C_5	3.2	1.4	3.9	0.0	3.8
	Sum	51.3	75.3	67.1	N/A	55.9
Liquid product yield (C%)	C_6	15.9	0.7	15.1	0.0	0.0
	C_7	15.3	1.9	8.5	0.0	0.0
	C_8	10.2	0.9	6.0	0.0	0.0
	C_9	2.6	0.0	2.3	0.0	0.0
	$\geq C_{10}$	0.0	0.0	0.0	0.0	0.0
	Sum	43.4	3.5	28.9	N/A	0.0
Coke yield (C%)		5.3	21.2	4.1	N/A	44.1
EE (mmol/kJ)		0.115	0.139	0.137	N/A	0.108

Table 6.6 The effect of methane/propane molar ratio on the conversion of methane and propane

Molar ratio		1:1	2:1	4:1	7:1
Conversion (%)	CH_4	3.3	5.1	14.7	14.7
	C_3H_8	36.4	33.3	37.4	32.9
Yield of H_2 (%)		23.6	35.3	36.0	47.5
Gas product yield (C%)	C_1	0.0	0.0	0.0	0.0
	C_2	25.9	27.4	28.8	37.5
	C_3	13.7	16.8	13.2	15.5
	C_4	8.4	11.1	10.0	11.4
	C_5	3.2	3.6	3.1	3.1
	Sum	51.3	58.9	55.2	67.6
Liquid product yield (C%)	C_6	15.9	14.5	13.7	7.3
	C_7	15.3	13.7	13.6	8.3
	C_8	10.2	6.6	7.8	5.7
	C_9	2.6	2.1	3.4	2.2
	$\geq C_{10}$	0.0	0.0	0.0	0.0
	Sum	43.4	33.4	37.9	23.0
Coke yield (C%)		5.3	7.7	6.9	9.4
EE (mmol/kJ)		0.115	0.103	0.118	0.091

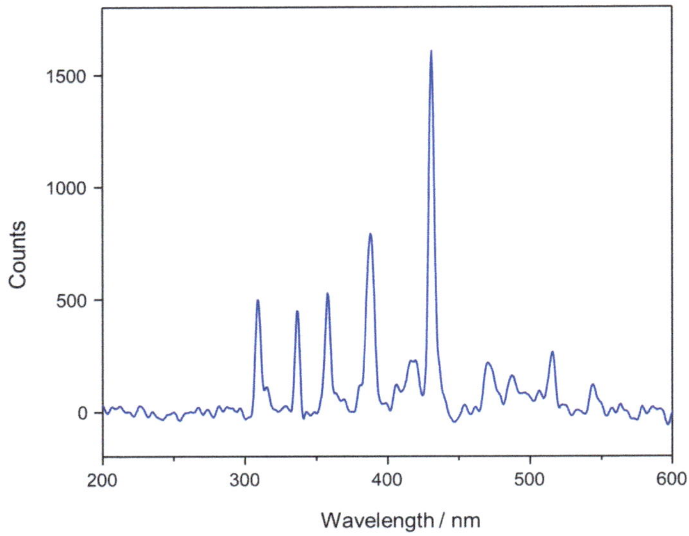

Fig. 6.8 UV/Vis spectrometer spectrum of light emitted from nonthermal plasma (Reprinted with permission from (Ref. [38]. Copyright (2020) American Chemical Society)

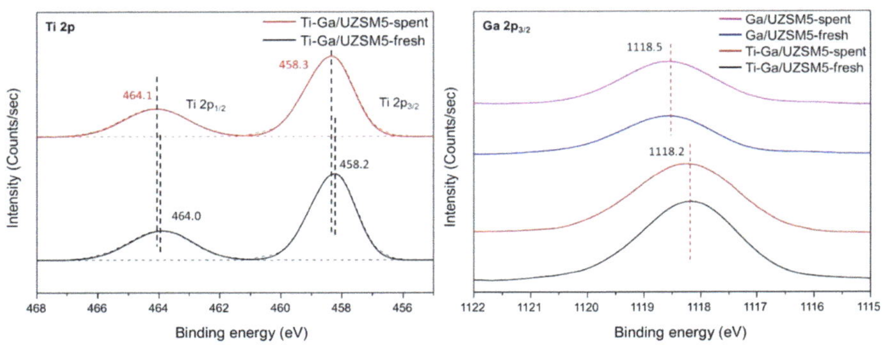

Fig. 6.9 Ti 2p and Ga $2p_{3/2}$ XPS spectra of fresh and spent catalysts (Reprinted with permission from (Ref. [38]. Copyright (2020) American Chemical Society)

visible light formed during plasma generation. In other words, more input energy is utilized for reactant conversion, particularly for propane conversion since propane is much easier to be activated compared to methane due to the aforementioned lower molecular activation energy. Higher propane conversion would provide more C_1–C_3 radicals for the formation of C_6–C_9 hydrocarbons. As a result, more liquid products are formed over Ti-Ga/UZSM-5. Additional control experiments with external UV source were also performed to prove this hypothesis (Table 6.7). With stronger external UV radiation, the conversion of propane further increases (by 5.4%), which leads to higher liquid yield (78.1 C%) and less coke formation

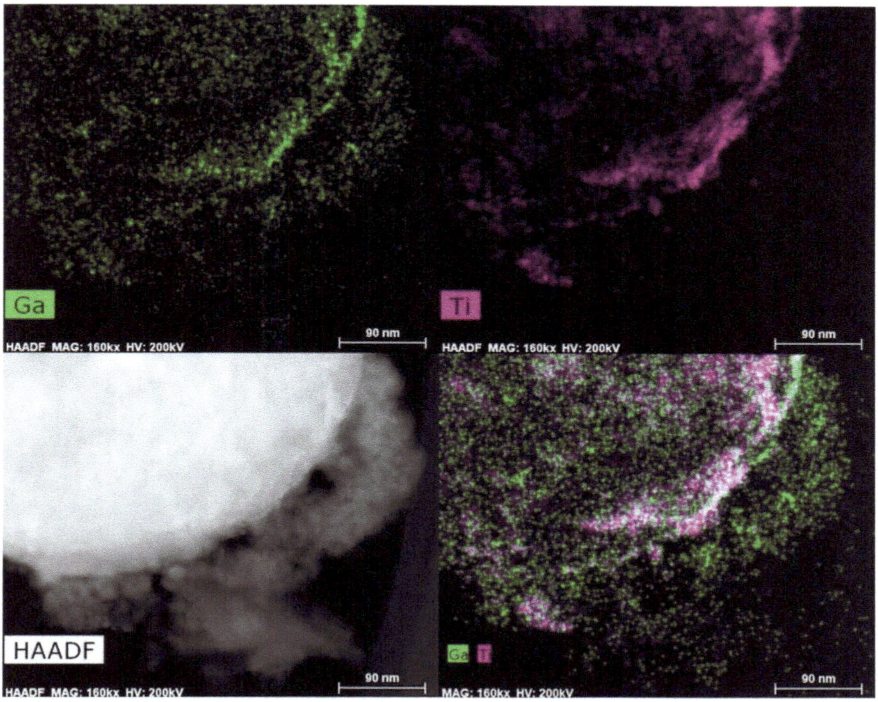

Fig. 6.10 HAADF and EDX images of fresh Ti-Ga/UZSM-5 (Reprinted with permission from (Ref. [38]. Copyright (2020) American Chemical Society)

(2.0 C%). However, without a plasma, there is no methane or propane conversion with the external UV light. These results suggest that the UV light can facilitate the conversion under a plasma but is unable to activate the reactant solely at low temperature.

Isotope-labeled CH_4 ($^{13}CH_4$ and CD_4) is employed to elucidate the evolution of C and H of methane in the reaction. As shown in Fig. 6.14a, compared to ^{13}C signal from the non-isotopic-labeled counterpart (CH_4 + C_3H_8), ^{13}C signal acquired from the $^{13}CH_4$ run ($^{13}CH_4$ + C_3H_8) is significantly enhanced (the peak area with respect to $CDCl_3$ reference increases from 0.3 to 11.0), which strongly supports methane incorporation into the final products. The signals in the region of 10–5 ppm correspond to the aliphatic resonances of ^{13}C in alkane. The peaks near 30.5, 23.0, and 19.9 ppm are ascribed to 2-^{13}C (CH_3-$^{13}CH_2$-) of paraffins, while the peaks at 14.7 and 12.0 ppm are attributed to 1-^{13}C ($^{13}CH_3$-) of paraffins. There are small peaks around 32.2, 34.6, 36.5, 39.2, and 41.7 ppm, which are derived from 3-^{13}C or 4-^{13}C of paraffins. The significant increase in the peaks at 12.0, 14.7, 19.9, 23.0, and 30.5 ppm suggests methane prefers to be incorporated into 1-C or 2-C of alkane as a terminal group [47, 48]. This is highly consistent with the observation in ^2H-NMR spectra as shown in Fig. 6.14b, where there is a broad deuterium peak (0–2 ppm)

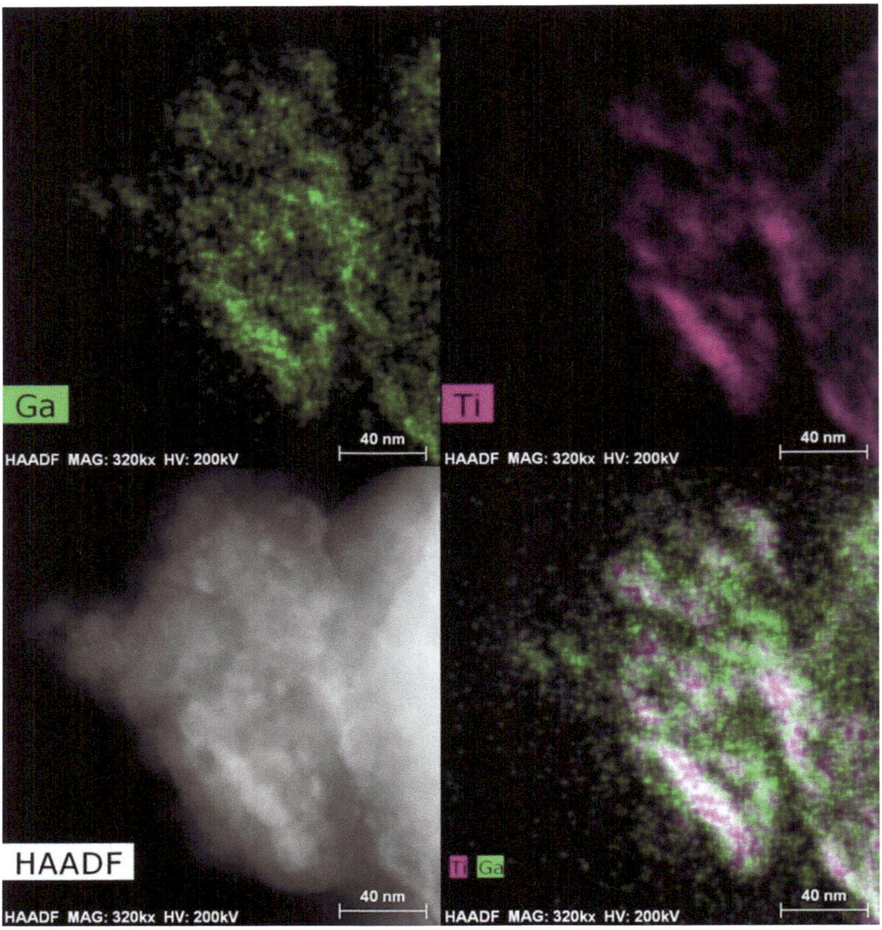

Fig. 6.11 HAADF and EDX images of spent Ti-Ga/UZSM-5 (Reprinted with permission from (Ref. [38]. Copyright (2020) American Chemical Society)

which can be deconvoluted into three peaks at 0.9, 1.3, and 1.6 ppm, corresponding to -CD_3, -CD_2, and -CD [49], respectively.

This proof-of-concept investigation highlights the synergy of plasma activation and photocatalysis in the upgrading of simulated natural gas. They demonstrate an efficient protocol for nonthermal plasma-assisted photocatalytic conversion of natural gas, which is cheap, abundant, and naturally available, to gasoline range chemicals at ambient conditions, thus greatly reducing the heavy dependence on finite other fossil fuels such as oil and coal.

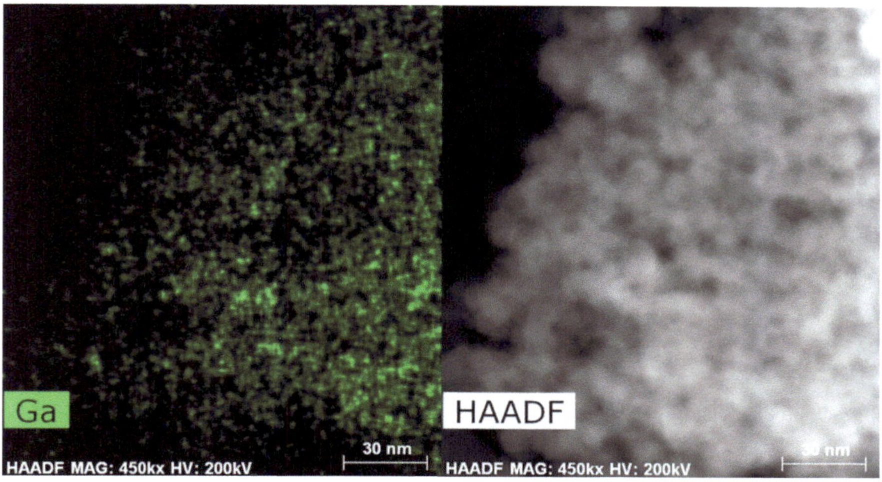

Fig. 6.12 HAADF and EDX images of fresh Ga/UZSM-5 (Reprinted with permission from (Ref. [38]. Copyright (2020) American Chemical Society)

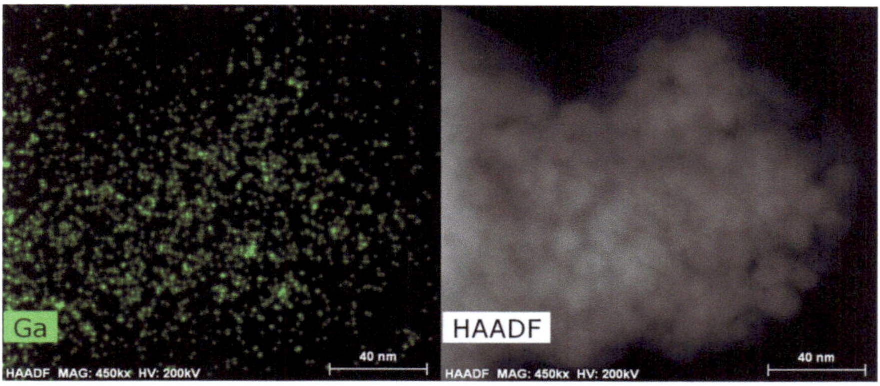

Fig. 6.13 HAADF and EDX images of spent Ga/UZSM-5 (Reprinted with permission from (Ref. [38]. Copyright (2020) American Chemical Society)

6.3 NTP-Assisted Catalytic Liquefaction of Biogas

Biogas is a promising renewable energy source, which primarily consists of methane (55–65%) and carbon dioxide (35–45%) [50] produced from the anaerobic digestion of biomass, landfill, and wastewater treatment. Extensive investigations on the conversion of biogas with NTP are mostly concentrated on producing syngas or higher hydrocarbons (C_2–C_4) [51–55], while very little effort is devoted to the formation of liquid products [21]. Another critical issue for a plasma-catalytic reaction is the formation of carbonaceous species, which are both a waste of raw

Table 6.7 The effect of external UV light on the conversion, the yields of products (gas, liquid, and coke), and EE over Ti-Ga/UZSM-5 catalyst

Reaction condition		Catalyst + plasma	Catalyst + plasma + external UV lamp	Catalyst + UV lamp
Conversion (%)	CH_4	15.0	16.8	0.0
	C_3H_8	48.6	54.0	0.0
Yield of H_2 (%)		28.0	19.2	N/A
Gas product yield (C%)	C_1	0.0	0.0	N/A
	C_2	21.7	14.7	N/A
	C_3	6.4	3.6	N/A
	C_4	5.2	1.6	N/A
	C_5	2.0	0.0	N/A
	Sum	35.3	19.9	N/A
Liquid product yield (C%)	C_6	15.1	20.7	N/A
	C_7	17.4	22.1	N/A
	C_8	19.7	27.3	N/A
	C_9	6.4	8.0	N/A
	$\geq C_{10}$	0.0	0.0	N/A
	Sum	58.4	78.1	N/A
Coke yield (C%)		6.3	2.0	N/A
EE (mmol/kJ)		0.134	0.148	N/A

materials and poison to the catalyst. Carbonaceous species is largely overlooked within plasma-assisted catalytic DRM reported in the literature and predominantly estimated to be higher than 10% [21, 51]. Actually, the coke yield is more severe. Importantly, the nature of carbonaceous species has a significant effect upon the plasma itself, with more stable carbon species (oxidized at $T > 650$ K) destabilizing the plasma, resulting in arcing and poorer methane conversion [56]. On this note, Wang et al. employed NTP to simultaneously convert biogas to liquid chemicals with controlled carbon deposition at low temperature and atmospheric pressure [57].

The conversion of methane and carbon dioxide was carried out in a coaxial DBD reactor at room temperature and atmospheric pressure (Fig. 6.15). As shown in Table 6.8, no conversion of both CH_4 and CO_2 (entry 1) indicates no reaction in the "catalyst-only" mode (without NTP and the used catalysts including CZSM-5, γ-Al_2O_3, UZSM-5, Pt/CZSM-5, and Pt/UZSM-5) due to such a low reaction temperature (<200 °C). However, the conversion of CH_4 and CO_2 increases significantly when applying NTP (entries 2–12) on the reaction and produces gas, liquid, and coke products. H_2 and CO are the major gas products detected by micro-GC, and the incondensable hydrocarbons (C_xH_y) are mainly composed of C_2H_6 with trace amounts of C_2H_4, C_3H_8, and C_4H_{10}. The collected liquid products are analyzed by GC-MS and ^{1}H NMR and identified to be C_1–C_3 oxygenates as shown in Fig. 6.16. The result of ^{1}H NMR (Fig. 6.16b) shows the presence of protons in the functional groups -CHO (~9.8 ppm, aldehydes), -OH or -O-CH_3 (3–5 ppm, alcohols/ethers), and -CO-CH_3 (2.0–2.3 ppm, ketones), which is highly consistent with GC-MS

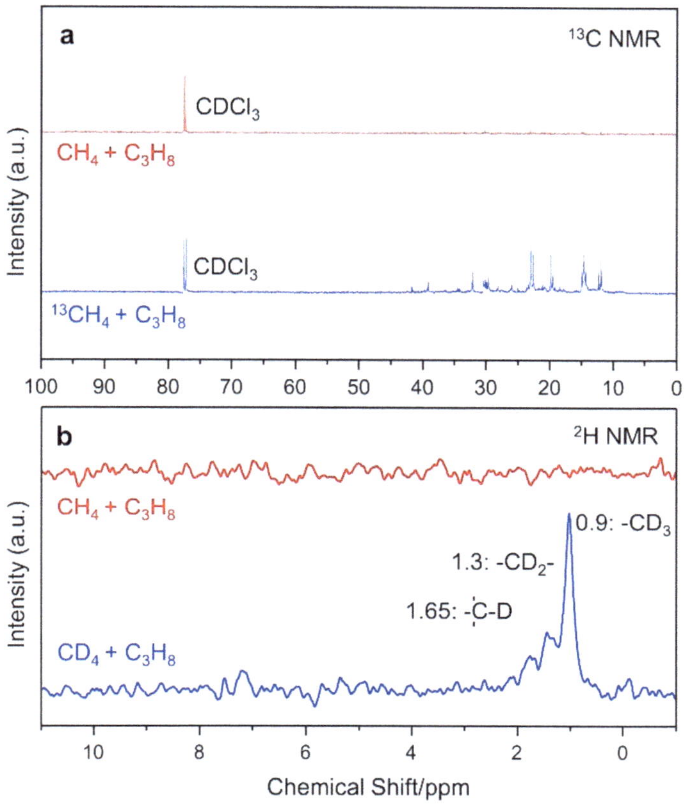

Fig. 6.14 Isotope labelling investigations on the conversion of methane and propane: (**a**) ^{13}C NMR and (**b**) ^{2}H NMR (Reprinted with permission from (Ref. [38]. Copyright (2020) American Chemical Society)

analysis (Fig. 6.16a) of the liquid products, where formaldehyde (HCHO), methanol (CH_3OH), ethanol (C_2H_5OH), and acetone (CH_3COCH_3) are detected.

Under the "plasma-only" mode (without packing catalysts), the conversion of CH_4 and CO_2 is 25.4% and 14.7%, respectively (Table 6.8, entry 2). As a result of the higher dissociation energy of CO_2 (5.52 eV) as compared to CH_4 (4.57 eV) [58], the conversion of CO_2 is lower than CH_4. However, no liquid products are condensed, and coke yield is as high as 57.7%. When placing the catalyst in the plasma zone, the formation of liquid products is observed (Table 6.8 except for SiO_2, entry 5), and carbon deposition is significantly reduced, which shows the great potential to convert biogas (CH_4 and CO_2) into valuable and renewable chemicals. It is also noticed that the conversion of CH_4 and CO_2 as well as energy efficiency (EE) for the reaction slightly drops under the plasma + catalyst mode, compared to the "plasma-only" mode, probably due to the change in discharge behavior induced by the catalyst [21].

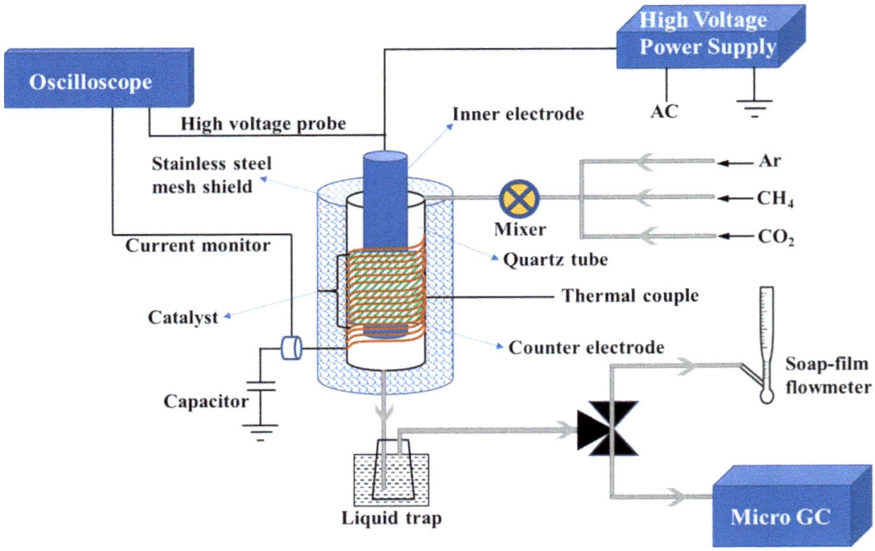

Fig. 6.15 A schematic diagram of the experimental setup (Reproduced from Ref. [57] with permission from Elsevier)

Table 6.8 The effect of operating modes and various catalysts on the reaction

	Entry	CH_4 conv. (%)	CO_2 conv. (%)	EE (mmol/kJ)	The yields of products (%)				
					H_2	CO	C_xH_y	Liquid	Coke
Catalyst only	1	0	0	N/A	0	0	0	0	0
Plasma only	2	25.4	14.7	0.23	40.8	22.9	19.4	0.0	57.7
CZSM-5	3	24.2	15.3	0.24	38.9	18.4	12.3	34.4	34.9
γ-Al_2O_3	4	19.4	10.7	0.17	50.3	21.1	21.2	25.8	31.9
SiO_2	5	17.0	10.9	0.14	49.4	26.5	24.5	0.0	49.0
Ag/CZSM-5	6	18.7	13.6	0.19	44.4	19.5	13.6	33.1	33.7
Pt/CZSM-5	7	21.5	12.6	0.21	32.7	18.2	11.8	60.7	9.3
Pd/CZSM-5	8	20.4	14.9	0.19	41.3	27.7	16.0	39.9	16.3
Re/CZSM-5	9	23.1	16.0	0.21	43.5	22.7	15.2	31.8	30.3
Ir/CZSM-5	10	21.2	12.8	0.19	43.6	23.2	14.6	25.0	37.2
UZSM-5	11	23.9	13.7	0.21	38.4	19.1	14.5	51.2	15.2
Pt/UZSM-5	12	22.1	16.2	0.23	37.8	20.2	15.7	59.0	5.1

The reforming of biogas to liquid chemicals in the presence of a plasma is likely to occur via radical chain mechanism [21, 51], where CO, CH_3, and OH radicals are considered to be the key species during the plasma-catalytic CH_4/CO_2 conversion [21]. As shown in Fig. 6.17, in a plasma environment, CO_2 is vibrationally and

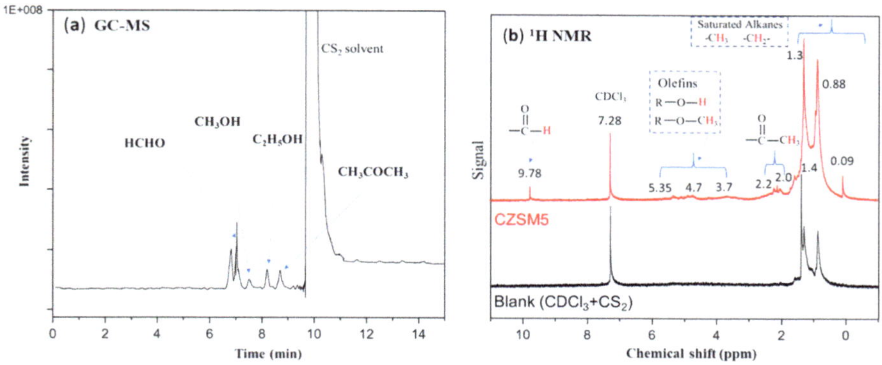

Fig. 6.16 GC-MS (**a**) and ^{1}H NMR (**b**) analyses of the liquid products formed under plasma + CZSM-5 catalyst mode (Reproduced from Ref. [57] with permission from Elsevier)

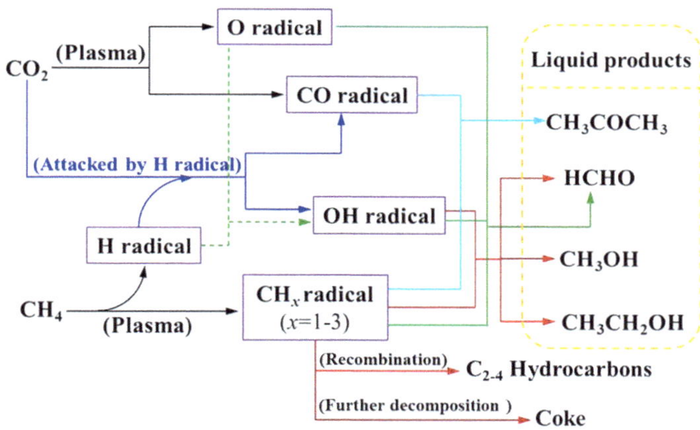

Fig. 6.17 The possible reaction pathway for the nonthermal plasma-catalytic conversion of biogas to liquid chemicals

electronically excited by electron impact reactions and dissociated into CO and O radicals. The electron impact dissociation of CH_4 generates H and CH_x (x = 1–3) radicals with CH_3 as the major radicals [59]. CH_x radicals could recombine to form C_2–C_4 hydrocarbons (C_xH_y) [4] as seen in Table 6.8. Meanwhile, further decomposition of CH_x radicals by highly energetic electrons leads to the coke formation on the catalyst surface. OH radicals could be derived from the recombination of H and O radicals. In addition, H radicals from methane dissociation may attack the vibrationally excited CO_2 species, proposed to be the main reaction pathway to form OH and CO radicals [21]. The coupling reaction between CH_3 and OH radicals directly gives methanol (CH_3OH). The recombination of CH_x radicals could form C_2H_5 radicals, which react with OH radicals to produce ethanol (C_2H_5OH). The oxidation of CH_x by O/OH radicals may be responsible for the formation of HCHO. While CO radicals react with CH_3 radicals to give acetyl radicals (CH_3CO), followed by recombination with CH_3 radicals to generate acetone as the final product. The alteration in plasma discharge behavior by placing the catalyst in DBD, which has been characterized as a shift from gas microdischarges to surface discharges and weaker microdischarges [60], may effectively inhibit the direct dissociation of CH_x to form carbonaceous species. Moreover, the porous structure of the catalyst could offer higher specific surface area for the radical reactions, thus enhancing liquid yield [61].

Three bare supports (zeolite, γ-Al_2O_3, SiO_2) are comparatively tested. Combining NTP with zeolite catalyst (CZSM-5, commercial) gives similar conversions of CH_4 and CO_2 with a slightly higher EE for the reaction, compared to the "plasma-only" mode. The conversion of CH_4 and CO_2 as well as EE are reduced when using γ-Al_2O_3 or SiO_2 catalyst. Moreover, the plasma reaction with CZSM-5 catalyst enhances the yield of liquid products to 34.4% and reduces the coke formation by 22.8% (entry 3), while using γ-Al_2O_3 catalyst leads to 25.8% of liquid product formation with an increase in H_2 yield by ca. 10% (entry 4) and packing SiO_2 catalyst in the NTP reactor leads to the increase in the yields of H_2, CO, and C_xH_y by 8.6%, 3.6%, and 5.1%, respectively. The modification of CZSM-5 is conducted to further improve the reaction performance by introducing metal species (Ag, Pt, Pd, Re, and Ir) and controlling morphology and acidity (UZSM-5, hydrothermal synthesized). Powder XRD patterns (Fig. 6.18a) show that a typical pattern of MFI structure ($2\theta = 8.3°$, $9.2°$, $23.5°$, and $24.2°$ corresponding to (101), (200), (501), and (303) crystal planes) exists in these fresh zeolite catalysts, indicating that the loaded metal species are highly dispersed on the zeolite support and cause insignificant distortion of the catalyst framework after the loading of various metal species on zeolite. The varying degrees of reduction in the intensity from (101) and (200) crystal planes over the supported metal (Ag, Pt, Pd, Re, Ir) CZSM-5 catalysts after reactions (Fig. 6.18b) might be due to the agglomeration of metal species on these two crystal surfaces. More importantly, the typical pattern of MFI structure is clearly observed in powder XRD patterns of spent catalysts (Fig. 6.18b), which indicates that the zeolite framework keeps intact during the plasma reaction. Compared to the performance over plasma + CZSM-5 (entry 3), the plasma reaction with Ag/CZSM-5 (entry 6) decreases the conversion of CH_4 and CO_2 and slightly enhances the gas yield. However, there is little effect on the formation of liquid products and coke reduction. Similar phenomenon is observed over Re/CZSM-5 (entry 9) and Ir/CZSM-5 (entry 10). Packing the Pd/CZSM5 catalyst in the DBD reactor (entry 8) results in higher yield of gas products (especially, CO yield increased by 9.3%) and lower coke formation (reduced by 18.6%) but has limited promotion effect on the liquid yield (increased by 5.5%). The highest liquid yield (60.7%) with the lowest coke formation (9.3%) is achieved over Pt/CZSM-5 (entry 7), which shows better plasma-catalytic performance over other catalysts.

DRIFT spectra (Fig. 6.19) of these fresh catalysts upon pyridine absorption exhibit similar distribution of Bronsted and Lewis acidic sites, implying that the distribution of Bronsted and Lewis acidic sites is not an important consideration for the plasma-assisted reaction. NH_3-TPD results in Table 6.9 suggest that the distribution of weak, medium, and strong acidic sites over the catalyst has little effect on the

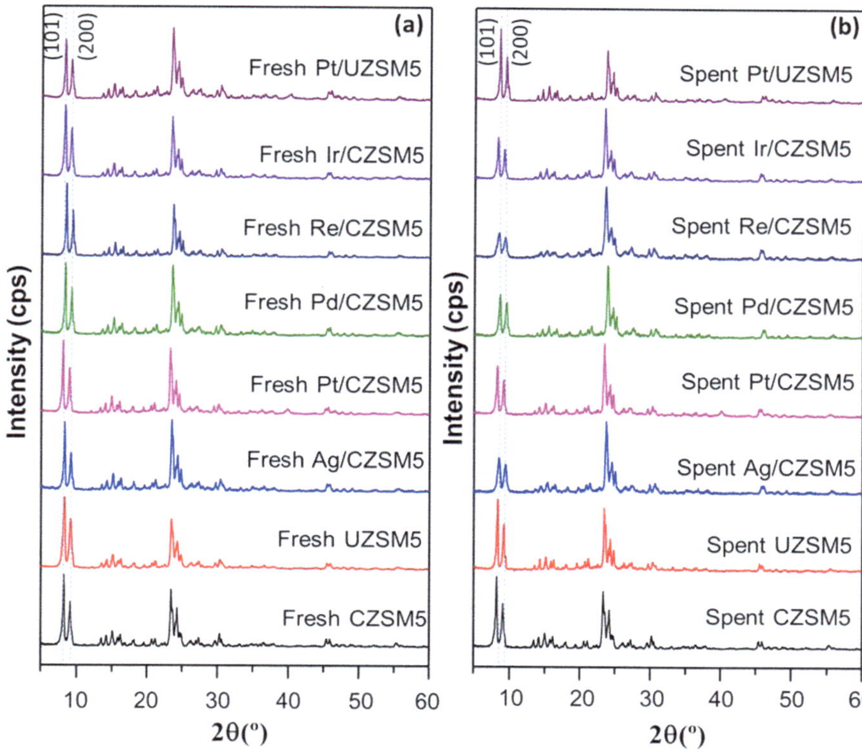

Fig. 6.18 Powder XRD patterns of various fresh (**a**) and spent (**b**) catalysts (Reproduced from Ref. [57] with permission from Elsevier)

conversion of CH_4 and CO_2, probably as both are exclusively plasma-activated at such a low reaction temperature. However, a higher ratio of weak acidic sites is beneficial for the formation of liquid products with less coke, as is evidenced by the high liquid yield (40–60%) and decreased coke yield (<20%) over Pd/CZSM-5 or Pt/CZSM-5 with more than 70% of weak acidic sites. In contrast, the absence of acidic sites (SiO_2) or higher ratio of medium and/or strong acidic sites (>30%, e.g., γ-Al_2O_3, CZSM-5, Ag/CZSM-5, and Re/CZSM-5) leads to increased coke formation (30–50%) accompanied by a reduction in liquid yield of 25–35%. Compared to the acidity distribution on CZSM-5, the loading of Pt, Pd, or Ir greatly increases the ratio of weak acidic sites, while the addition of Ag or Re reduces the ratio of weak acidic sites. Acidic sites (particularly, weak acidic sites) over the catalyst could promote the absorption and stabilization of the generated radicals to form more liquid products, probably through a Langmuir-Hinshelwood-type mechanism. However, the strong interaction between strong/medium acidic sites and the radicals could trap the formed products on the catalyst surface, suffering further attack from the plasma to form carbonaceous species as the final products. A possible explanation for the higher liquid yield achieved over Pt/CZSM-5, compared to Pd/CZSM-5

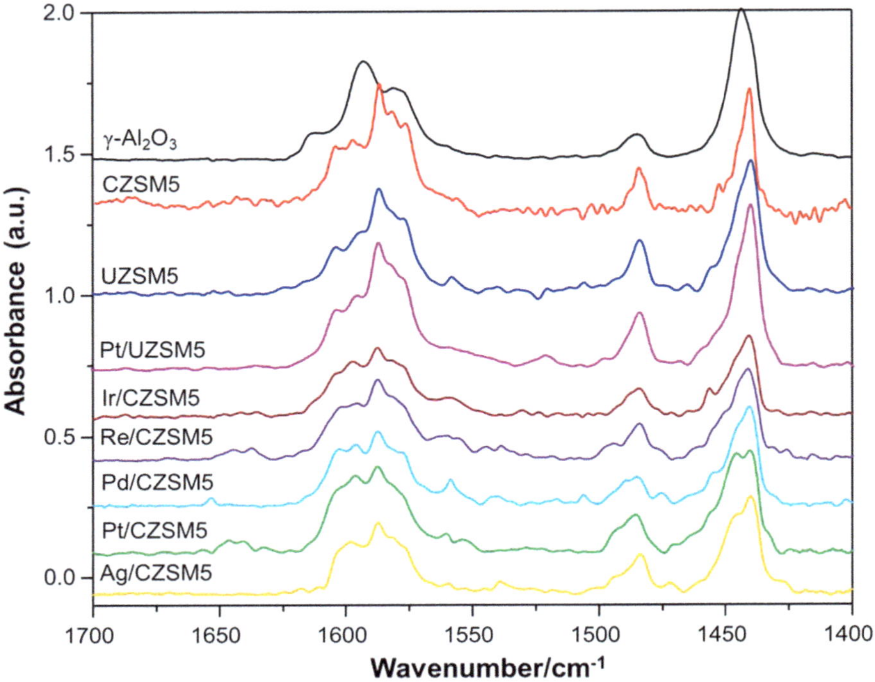

Fig. 6.19 DRIFT spectra collected over various fresh catalysts upon pyridine adsorption

Table 6.9 NH_3-TPD results of bare supports and various metal supported catalysts

Catalysts	Temperature range-wise desorption of NH_3 (μmol/g)			The ratio of weak acidic sites (%)	Total acid sites (μmol NH_3/g)
	Weak (<200 °C)	Medium (200–400 °C)	Strong (>400 °C)		
Al_2O_3	152	138	86	40	376
SiO_2	0	0	0	N/A	0
CZSM-5	197	0	166	54	363
Ag/CZSM-5	118	129	0	48	247
Pt/CZSM-5	287	0	87	77	375
Pd/CZSM-5	288	18	99	71	405
Re/CZSM-5	166	104	130	41	400
Ir/CZSM-5	302	55	77	70	435
UZSM-5	91	38	0	71	129
Pt/UZSM-5	84	0	0	100	84

(both have a similar ratio of weak acidic sites), may be that the generated radicals (e.g., CO, OH, CH_3, H) have a greater affinity towards the active Pt species on the catalyst surface. Indeed, such an interaction may stabilize the radical species and contribute to the formation of more liquid chemicals. Interestingly, coke characterization by TGA (Fig. 6.20) shows that the active carbonaceous species [51] (in the form of amorphous carbon with a small peak in the range of 300–400 °C) is observed on Pt/CZSM5 (Fig. 6.20b) and Pd/CZSM-5 (Fig. 6.20f) after reactions, while less active carbon deposition [51] (in the mixed form of amorphous and filamentous carbon with a peak in the range of 400–600 °C) is observed on CZSM-5 (Fig. 6.20a), Ag/CZSM-5 (Fig. 6.20e), Re/CZSM-5 (Fig. 6.20g), and Ir/CZSM-5 (Fig. 6.20h). It is reported that the active carbon species could react with O radicals (generated from CO_2 by plasma), thus regenerating the active sites and reducing the coke formation [62]. Based on these observations, it is speculated that the presence of a solid acid catalyst with a greater ratio of weak acidic sites and Pt species may benefit the adsorption and stabilization of generated radicals on the catalyst surface and prevent the formation of inactive carbonaceous species, thus yielding more liquid products with less coke.

This hypothesis is further verified by the acidity manipulation of zeolite catalyst. UZSM-5 with a higher ratio of weak acidic sites (higher SiO_2/Al_2O_3 ratio) is hydrothermally synthesized. Powder XRD patterns in Fig. 6.18 show that both UZSM-5 and CZSM-5 have the same MFI structure. DRIFT spectra in Fig. 6.19 indicate UZSM5 (including Pt/UZSM-5), and CZSM-5 exhibits the identical distribution of Bronsted and Lewis acidic sites. As expected, UZSM-5 has lower acidity (129 μmol NH_3/g) and higher ratio of weak acidic sites (71%) than CZSM-5 (363 μmol NH_3/g, 54% of weak acidic sites) as determined by NH_3-TPD in Table 6.9. Coke yield on UZSM-5 decreases significantly to 15.2% (entry 11, Table 6.8), compared to CZSM-5 (34.9%). Furthermore, the liquid yield is improved greatly from 34.4% (CZSM-5) to 51.2% (UZSM-5). Pt loading on UZSM-5 leads to the disappearance of medium acidic sites over UZSM-5 as shown in Table 6.9, which further reduces the coke formation to 5.1% and increases the liquid yield by 7.8% (entry 12, Table 6.8). Compared with Pt/CZSM-5 (entry 7), the plasma reaction with Pt/UZSM-5 produces less coke with similar liquid yield due to higher ratio of weak acidic sites. TGA profiles of UZSM5 (Fig. 6.20c) and Pt/UZSM-5 (Fig. 6.20d) indicating carbon species on these two catalysts show a similar feature as that on Pt/CZSM-5 (Fig. 6.20b). These observations support the previous hypothesis that coke formation could be controlled by increasing the ratio of weak acidic sites and loading Pt species over the catalyst. Extended time runs (Fig. 6.21) show that the activity of Pt/UZSM5 is stable following 3 h of reaction. After the reaction of 90 minutes, the conversion of CH_4 and CO_2 changes insignificantly. The yields of gas (H_2, CO, C_xH_y) and liquid products are stable, and the effect of carbonaceous coke on the catalyst performance is negligible. However, the conversion of CH_4 and CO_2 decreases greatly after the reaction of 180 minutes, and the liquid yield decreases significantly to 15.4% after 6 h. The poor performance after 3 h may be ascribed to more carbon deposition on the active sites (metal species and acidic sites), deactivating the catalyst. The coke characterized by TGA shows the formation of less active carbon species (with a peak in the range of 400–600 °C) after 3 h. As discussed previously, the presence of acidic sites with higher ratio of weak acidic sites and Pt species are thought to be fundamental for the formation of liquid products with less coke. More inactive carbon deposition on the catalyst leads to the loss of weak acidic sites and Pt species, thus reducing the formation of liquid chemicals.

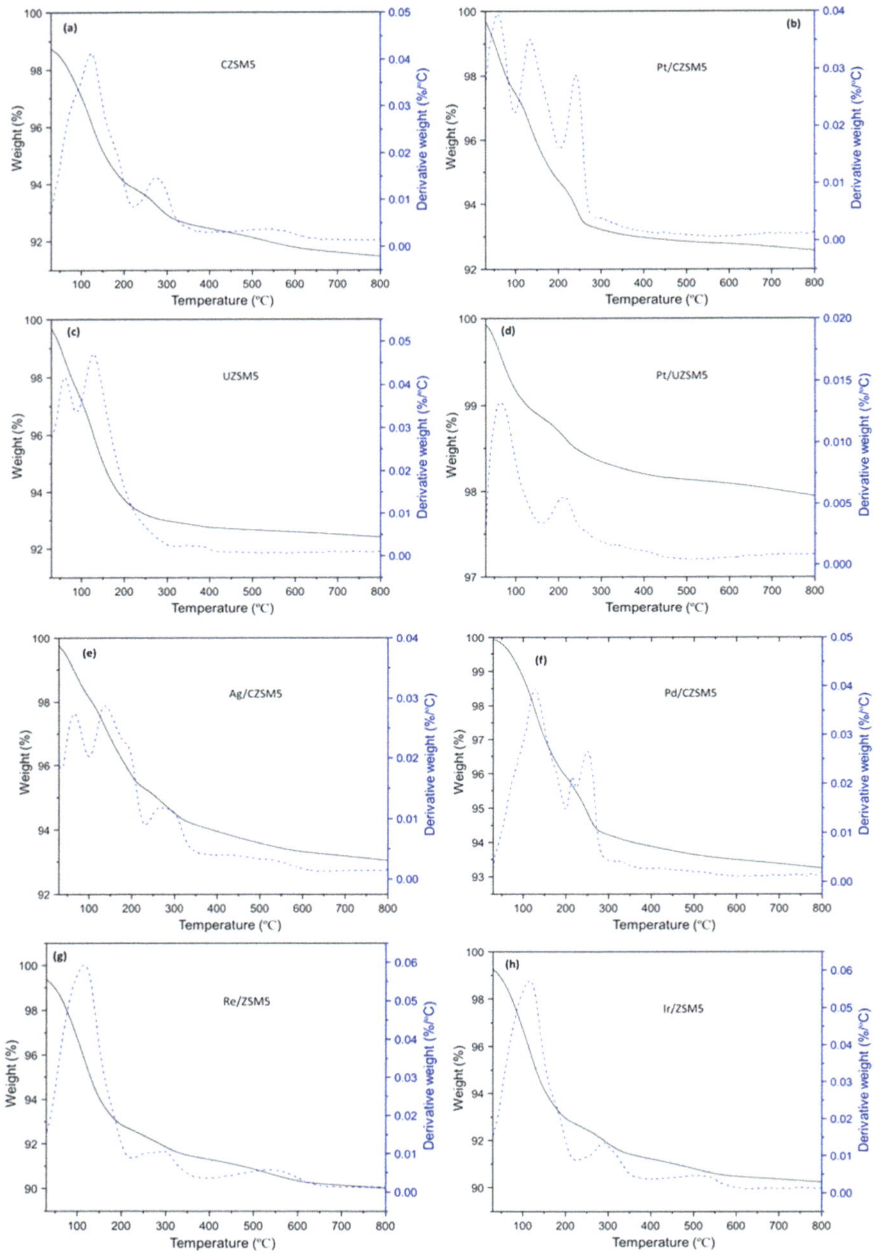

Fig. 6.20 TGA profiles of the spent catalysts (a: CZSM-5, b: Pt/CZSM-5, c: UZSM-5, d: Pt/UZSM-5, e: Ag/CZSM-5, f: Pd/UZSM-5, g: Re/ZSM-5, h: Ir/ZSM-5) after the reaction with plasma (Reproduced from Ref. [57] with permission from Elsevier)

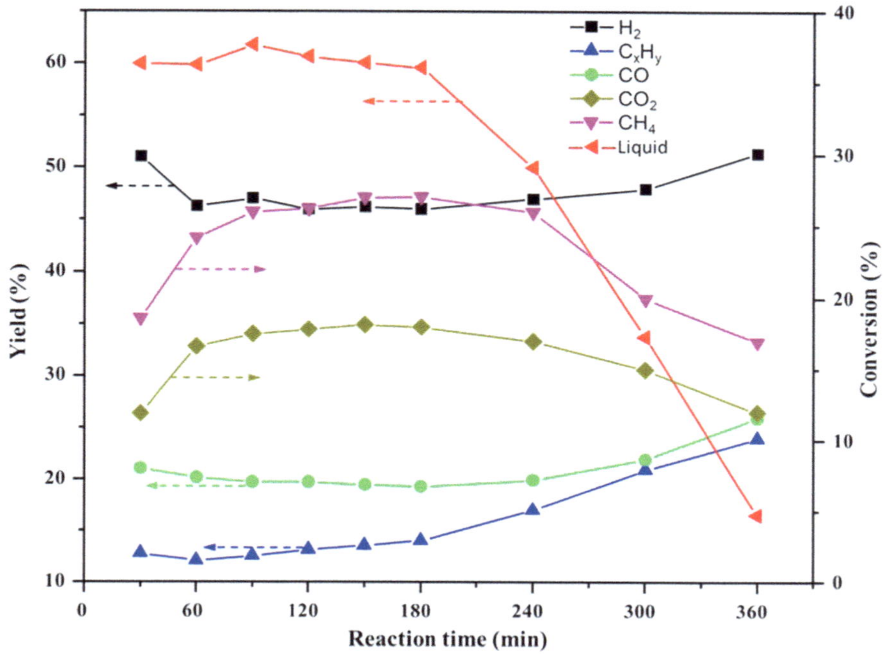

Fig. 6.21 The stability test of Pt/UZSM5 on the conversion of CH_4 and CO_2 under plasma (Reproduced from Ref. [57] with permission from Elsevier)

The physicochemical properties of Pt/UZSM5 catalyst are investigated as it showed the most promising performance on the plasma-catalytic conversion of CH_4 and CO_2. HAADF (high-angle annular dark field) TEM images (Fig. 6.22a and c) and Pt elemental mapping (Fig. 6.22b and d) of fresh and spent catalysts show that the spatial distribution and the particle size of Pt species are well maintained during the reaction. The MFI structure of Pt/UZSM-5 was kept intact as evidenced by XRD (Fig. 6.18). Unlike Pt/CZSM-5, no decrease in the peak intensity of (101) and (200) crystal planes is observed over Pt/UZSM-5, which is in agreement with TEM results. XANES spectra at Pt L3 edge (Fig. 6.22e) show a decrease in the intensity of fresh and spent Pt/UZSM-5 relative to the foil at Pt L3 edge, which is a characteristic of metal nanoparticles [63], consistent with TEM observations (~8.5 nm in Fig. 6.22a and c). Interestingly, the edge energy for fresh Pt/UZSM-5 (11564.63 eV) is identical to Pt foil (11564.63 eV), but higher than the spent catalyst (11564.29 eV). The 0.34 eV shift in the edge position for Pt/UZSM-5 after reaction may be attributed to the mild reduction of Pt nanoparticles in the plasma process. This speculation is reasonable and supported by EXAFS analysis of Pt L3 edge (Fig. 6.22f). The intensity of the main peaks at the bond distance of 2.60 and 2.75 Å (corresponding to Pt-O and Pt-Pt coordination [63–65]) in spent Pt/UZSM-5 is slightly weaker than that in the fresh one. A small peak with the distance of 1.65 Å appears in both Pt foil and spent Pt/UZSM-5, not in the fresh catalyst. These observations are in line with the previous hypothesis on the mild reduction of Pt species.

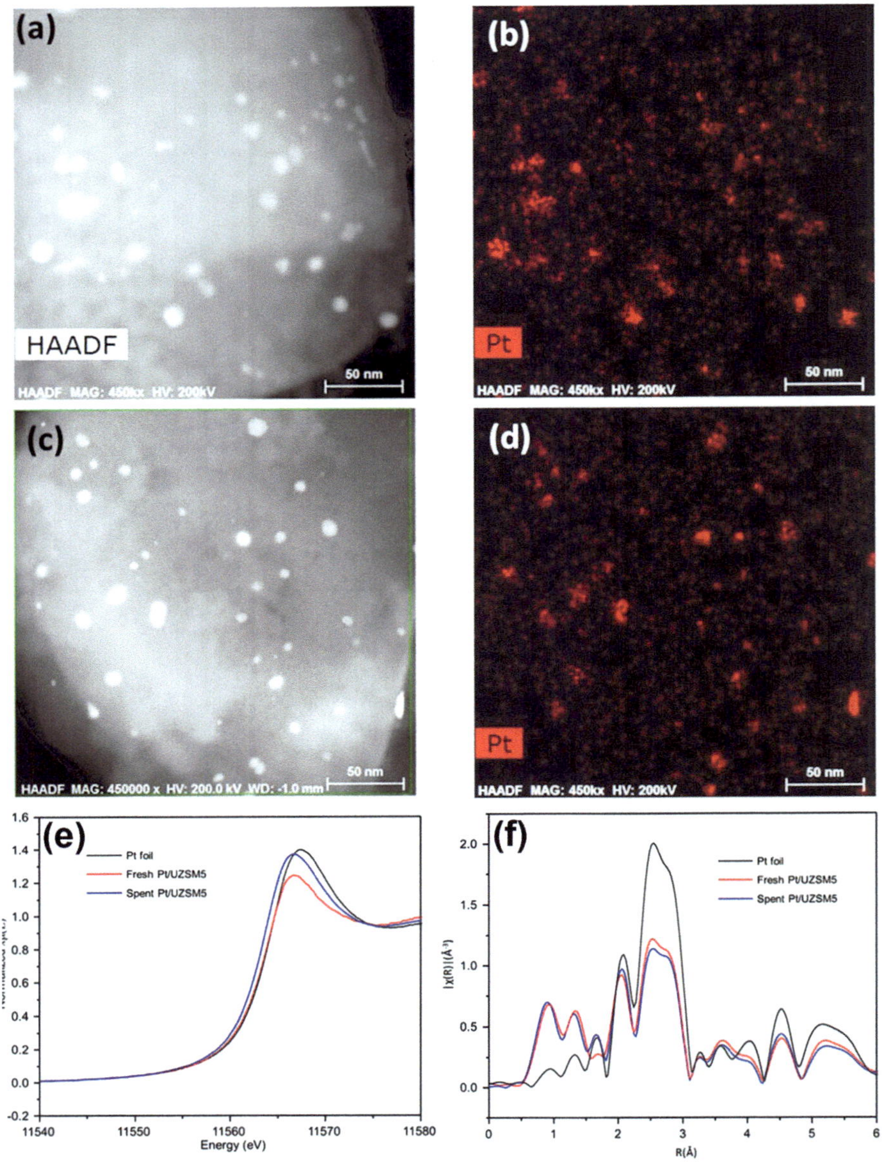

Fig. 6.22 High-angle annular dark field TEM images (HAADF) and Pt elemental mapping of fresh (**a**, **b**) and spent (**c**, **d**) Pt/UZSM5 catalysts. The particle sizes of fresh and spent Pt/UZSM-5 catalysts are calculated to be 8.5 ± 2.3 nm and 8.6 ± 3.7 nm, respectively. The normalized XANES spectra (**e**) and Fourier transform magnitudes of EXAFS (**f**) at Pt L3 edge for fresh and spent Pt/UZSM-5 catalysts (Reproduced from Ref. [57] with permission from Elsevier)

In summary, our work demonstrates the direct plasma-catalytic conversion of biogas to valuable liquid chemicals with limited coke formation at low temperature and ambient pressure. Combining NTP with the catalyst enhances the formation of liquid products while significantly inhibiting coke formation. High liquid yield with low carbon deposition is achieved over a unique Pt/UZSM5 catalyst. The total yield of liquid chemicals reaches approximately 60%, with the composition of formaldehyde, methanol, ethanol, and acetone. Coke formation is limited to 5.1%. The liquid yield and coke formation can be controlled by the modification of acidic properties and Pt species. Acidity characterizations upon pyridine absorption and NH_3-TPD show that the distribution of Bronsted and Lewis acidic sites and the number of acidic sites may not be critical factors for the reaction performance. High ratio of weak acidic sites benefits the formation of liquid chemicals and coke reduction. These findings suggest that the combination of NTP and heterogeneous catalysis has the great potential to produce renewable oxygenates from the biogas conversion with limited carbon deposition.

6.4 Conclusion

Several works demonstrate the direct plasma-catalytic conversion of hydrocarbons to valuable liquid chemicals with limited coke formation at low temperature and ambient pressure. Combining NTP with the catalyst enhances the formation of liquid products while significantly inhibiting coke formation. High liquid yield with low carbon deposition is achieved over zeolite catalyst. The liquid yield and coke formation can be controlled by the modification of acidic properties and loaded metal species. High ratio of weak acidic sites benefits the formation of liquid chemicals and coke reduction. These findings suggest that the combination of NTP and heterogeneous catalysis has a great potential in efficient methane conversion to liquid products and renewable oxygenates with limited carbon deposition and the plasma-generated light could be fully utilized by combining plasma with photocatalysis.

References

1. Y.-R. Luo, *Handbook of Bond Dissociation Energies in Organic Compounds* (CRC Press, Boca Raton, FL, 2002)
2. S. Kado, K. Urasaki, Y. Sekine, K. Fujimoto, Direct conversion of methane to acetylene or syngas at room temperature using non-equilibrium pulsed discharge. Fuel **82**(11), 1377–1385 (2003)
3. T. Hammer, T. Kappes, M. Baldauf, Plasma catalytic hybrid processes: Gas discharge initiation and plasma activation of catalytic processes. Catal. Today **89**(1–2), 5–14 (2004)
4. A.H. Khoja, M. Tahir, N.A.S. Amin, Dry reforming of methane using different dielectric materials and DBD plasma reactor configurations. Energy Convers. Manag. **144**, 262–274 (2017)
5. M. Usman, W.W. Daud, H.F. Abbas, Dry reforming of methane: Influence of process parameters—A review. Renew. Sust. Energ. Rev. **45**, 710–744 (2015)

6. S. Sadrameli, Thermal/catalytic cracking of hydrocarbons for the production of olefins: A state-of-the-art review I: Thermal cracking review. Fuel **140**, 102–115 (2015)
7. G. Keller, M. Bhasin, Synthesis of ethylene via oxidative coupling of methane: I. determination of active catalysts. J. Catal. **73**(1), 9–19 (1982)
8. A. Maitra, Critical performance evaluation of catalysts and mechanistic implications for oxidative coupling of methane. Appl. Catal. A Gen. **104**(1), 11–59 (1993)
9. L.D. Nguyen, S. Loridant, H. Launay, A. Pigamo, J.L. Dubois, J.M.M. Millet, Study of new catalysts based on vanadium oxide supported on mesoporous silica for the partial oxidation of methane to formaldehyde: Catalytic properties and reaction mechanism. J. Catal. **237**(1), 38–48 (2006)
10. A.T. Ashcroft, A.K. Cheetham, M. Green, Partial oxidation of methane to synthesis gas using carbon dioxide. Nature **352**(6332), 225–226 (1991)
11. A. De Vekki, S. Marakaev, Catalytic partial oxidation of methane to formaldehyde. Russ. J. Appl. Chem. **82**(4), 521–536 (2009)
12. X. Guo, G. Fang, G. Li, H. Ma, H. Fan, L. Yu, et al., Direct, nonoxidative conversion of methane to ethylene, aromatics, and hydrogen. Science **344**(6184), 616–619 (2014)
13. J.J. Spivey, G. Hutchings, Catalytic aromatization of methane. Chem. Soc. Rev. **43**(3), 792–803 (2014)
14. U. Zavyalova, M. Holena, R. Schlögl, M. Baerns, Statistical analysis of past catalytic data on oxidative methane coupling for new insights into the composition of high-performance catalysts. ChemCatChem **3**(12), 1935–1947 (2011)
15. A. Rutscher, Characteristics of low-temperature plasmas under nonthermal conditions–a short summary. Low Temp. Plasmas: Fundam. Technol. Techn. **1**, 1–14 (2008)
16. Bromberg, L.; Cohn, D.; Rabinovich, A.; Alexeev, N., Hydrogen Manufacturing Using Low Current, Non-thermal Plasma Boosted Fuel Converters. American Chemical Society, San Diego, CA, United States 2001
17. A. Zille, F.R. Oliveira, A.P. Souto, Plasma treatment in textile industry. Plasma Process. Polym. **12**(2), 98–131 (2015)
18. J.C. Devins, M. Burton, Formation of hydrazine in electric discharge decomposition of Ammonia1, 2. J. Am. Chem. Soc. **76**(10), 2618–2626 (1954)
19. T. Nozaki, K. Okazaki, Non-thermal plasma catalysis of methane: Principles, energy efficiency, and applications. Catal. Today **211**, 29–38 (2013)
20. W.-C. Chung, M.-B. Chang, Review of catalysis and plasma performance on dry reforming of CH4 and possible synergistic effects. Renew. Sust. Energ. Rev. **62**, 13–31 (2016)
21. L. Wang, Y. Yi, C. Wu, H. Guo, X. Tu, One-step reforming of CO2 and CH4 into high-value liquid chemicals and fuels at room temperature by plasma-driven catalysis. Angew. Chem. Int. Ed. Engl. **56**(44), 13679–13683 (2017)
22. M. Shareei, H. Taghvaei, A. Azimi, A. Shahbazi, M. Mirzaei, Catalytic DBD plasma reactor for low temperature partial oxidation of methane: Maximization of synthesis gas and minimization of CO2. Int. J. Hydrog. Energy **44**(60), 31873–31883 (2019)
23. L. Song, Y. Kong, X. Li, Hydrogen production from partial oxidation of methane over dielectric barrier discharge plasma and NiO/γ-Al2O3 catalyst. Int. J. Hydrog. Energy **42**(31), 19869–19876 (2017)
24. H.L. Chen, H.M. Lee, S.H. Chen, Y. Chao, M.B. Chang, Review of plasma catalysis on hydrocarbon reforming for hydrogen production—Interaction, integration, and prospects. Appl. Catal. B Environ. **85**(1–2), 1–9 (2008)
25. H. Long, S. Shang, X. Tao, Y. Yin, X. Dai, CO2 reforming of CH4 by combination of cold plasma jet and Ni/γ-Al2O3 catalyst. Int. J. Hydrog. Energy **33**(20), 5510–5515 (2008)
26. A. Aziznia, H.R. Bozorgzadeh, N. Seyed-Matin, M. Baghalha, A. Mohamadalizadeh, Comparison of dry reforming of methane in low temperature hybrid plasma-catalytic corona with thermal catalytic reactor over Ni/γ-Al2O3. J. Nat. Gas Chem. **21**(4), 466–475 (2012)
27. A. Ghorbanzadeh, S. Norouzi, T. Mohammadi, High energy efficiency in syngas and hydrocarbon production from dissociation of CH4–CO2 mixture in a non-equilibrium pulsed plasma. J. Phys. D. Appl. Phys. **38**(20), 3804 (2005)

28. X. Tu, J. Whitehead, Plasma-catalytic dry reforming of methane in an atmospheric dielectric barrier discharge: Understanding the synergistic effect at low temperature. Appl. Catal. B Environ. **125**, 439–448 (2012)
29. J. Sentek, K. Krawczyk, M. Młotek, M. Kalczewska, T. Kroker, T. Kolb, A. Schenk, K.-H. Gericke, K. Schmidt-Szałowski, Plasma-catalytic methane conversion with carbon dioxide in dielectric barrier discharges. Appl. Catal. B Environ. **94**(1), 19–26 (2010)
30. D. Kang, N. Rahimi, M.J. Gordon, H. Metiu, E.W. McFarland, Catalytic methane pyrolysis in molten MnCl2-KCl. Appl. Catal. B Environ. **254**, 659–666 (2019)
31. J.R. Fincke, R.P. Anderson, T. Hyde, B.A. Detering, R. Wright, R.L. Bewley, D.C. Haggard, W.D. Swank, Plasma thermal conversion of methane to acetylene. Plasma Chem. Plasma Process. **22**(1), 105–136 (2002)
32. D.H. Lee, Y.-H. Song, K.-T. Kim, J.-O. Lee, Comparative study of methane activation process by different plasma sources. Plasma Chem. Plasma Process. **33**(4), 647–661 (2013)
33. X.-S. Li, C. Shi, Y. Xu, K.-J. Wang, A.-M. Zhu, A process for a high yield of aromatics from the oxygen-free conversion of methane: Combining plasma with Ni/HZSM-5 catalysts. Green Chem. **9**(6), 647–653 (2007)
34. L. Chen, X.-W. Zhang, L. Huang, L.-C. Lei, Partial oxidation of methane with air for methanol production in a post-plasma catalytic system. Chem. Eng. Process. Process Intensif. **48**(8), 1333–1340 (2009)
35. P. Chawdhury, D. Ray, T. Vinodkumar, C. Subrahmanyam, Catalytic DBD plasma approach for methane partial oxidation to methanol under ambient conditions. Catal. Today **337**, 117–125 (2019)
36. I. Belov, S. Paulussen, A. Bogaerts, Appearance of a conductive carbonaceous coating in a CO2 dielectric barrier discharge and its influence on the electrical properties and the conversion efficiency. Plasma Sources Sci. Technol. **25**(1), 015023 (2016)
37. S. Faramawy, T. Zaki, A.A.E. Sakr, Natural gas origin, composition, and processing: A review. J. Nat. Gas Sci. Eng. **34**, 34–54 (2016)
38. S. Meng, A. Wang, P. He, H. Song, Nonthermal plasma-assisted photocatalytic conversion of simulated natural gas for high-quality gasoline production near ambient conditions. J. Phys. Chem. Lett. **11**(10), 3877–3881 (2020)
39. T.M. Lovestead, T.J. Bruno, Application of the advanced distillation curve method to the aviation fuel avgas 100LL. Energy Fuel **23**(4), 2176–2183 (2009)
40. P. He, R. Gatip, M. Yung, H.B. Zeng, H. Song, Co-aromatization of olefin and methane over Ag-Ga/ZSM-5 catalyst at low temperature. Appl. Catal. B-Environ. **211**, 275–288 (2017)
41. W. Huang, S. Zhang, Y. Tang, Y. Li, L. Nguyen, Y. Li, J. Shan, D. Xiao, R. Gagne, A.I. Frenkel, Low-temperature transformation of methane to methanol on Pd1O4 single sites anchored on the internal surface of microporous silicate. Angew. Chem. Int. Ed. **55**(43), 13441–13445 (2016)
42. T.E. Tshabalala, N.J. Coville, M.S. Scurrell, Dehydroaromatization of methane over doped Pt/Mo/H-ZSM-5 zeolite catalysts: The promotional effect of tin. Appl. Catal. A Gen. **485**, 238–244 (2014)
43. A.J.K.A.W.R. Parrish, *Fundamentals of Natural Gas Processing* (CRC Press, Boca Raton, FL, 2006), p. 9
44. L. Yuliati, H. Yoshida, Photocatalytic conversion of methane. Chem. Soc. Rev. **37**(8), 1592–1602 (2008)
45. R. Carman, R. Mildren, B. Ward, D. Kane, High-pressure (>1 bar) dielectric barrier discharge lamps generating short pulses of high-peak power vacuum ultraviolet radiation. J. Phys. D. Appl. Phys. **37**(17), 2399 (2004)
46. A. El-Dakrouri, J. Yan, M. Gupta, M. Laroussi, Y. Badr, VUV emission from a novel DBD-based radiation source. J. Phys. D. Appl. Phys. **35**(21), L109 (2002)
47. D.M. Grant, E.G. Paul, Carbon-13 magnetic resonance. II. Chemical shift data for the alkanes. J. Am. Chem. Soc. **86**(15), 2984–2990 (1964)
48. E. Pretsch, P. Buehlmann, C. Affolter, E. Pretsch, P. Bhuhlmann, C. Affolter, *Structure Determination of Organic Compounds* (Springer, 2000)

49. D.A. Skoog, F.J. Holler, S.R. Crouch, *Principles of Instrumental Analysis* (Cengage Learning, Belmont, CA, 2017)
50. S.E. Hosseini, M.A. Wahid, Biogas utilization: Experimental investigation on biogas flameless combustion in lab-scale furnace. Energy Convers. Manag. **74**, 426–432 (2013)
51. Y.X. Zeng, L. Wang, C.F. Wu, J.Q. Wang, B.X. Shen, X. Tu, Low temperature reforming of biogas over K-, Mg- and Ce-promoted Ni/Al2O3 catalysts for the production of hydrogen rich syngas: Understanding the plasma-catalytic synergy. Appl. Catal. B Environ. **224**, 469–478 (2018)
52. S. Mao, Z. Tan, L. Zhang, Q. Huang, Plasma-assisted biogas reforming to syngas at room temperature condition. J. Energy Inst. **91**(2), 172–183 (2018)
53. Y. Zeng, X. Zhu, D. Mei, B. Ashford, X. Tu, Plasma-catalytic dry reforming of methane over γ-Al2O3 supported metal catalysts. Catal. Today **256**, 80–87 (2015)
54. H.H. Nguyen, K.-S. Kim, Combination of plasmas and catalytic reactions for CO2 reforming of CH4 by dielectric barrier discharge process. Catal. Today **256**, 88–95 (2015)
55. B. Wang, G. Xu, Conversion of methane to C2 hydrocarbons via cold plasma reaction. J. Nat. Gas Chem. **12**(3), 178–182 (2003)
56. C.J. Liu, L.L. Lobban, R.G. Mallinson, Experimental investigations on the interaction between plasmas and catalyst for plasma catalytic methane conversion (PCMC) over zeolites. Stud. Surf. Sci. Catal. **119**, 361–366 (1998)
57. A. Wang, J.H. Harrhy, S. Meng, P. He, L. Liu, H. Song, Nonthermal plasma-catalytic conversion of biogas to liquid chemicals with low coke formation. Energy Convers. Manag. **191**, 93–101 (2019)
58. Y.R. Luo, *Comprehensive Handbook of Chemical Bond Energies* (CRC Press, Boca Raton, FL, 2007), pp. 29–38
59. C. De Bie, B. Verheyde, T. Martens, J. van Dijk, S. Paulussen, A. Bogaerts, Fluid modeling of the conversion of methane into higher hydrocarbons in an atmospheric pressure dielectric barrier discharge. Plasma Process. Polym. **8**(11), 1033–1058 (2011)
60. X. Tu, J.C. Whitehead, Plasma-catalytic dry reforming of methane in an atmospheric dielectric barrier discharge: Understanding the synergistic effect at low temperature. Appl. Catal. B Environ. **125**, 439–448 (2012)
61. D. Park, J. Kim, T. Kim, Nonthermal plasma-assisted direct conversion of methane over NiO and MgO catalysts supported on SBA-15. Catal. Today **299**, 86–92 (2018)
62. A.D. Ballarini, S.R. de Miguel, E.L. Jablonski, O.A. Scelza, A.A. Castro, Reforming of CH4 with CO2 on Pt-supported catalysts. Catal. Today **107-108**, 481–486 (2005)
63. S.L. Yohe, H.J. Choudhari, D.D. Mehta, P.J. Dietrich, M.D. Detwiler, C.M. Akatay, et al., High-pressure vapor-phase hydrodeoxygenation of lignin-derived oxygenates to hydrocarbons by a PtMo bimetallic catalyst: Product selectivity, reaction pathway, and structural characterization. J. Catal. **344**, 535–552 (2016)
64. J. de Graaf, A.J. van Dillen, K.P. de Jong, D.C. Koningsberger, Preparation of highly dispersed Pt particles in zeolite Y with a narrow particle size distribution: Characterization by hydrogen chemisorption, TEM, EXAFS spectroscopy, and particle modeling. J. Catal. **203**(2), 307–321 (2001)
65. J. Mao, W. Chen, D. He, J. Wan, J. Pei, J. Dong, et al., Design of ultrathin Pt-Mo-Ni nanowire catalysts for ethanol electrooxidation. Sci. Adv. **3**(8), e1603068 (2017)

Chapter 7
Biomass Valorization Under Methane Environment

7.1 Introduction

Energy consumption continues to increase, while the use of unrenewable energy resources such as fossil fuels brings a myriad of problems such as energy crises, environment pollution, greenhouse gas accumulation, deglaciation, and ecosystem destruction [1]. Therefore, developing renewable energy resources is of great urgency. Biomass refers to faunal and floral materials used for energy generation or the production of a range of chemicals, which can be regarded as a renewable substituent of conventional resources. Biomass also covers a variety of agricultural and domestic waste materials such as wheat straw, corn cobs, manure, and household garbage [2]. Therefore, the utilization of biomass not only compensates the shortage of energy supply but also converts various intractable wastes into valuable products, namely, turning "trash" into "treasure" [3]. As a result, corresponding studies are critical for environment protection and sustainable development.

Current biomass utilization techniques include biological conversion, gasification, liquefaction, pyrolysis, and hydrodeoxygenation [4]. Corresponding research of these pathways provides solutions for using biomass effectively, while several intrinsic drawbacks are still hard to overcome. Biological processes (Fig. 7.1) are often operated at mild conditions, and the selectivity towards valuable products is often favorable. However, the product yield and efficiency are still undesirable, leading to high operational cost and low profitability of the whole process [5]. Gasification (Fig. 7.2) is less costly and can handle biomass with high water content, but the operation temperature is relatively high, and the product distribution is hard to control, resulting in extra energy consumption and greenhouse gas emissions [6]. Liquefaction (Fig. 7.3) occurs at much milder temperatures and produces more liquid products, while a solvent is usually needed, and the intricacy and high cost limit its practical applications [7]. Pyrolysis (Fig. 7.4) also adopts relatively mild reaction temperatures and pressures. In this process, gas, liquid, and solid

H. Song et al., *Methane Activation and Utilization in the Petrochemical and Biofuel Industries*, https://doi.org/10.1007/978-3-030-88424-6_7

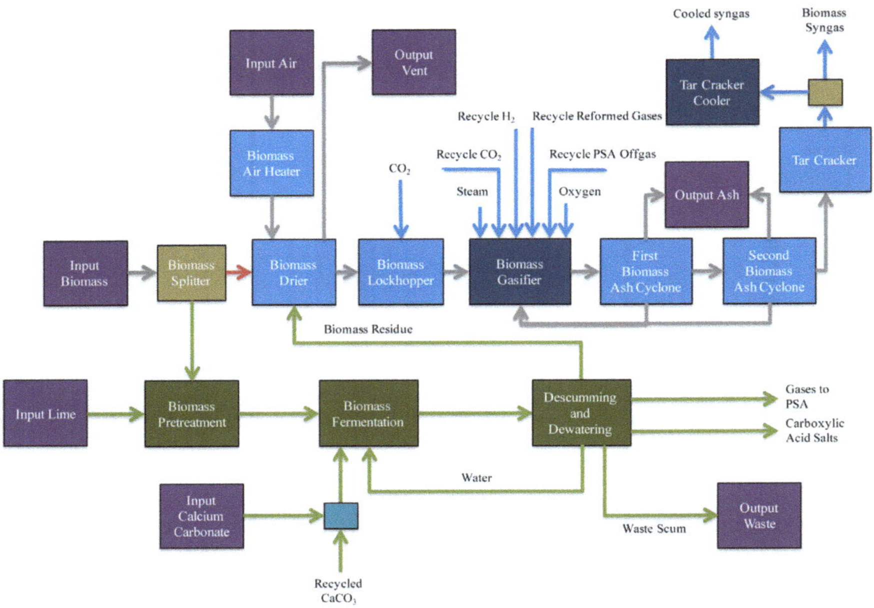

Fig. 7.1 Biomass feed to the combined process superstructure for thermochemical and biological conversion of biomass. Reproduced with permission from [5]. Copyright (2014) American Chemical Society

Fig. 7.2 Main stages and the reactions involved in the gasification process. Reproduced with the permission from Elsevier

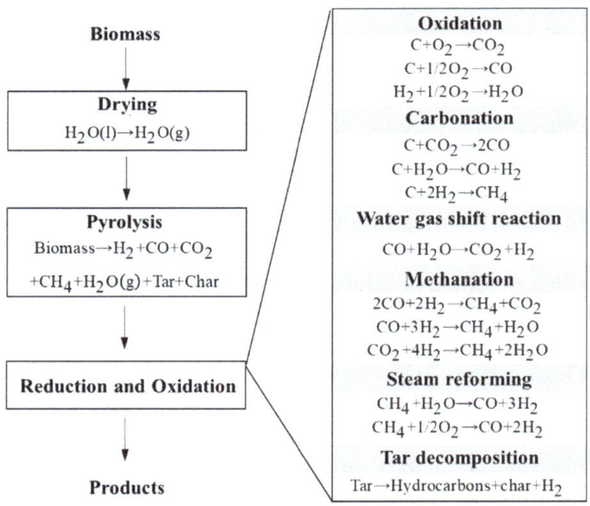

products are generated simultaneously, and the yield of each part can be tuned by reaction conditions such as temperature and residence time. However, the quality of products strongly depends on the elemental composition in feedstocks, which often

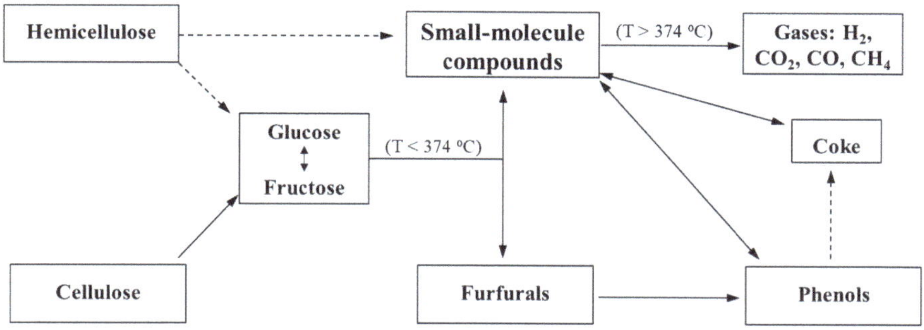

Fig. 7.3 Simplified reaction mechanism for the degradation of biomass during liquefaction process. Reproduced with the permission from Elsevier

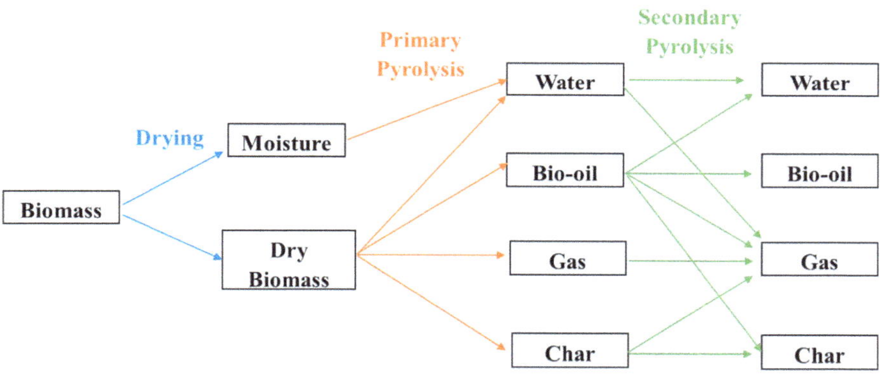

Fig. 7.4 Generalized pathways of biomass pyrolysis process; the yield and properties of bio-oil, gas, and char might vary significantly according to the type of biomass and reaction conditions. Reproduced with the permission from Elsevier

suffer from high oxygen content and low hydrogen-to-carbon ratio, leading to undesired products [8]. Meanwhile, the detailed mechanism still remains unknown due to the complexity of the thermal driving process [9]. Hydrodeoxygenation (Fig. 7.5) often deals with highly oxygenated biomass feedstocks, which significantly improve the value of the oil by removing oxygen atoms and producing valuable hydrocarbons [10, 11]. However, hydrogen resources are not naturally available and are generally produced from the methane reforming reaction, leading to high cost and inevitable CO_2 emission [12]. Therefore, the development of new techniques with better energy utilization, higher economic efficiency, simpler process, and improved quality control is of great importance.

Methane is the main component in natural gas resources, with abundant reserves and an underestimated value. It is evaluated that the cost of natural gas for producing a certain amount of energy is only one third of that of hydrogen [13]. Besides, it is estimated that the production of natural gas is still outpacing consumption, leading to the decline of its price from 8.85 USD/MMBtu (million British thermal units) at

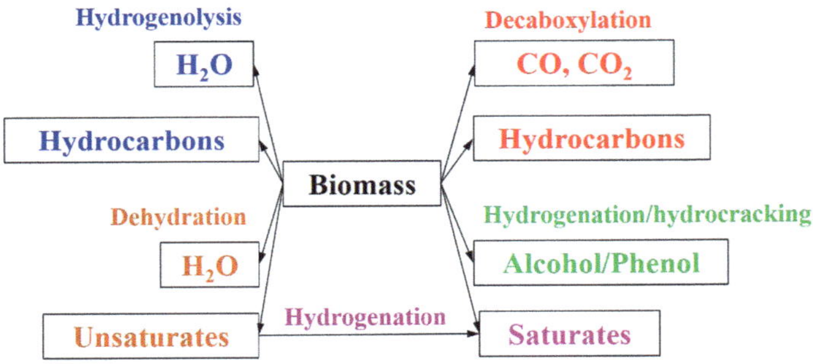

Fig. 7.5 Generalized pathways of biomass hydrodeoxygenation process. Drawn with reference to [11]

Henry Hub, Unites States, in 2008 to 2.57 in 2019 and a projection of 2.33 in 2020 [14–16]. Thanks to the natural occurrence and low price of natural gas, the substitution of hydrogen by methane facilitates an immediate reduction of operational costs in upgrading processes. Furthermore, the hydrogen-to-carbon ratio is 4:1 in the formula CH_4, making it a competent hydrogen donor to replace H_2 for upgrading processes, where a reductive atmosphere and a high hydrogen-to-carbon ratio in products are generally required. It should also be noted that the carbon atom in the methane molecule can also be fully utilized by incorporating into final products, improving the yield of desired product and reducing the emission of CO_2 by avoiding the steam methane reforming process. Therefore, biomass upgrading under methane environment is fully supported by economic and environmental advantages.

In this chapter, a collection and brief introduction will be firstly presented regarding the existing catalysts for biomass conversion under methane environment. Then, the mechanism of methane activation will be introduced, which is a crucial primary step for its utilization in biomass upgrading processes. Next, the property-performance relationship of catalysts for biomass valorization under methane environment will be discussed, which helps to get better understanding of the valorization process for catalyst design and optimization. After reviewing the relevant work, the gap between laboratory research and real industrialization application will be proposed. Finally, biomass valorization under methane environment will be briefly summarized, during which the status quo and perspectives for this promising process will be proposed.

7.2 Catalysts for Biomass Valorization Under Methane Environment

The early work of catalytic upgrading process using methane can be dated back to late twentieth century. In 1997, Choudhary et al. [17] found that methane can be activated on an MFI zeolite at a low temperature, which can be regarded as the pioneering work for new paths of methane utilization. Subsequent studies further revealed that despite the inert nature of methane, if it was co-fed with other compounds such as hydrocarbons or oxygenated species, the interactive reaction could occur at much milder reaction conditions with the assistance of zeolite catalysts, and the hydrogen atoms in methane can be fully used as hydrogen donors to produce valuable products [18–21]. After that, zeolites have become the main choice as the catalyst supports for the upgrading process in the presence of methane for decades.

In 2014, He et al. [22] performed catalytic biomass conversion under methane environment with phosphorus- and cerium-modified Ag/ZSM-5 catalysts utilizing raw biomass feedstocks including sawdust and flex straw. They found that the introduction of methane led to increased oil yield and dramatically enhanced hydrogen-to-carbon ratio in collected oil products. Besides, Ag exhibited the best performance as a loading metal among a series of transitional metals including Ir, In, Zn, Ce, Mo, and Ga. In 2015, He et al. [23] compared a series of zeolite catalysts loading with non-noble metals including iron, cobalt, copper, nickel, manganese, zirconium, cerium, and zinc in biomass upgrading by natural gas and found that Zn/ZSM-5 provided favorable quality improvement of bio-oil in terms of high H/C ratio, low O/C ratio, and low total acid number (TAN) as well as enhanced oil yield with the assistance of methane-enriched environment. In 2016, Yang et al. [24] developed the valorization of biomass constituents including cellulose, hemicellulose, and lignin under methane environment with Mo/HZSM-5 catalyst. They found that the aromatic yield can be dramatically improved when lignin was used as the feedstock, while no significant enhancement was observed by feeding cellulose and hemicellulose. They also suggested that the introduction of methane triggered the deoxygenation reaction of phenol groups, facilitating the production of aromatics. Wang et al. [25] studied the co-conversion of ethanol and methane with Ag/ZSM-5 catalysts. Comparing with N_2 control experiments, they confirmed that methane successfully and dramatically intensified the aromatization process of ethanol by taking part in the reaction process. In 2017, Wang et al. [26] investigated the upgrading process of acetic acid as a model compound for a biomass-derived liquid. The reaction was operated at a relatively mild temperature of 400 °C, benefiting high liquid product yield and low energy consumption. They prepared a series of Ag-, Zn-, and Ga-based catalysts and found that ZnGa/ZSM-5 catalyst demonstrated excellent performance, which was indicated by higher methane conversion and aromatic yield as well as lower yields of phenol derivatives and coke. Besides, they proved that the introduction of methane led to higher aromatic selectivity and lower phenol selectivity by carrying out control experiments under N_2 atmosphere, indicating the positive influence of methane environment. Yang et al. [27] explored the pyrolysis process of torrefied switchgrass under methane environment with MoZn/HZSM-5 catalyst. Comparing with the results in helium atmosphere, methane was proved to significantly enhance the aromatic yield. Gunawardena et al. [28] found that the yield of aromatics, especially BTEX, can be increased with the presence of methane, using glucose as the reactant and Mo/HZSM-5 as the catalyst. In 2018, Austin et al. [29] studied the aromatization of acetone under methane atmosphere. They found that zinc and gallium loaded on HZSM-5 support demonstrated excellent catalytic performance in terms of BTEX yield. Shafaghat et al. [30] used real biomass material including lignin and yellow poplar to study the effect of methane feeding on the catalytic pyrolysis process with Ni/HZSM-5 catalyst. Comparing with N_2 atmosphere, the feeding of CH_4 led to significant increase in the yields of BTEX as well as other aromatics. Austin et al. [31] carried out glycerol upgrading experiments under methane environment using ZnCuGa/ZSM-5 catalyst. They compared a serious of catalyst preparation methods including incipient wetness impregnation, deposition precipitation, coprecipitation, ion exchange, and hydrothermal synthesis. After the comparison, they concluded that deposition precipitation led to the most promising catalytic performance. Besides, they confirmed the introduction of methane not only facilitated the dispersion of active metals but also participated in the reaction to produce more valuable liquid products, especially aromatics. Wang et al. [32] studied the reaction between cellulose and methane with a series of Zn/ZSM-5 catalysts and suggested that the addition of methane could mitigate coke formation and improve oxygen removal efficiency, producing more liquid products with high quality. Wang et al. [33] explored the co-conversion of methane and lignin with a serious of zeolite supports including HZSM-5, Hbeta, and HY. They found that ZSM-5 performed best among zeolites and ZnGa/ZSM-5 catalysts resulted in favorable lignin conversion, aromatic yield, and BTEX selectivity. Wang et al. [34] developed a valorization process of furfural under methane environment using Zn-Ga/ZSM-5 catalysts and proved that methane could be used in an effective way to produce more valuable aromatic products. In 2019, Lalsare et al. [35] used FeMo/ZSM-5 catalyst for methane-activated gasification of hardwood and found that the introduction of methane led to high hydrogen yield, in contrast to high methane yield with the absence of methane in feed gas, making the product gas more profitable. Wang et al. [36] studied the deoxygenation process of phenol under methane environment with Zn/ZSM-5 catalyst. They demonstrated that the liquid yield and selectivity towards BTEX could be considerably enhanced with the presence of methane. Besides, the catalyst could also perform better due to the enhanced activity and inhibited coke formation process after methane introduction. They also did a comprehensive screening of active metals and found that Zn, Pt, and Ir were effective for improving methane conversion, while Ag, Ga, Pd, and Ru led to increased coke formation.

Meanwhile, non-zeolitic catalysts also begin to emerge for biomass upgrading under methane environment in recent years. In 2009, Sadykov et al. [37] developed a series of monolithic materials for steam reforming of co-fed methane and oxygenates (e.g., ethanol, acetone). In 2015, Xiao et al. [38] developed Pt-Bi catalyst loaded on activated carbon for the deoxygenation of guaiacol using methane as the reductant. In 2018, Huang et al. [39] studied the steam reforming of glycerol and methane with Ni/alumina catalyst. In order to improve the performance of catalyst, they modified the support with alkaline-earth metal oxides including CaO and MgO and introduced transitional metals including lanthanum, cerium, and zirconium. They found the addition of methane in the steam reforming system led to reduced CO_2 emission and increased syngas yield. Besides, they verified the stability of the catalyst by a 500-h long-term experiment. Shen et al. [40] found that the introduction of methane and the application of Ni/Al_2O_3 catalyst significantly improved the biomass gasification process by increasing the ratio of hydrogen to carbon monoxide as well as the biomass conversion while decreasing the tar yield.

The basic information of recent studies concerning biomass upgrading under methane environment is summarized in Table 7.1 in a chronological way. It can be seen that there is a decreasing trend of reaction temperature and gas yield, while the liquid yield is increasing over time, indicating the progress of biomass upgrading under methane environment represented by more desirable products and milder reaction conditions.

7.3 Mechanism of Methane Activation

As discussed in previous chapters, methane is the most stable hydrocarbon compound, evidenced by its symmetrical tetrahedron structure and high C-H bond energy of 425 kJ mol^{-1}. Therefore, the intrinsic inertness could be the biggest challenge for its utilization in chemical processes. Despite the difficulties, methane conversion has been realized through a variety of pathways. Among different conversion strategies, direct methane conversion, which refers to one-step methane conversion to useful chemicals, deserves extra attention. Corresponding examples include methane to hydrogen cyanide, methane oxidative coupling, methane to methanol and formaldehyde, methane halogenation and oxyhalogenation, and methane aromatization [43]. It can be seen from these processes that methane conversion is inextricably connected with C-H bond activation and cleavage, which is of paramount importance in all relevant researches.

Depending on the extent of C-H bond removal, direct methane conversion can be classified from another criterion into complete C-H removal and partial C-H removal [44]. Complete C-H removal is generally realized in drastic reaction conditions such as methane reforming and methane pyrolysis processes, which produce valuable H_2 resources. However, the adopted high temperature and pressure make the processes highly energy-intensive and the control of product selectivity difficult. Partial C-H bond removal and subsequent formation of other chemical bonds can be achieved in much milder conditions, and the process can be finely tuned to produce desired products, leading to better versatility. Biomass valorization under methane environment generally leads to partial C-H bond removal, and the corresponding mechanism is thus exclusively discussed here.

Two pathways have been proposed for partial C-H bond activation depending on the charge of methyl group after methane molecule is activated, which can be called carbenium (CH_3^+) pathway and alkyl (CH_3^-) pathway. Figure 7.6 provides a proposed mechanism of methane activation over Mo/HZSM-5 for aromatization [45]. First, methane is adsorbed on the active molybdenum oxide component over the zeolite catalyst. Then, it is unsymmetrically polarized to form carbenium CH_3^+ and anion H^-. H^- can be further combined with H^+ from the Bronsted acid site of the

Table 7.1 Recent studies of catalytic biomass upgrading processes under methane environment. GY, LY, and SY represent gas yield, liquid yield, and solid yield, respectively

Year	Author	Catalyst	Biomass feedstock	Desired products	Reaction temp. (°C)	GY (wt.%)	LY/ (wt.%)	SY(wt.%)	Ref.
2009	Sadykov	Ru/(PrCeSmZr)O_2/Cr-Al composite	Ethanol	H_2, CO	860	100			[37]
2009	Sadykov	Pt/Ni/La-Ce-Zr-O/ corundum	Acetone	H_2, CO	850	100			[37]
2011	Mendiara	NiO/NiAl_2O_4	Biomass tar	H_2, CO	600–800	99	0	1	[41]
2012	Chattanathan	Reformax 250	Biomass synthesis gas	H_2, CO	800	99	0	1	[42]
2014	He	Ag/ZSM-5	Flex straw	Oil	500	23	28	45	[22]
2015	He	Zn/ZSM-5	Pyrolysis oil	Bio-oil	300–600		20		[23]
2015	Xiao	Pt-Bi/C	Guaiacol	Deoxygenated organics	400	30	70	0	[38]
2016	Yang	Mo/HZSM-5	Cellulose, hemicellulose, lignin	Aromatics	700		15–35		[24]
2016	Wang	Ag/ZSM-5	Ethanol	Aromatics		60	35	0.8	[25]
2017	Wang	ZnGa/ZSM-5	Acetic acid	Aromatics	400	60	35	5	[26]
2017	Yang	MoZn/HZSM-5	Switchgrass	Aromatics	700	33	42	25	[27]
2017	Gunawardena	Mo/HZSM-5	Glucose	Aromatics	700	20	2	57	[28]
2018	Huang	NiLaCeZr/ CaO-MgO-Al_2O_3	Glycerol	H_2, CO	850	97	2	1	[39]
2018	Shen	Ni/γ-Al_2O_3	Pine	H_2, CO	1000	70	0	30	[40]
2018	Austin	ZnGa/HZSM-5	Acetone	Aromatics	400	31	70	0.5	[29]

2018	Shafaghat	Ni/HZSM-5	Lignin	Aromatics	600	21	27	52	[30]
2018	Shafaghat	Ni/HZSM-5	Yellow poplar	Aromatics	600	47	22	31	[30]
2018	Austin	ZuCuGa/ZSM-5	Glycerol	Aromatics	400	34	52	4	[31]
2018	Wang	Zn/HZSM-5	Cellulose	Aromatics	450	43	33	21	[32]
2018	Wang	ZnGa/ZSM-5	Lignin	Aromatics	400	17	37	52	[33]
2018	Wang	ZnGa/ZSM-5	Furfural	Aromatics	400	77	30	13	[34]
2019	Lalsare	FeMo/ZSM-5	Hardwood	H_2	850	90	0	10	[35]
2019	Wang	Zn/ZSM-5	Phenol	Aromatics	400	6	85	9	[36]

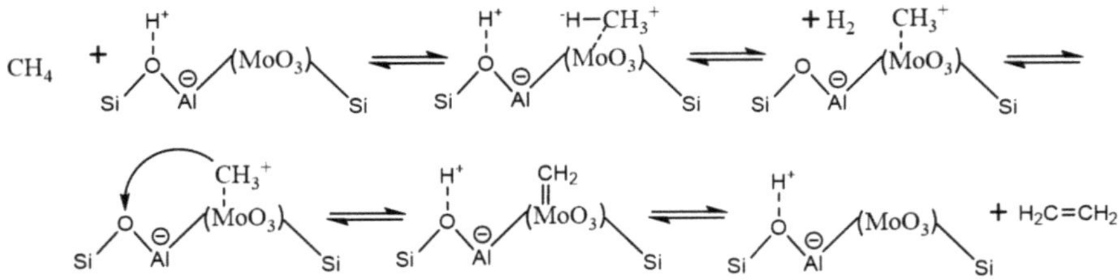

Fig. 7.6 The mechanism of methane activation over Mo/HZSM-5. Reproduced with the permission from Elsevier

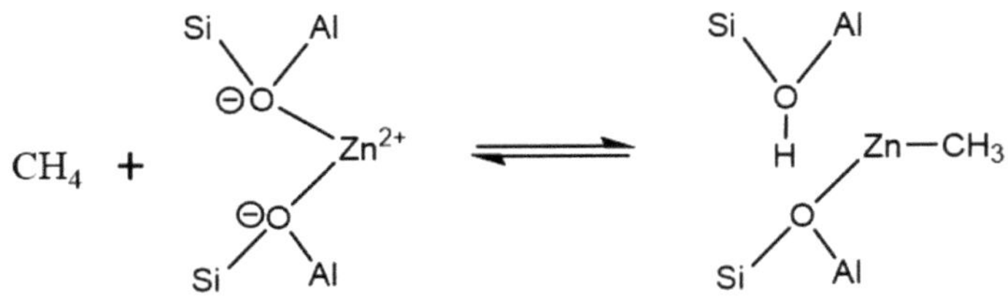

Fig. 7.7 Methane activation reaction over Zn^{2+}/ZSM-5 through alkylation pathway. Reproduced with the permission from Elsevier

zeolite support to produce H_2, and the exposed O atom with lone pair electrons can thus receive another proton from carbenium CH_3^+ to from a carbene intermediate (CH_2=). These carbene species can be further combined through dimerization to form ethylene, and the latter can be further converted into aromatics with the assistance of zeolite structure through hydrogen transfer and dehydrogenation processes. Figure 7.7 illustrates another possibility of methane activation over HZSM-5 zeolite exchanged by zinc cations [46]. In this process, CH_4 can firstly interact with Zn^{2+} cation, leading to the formation of metal-methyl species and proton. The proton can be further deprived by the unoccupied oxygen vacancies over the zeolite structure, which is proved to be crucial for the realization of methane activation through alkyl pathway.

Interestingly, it is also observed that these two methane activation patterns can occur simultaneously over one catalyst if the corresponding active sites are present. For example, it is observed that not only zinc-methyl species can be formed over Zn/ZSM-5 through its interaction with zinc cation, but also methane can be adsorbed on the oxygen atom of zinc oxide component, generating the methoxy group (Fig. 7.8) [47]. The presence of these species has been confirmed by ^{13}C spectra of the solid-state NMR technology.

According to the results, it is promising that methane activation is feasible over a variety of catalytic active sites regardless of its inert nature. To take advantage of methane introduction, the selection of catalyst which could produce appropriate species upon methane activation and effective utilization of these intermediates for desired products should be taken into careful consideration for successful biomass valorization processes.

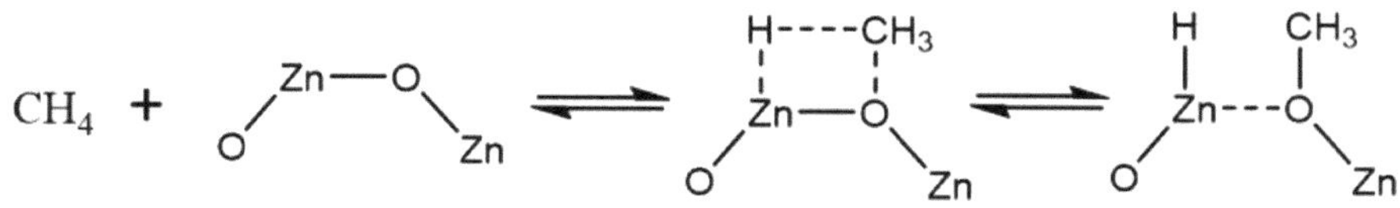

Fig. 7.8 Possible pathways of methane activation over Zn/ZSM-5. Reproduced with the permission from Elsevier

7.4 Property-Performance Relationship of Catalytic Biomass Valorization Under Methane Environment

7.4.1 Catalytic Activity

As one of the pioneering studies for methane activation, Choudhary et al. [17] discovered that methane could be evidently activated and converted at relatively low temperatures (400–600 °C) with the presence of other hydrocarbon species such as higher alkanes or alkenes. This synergistic effect could be explained by extra hydrogen transfer processes between methane and other hydrocarbons, making the reaction from methane to other hydrocarbons (e.g., benzene) more thermodynamically favorable. This characteristic has been widely utilized for methane utilization at mild conditions, including biomass upgrading process operated under methane environment. The catalyst used in this study (HGaAl-MFI zeolite) also became typical for methane activation. From then on, other metals have also been found effective for facilitating advantageous effects from methane in biomass upgrading processes, which is discussed in the following part.

In 2014, He et al. [22] suggested that the loading of Ag with a subsequent calcination process resulted in well-dispersed Ag^+ in the inner channel of ZSM-5 support evidenced by XRD patterns, which changed the structure and acidity of the support and facilitated methane activation (Fig. 7.9). They also observed in TEM images that the addition of P and Ce atoms led to the formation of a new P-doped ceria phase with needle shape, which was effective for mitigating the over-cracking of reactants through surface acidity modification of zeolite support (Fig. 7.10). Then, they systematically investigated the influence of initial silver nitrate concentration and calcination temperature on the dispersion of Ag particles, and the latter was found to be closely connected to the H/C ratio of upgrading products (Figs. 7.11 and 7.12).

In 2015, He et al. [23] validated that well-dispersed ZnO on zeolite attributed to favorable oil quality improvement in natural gas-assisted bio-oil upgrading process. However, XRD, STEM, and EDX characterizations suggested that the preferable zinc dispersion could only be achieved with low Zn loading (<5 wt.%). Higher Zn loading led to severe particle agglomeration and lowered catalytic performance. Xiao et al. [38] proposed that the main products of methane deoxygenation could be H_2O, H_2 and C_2H_6, indicating methane was activated to form methyl and hydrogen free radicals. In 2016, Yang et al. [24] found that molybdenum atoms can be evenly

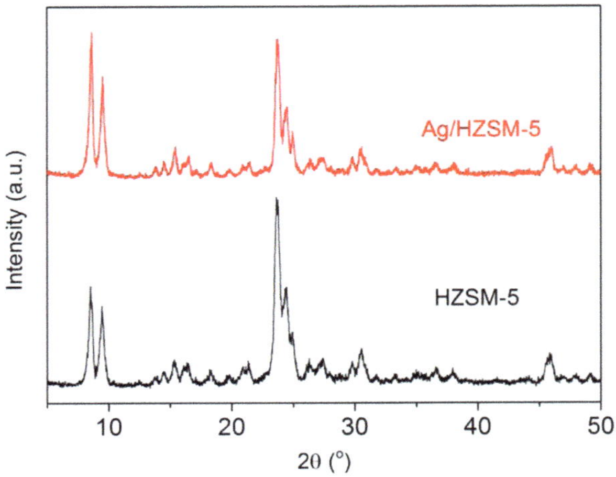

Fig. 7.9 XRD patterns of fresh catalysts of HZSM-5 and Ag/ZSM-5 for catalytic biomass conversion under methane environment. No Ag crystalline peak can be observed, indicating Ag is dispersed as fine particles with the size below detection limit (~4 nm). Reproduced with permission from [22]. Copyright (2014) American Chemical Society

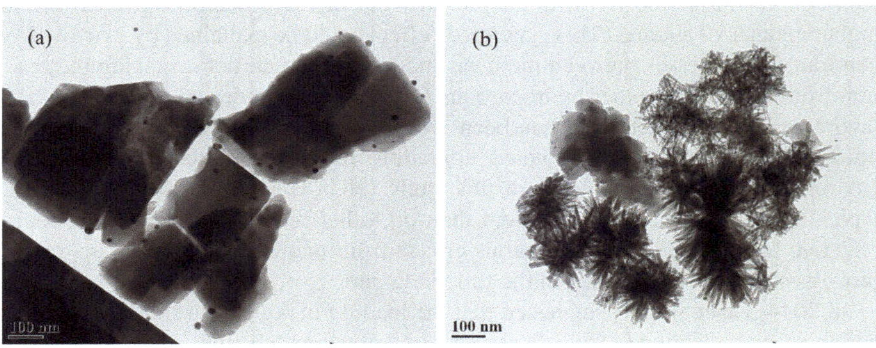

Fig. 7.10 TEM images of Ag/ZSM-5 (**a**) and Ag/P-Ce-ZSM-5 (**b**) for catalytic biomass conversion under methane environment. Reproduced with permission from [22]. Copyright (2014) American Chemical Society

distributed on HZSM-5 support by a simple wet impregnation method, which are responsible for the activation of methane as a hydrogen donor in deoxygenation process. However, the low yield indicated that methane did not get fully activated. Wang et al. [25] performed $^{13}CH_4$ isotope-labelling experiment along with carbon number distribution analysis to confirm the participation of methane in the co-conversion of ethanol and proved that methane could be successfully incorporated into alkyl, benzylic, and oxygenated groups but not favorably into benzene ring directly linked with alkyl groups. They further proved the good dispersion of active metals Ag and Zn and acceptable stability of the catalyst by XRD, TEM, and EDX characterizations. However, Ag formed bigger particles with a diameter of 10–20 nm, while Zn was evenly distributed on zeolite supports. Subsequent DRIFT and NH_3-TPD analyses indicated that the strong acid sites were responsible for aromatization, while the coke amount could increase correspondingly. In 2017,

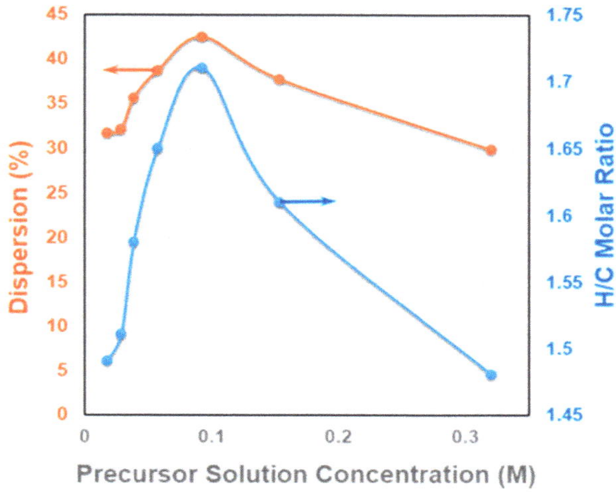

Fig. 7.11 Ag dispersion and H/C atomic ratio as a function of $AgNO_3$ concentration used for the synthesis of Ag/ZSM-5 for biomass conversion under methane environment. Reproduced with permission from [22]. Copyright (2014) American Chemical Society

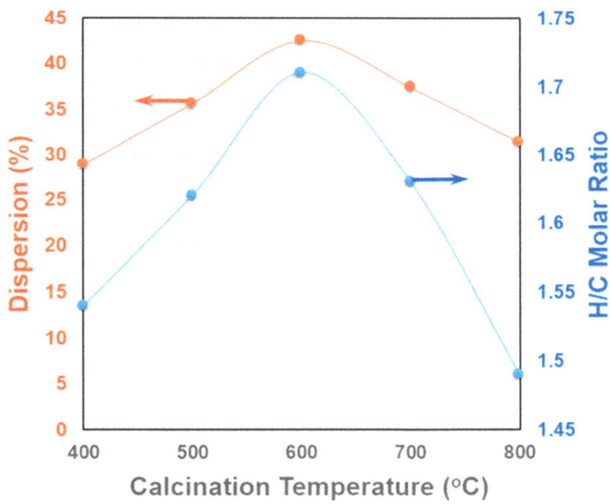

Fig. 7.12 Ag dispersion and H/C atomic ratio as a function of calcination temperature used for the synthesis of Ag/ZSM-5 for biomass conversion under methane environment. Reproduced with permission from [22]. Copyright (2014) American Chemical Society

Wang et al. [26] suggested that evenly distributed zinc and gallium species (proved by TEM and XRD results) and moderate acidity of zeolite (proved by DRIFT and ammonia-TPD) supports are key points for catalytic upgrading of acetic acid under methane environment. XPS further illustrated that gallium and zinc were mainly in +3 and + 2 valence, respectively. Zn in the form of zinc oxide was responsible for the aromatization of intermediates generated from acetic acid, while Ga facilitated the activation of methane in the reaction process. Furthermore, they fully investigate the reaction pathway by a series of control experiments. Firstly, the reaction was quenched with different reaction times to illustrate the evolution of acetic acid. They found that acetic acid was consumed in a short period of time with the production of acetone and CO_2 through ketonization. Then, mesityl oxide and isobutene were produced from acetone. Both of them can be further converted to aromatics through condensation, cracking, cyclization, and hydrogen transfer processes. Finally, experiments were performed using acetic acid with ^{13}C label in methyl and carboxyl group, respectively, to investigate the contribution of each carbon to final product. It was found that these two carbons contributed equally to benzene rings in final aromatic products, while carboxyl carbon was preferably incorporated in phenyl and benzylic groups. Yang et al. [27] suggested that MoZn/HZSM-5 catalyst resulted in better aromatic yield than MoAg/HZSM-5 for torrefied switchgrass under methane environment, which might be due to the lower intensity of carbon loss in benzene derivatives through dealkylation reaction. They also concluded that the aromatic yield of different components in biomass followed the sequence of cellulose > hemicellulose > lignin. Gunawardena et al. [28] investigated the pathway of glucose pyrolysis process under methane environment through isotope labelling. First, they confirmed the hydrogen donation from methane to aromatic products by reaction with deuterium labelled methane, where loaded metals played an indispensable role. Then, they used ^{13}C-labelled glucose to track the evolution of biomass molecules in the process and concluded that carbon coupling reaction occurred between methane and oxygenated species to form hydrocarbons and water molecules.

In 2018, Austin et al. [29] found that the silica-to-alumina (STA) value of zeolite support was essential for the catalytic upgrading of acetone under methane environment. A high STA value led to lower surface acidity and thus improved the yield of aromatics. However, extremely low BTEX yield was found with a Si-MFI (silicalite) zeolite, indicating acidity was still critical for the reaction despite the fact that low acidity was preferred (Fig. 7.13). Through XPS and XRD characterization, they suggested that the improved catalytic performance of ZnGa/HZSM-5 catalyst could be attributed to well-dispersed Zn^{2+} and Ga^{3+} species (Figs. 7.14 and 7.15). The difference of gas atmosphere was further interpreted as follows: as a reactant,

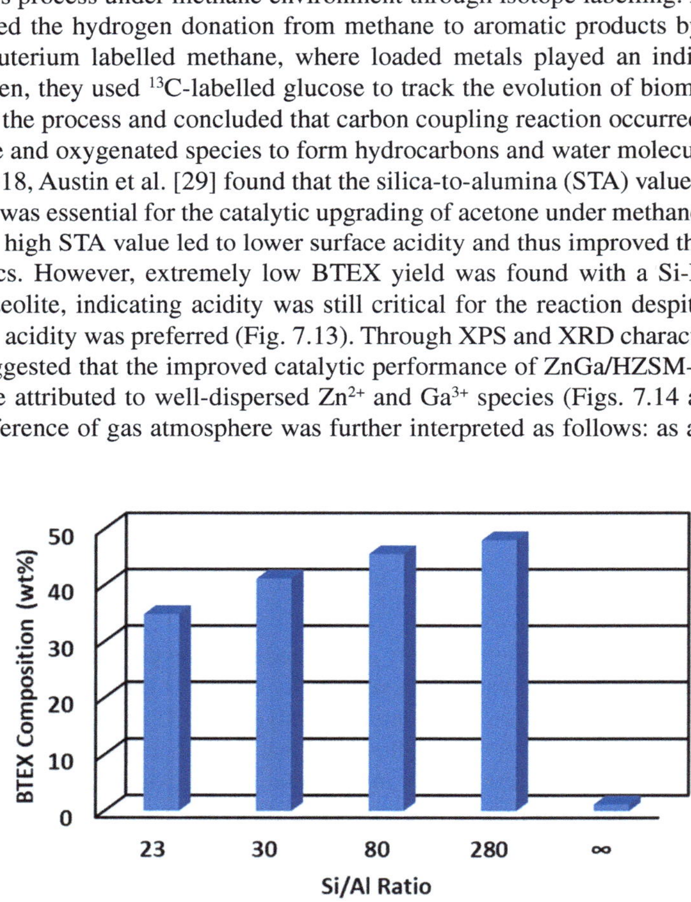

Fig. 7.13 Liquid product composition obtained over HZSM-5 with varying Si/Al ratios for catalytic acetone upgrading under CH_4 environment. Reproduced with permission from the Royal Society of Chemistry

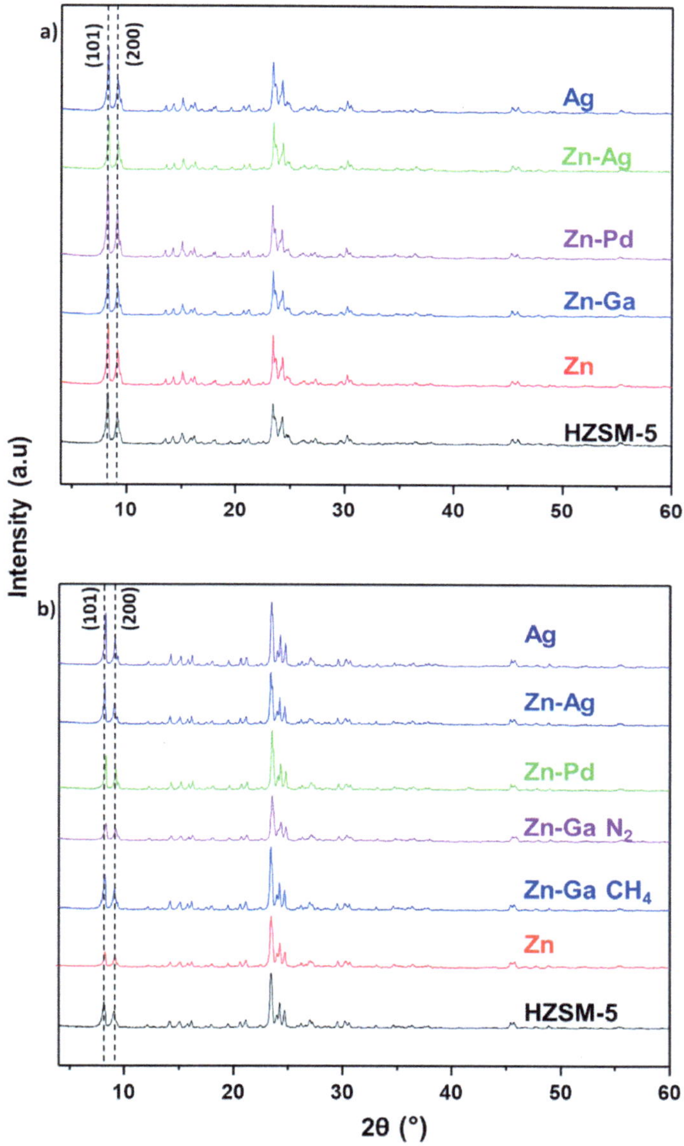

Fig. 7.14 XRD patterns collected for fresh (**a**) and spent (**b**) catalysts for catalytic acetone upgrading under methane environment. Reproduced with permission from the Royal Society of Chemistry

methane also played an important role for better dispersion of metal species, while N_2 led to metal agglomeration, as proved by EDS mapping results (Fig. 7.16). They also performed isotope-labelling experiments with $^{13}CH_4$ and CD_4 to validate the participation of methane in the reaction and found that methane was preferably

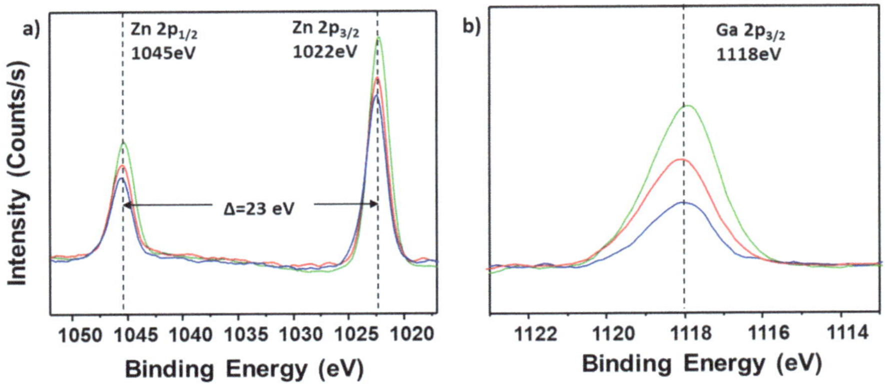

Fig. 7.15 XPS spectra of Zn-Ga/HZSM-5 catalysts for catalytic acetone upgrading under methane environment before reaction (red) and after reaction in CH_4 (blue) and N_2 (green) environments at the (**a**) Zn 2p and (**b**) Ga 2p regions. Reproduced with permission from the Royal Society of Chemistry

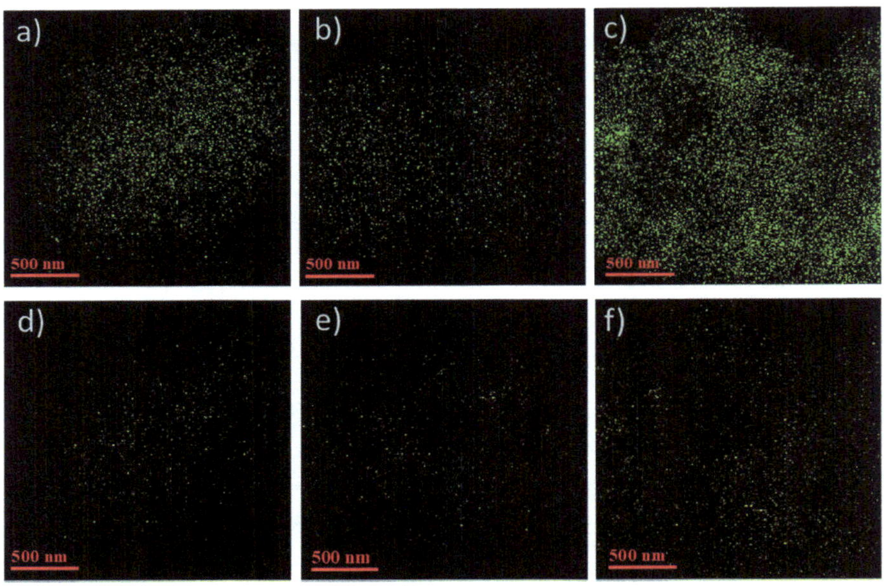

Fig. 7.16 TEM-EDS mapping of Zn for (**a**) fresh, (**b**) spent in CH_4, and (**c**) spent in N_2 samples and Ga for (**d**) fresh, (**e**) spent in CH_4, and (**f**) spent in N_2 samples over Zn-Ga/HZSM-5 for catalytic acetone upgrading under methane environment. Reproduced with permission from the Royal Society of Chemistry

incorporated into the aromatic ring rather than side chain substituents (Figs. 7.17 and 7.18). Through quenching the experiment at different reaction times, they proposed the reaction mechanism as follows. First, acetone was converted into acetic acid and isobutene through and condensation and β-scission. Then, acetic acid could

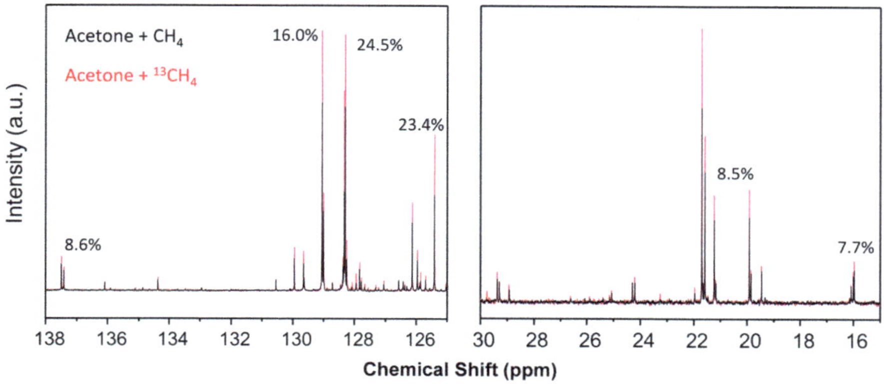

Fig. 7.17 ^{13}C liquid NMR spectra of the products collected from the aromatization of acetone over Zn-Ga/HZSM-5 under $^{13}CH_4$ and CH_4 environment, including the percent increase in peak area under $^{13}CH_4$ environment. Reproduced with permission from the Royal Society of Chemistry

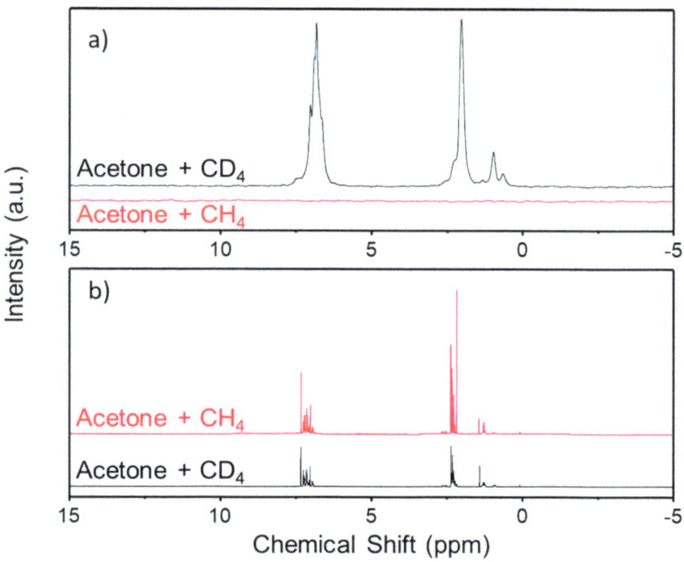

Fig. 7.18 ^{2}H (**a**) and ^{1}H (**b**) liquid NMR spectra of the products collected from the aromatization of acetone over Zn-Ga/HZSM-5 under CH_4/CD_4 environment. Reproduced with permission from the Royal Society of Chemistry

be converted back to acetone through ketonization with the release of CO_2, while isobutene together with other short-chain olefin intermediates could form higher olefins. Finally, the olefin intermediates with 6–8 carbon numbers reacted with methane to form cycloalkanes, and the latter could be converted into aromatics through hydrogen transfer process, in which excessive hydrogen could be used for olefin saturation to produce alkanes. Meanwhile, the produced BTEX species also reacted with activated methane to form heavier aromatics in the final stage, and thus the control of reaction time was essential for higher BTEX yields (Figs. 7.19 and 7.20).

In the same year, Huang et al. [39] confirmed that methane activated by modified Ni/alumina catalyst was responsible for the inhibition of CO_2 emission in steam reforming process. Comparing with non-modified Ni/alumina catalyst, the enhancement of catalytic performance by alkaline-earth metal and transition metal oxide promoters could be attributed to the formation of spinel phase, which helped the nickel species get better dispersed and thus increased the number of active sites. They also suggested that the improved catalytic performance could also be related to decreased surface acidic sites by NH_3-TPD analysis. Shen et al. [40] concluded that nickel(II) oxide could be evenly dispersed on γ-alumina support, which was highly active for the reforming of methane and gasified biomass. They also studied the deactivation process and suggested the carbon deposition should be the main reason for decreasing activity, which was highly dependent on the stream-to-carbon ratio of the process. Shafaghat et al. [30] indicated that the introduction of Ni resulted in the formation of new acid sites on HZSM-5 by NH_3-TPD. They also proved that the introduced Ni particles were highly distributed on ZSM-5 support according to XRD and TEM observations, which led to increased acid sites and better methane activation in the pyrolysis process. Austin et al. [31] suggested that methane led to side chain substitution reaction when it was co-fed with glycerol, which was proved by higher yield of aromatics and lower yield of naphthalene and phenol (Fig. 7.21). They also evaluated the surface acidity of various catalysts by pyridine adsorption in DRIFT analysis and NH_3-TPD, concluding that the favorable aromatization ability of ZnCuGa/ZSM-5 catalyst prepared by deposition

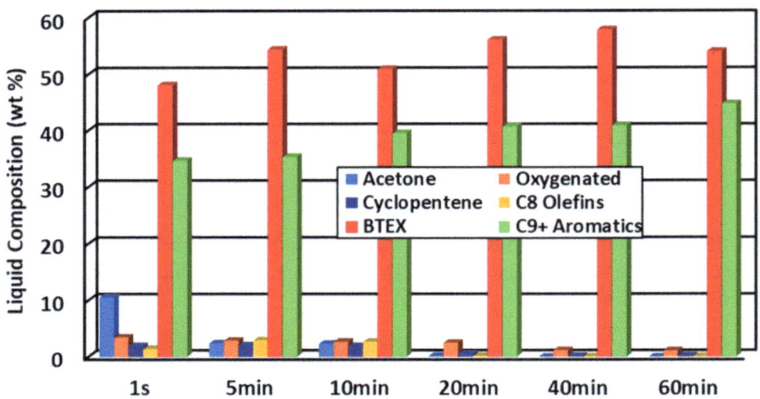

Fig. 7.19 Liquid product compositions obtained after varying reaction times over ZnGa/HZSM-5 for catalytic acetone upgrading under methane environment. Reproduced with permission from the Royal Society of Chemistry

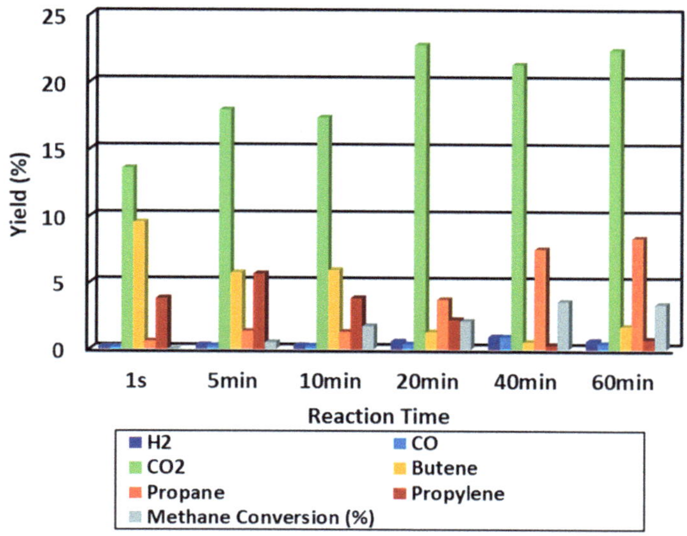

Fig. 7.20 Gas product yields and methane conversion obtained after varying reaction times over Zn-Ga/HZSM-5 for catalytic acetone upgrading under methane environment. Reproduced with permission from the Royal Society of Chemistry

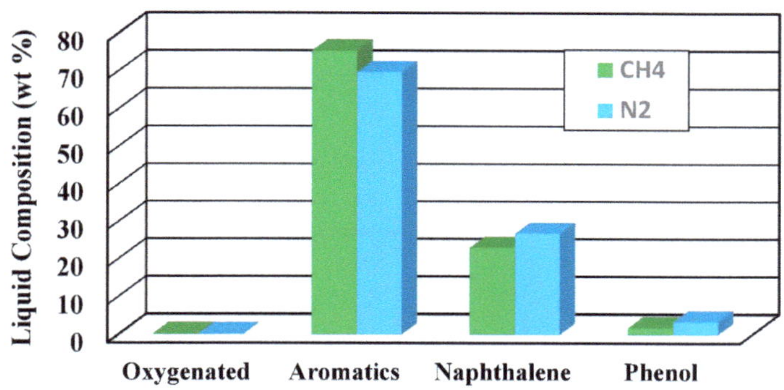

Fig. 7.21 A comparison of liquid product distribution obtained over ZnCuGa/ZSM-5 catalyst under CH_4 and N_2 environment for glycerol valorization. Reproduced with permission from Elsevier

precipitation could be due to the sufficient medium acidity sites (Figs. 7.22 and 7.23). They attributed the increased surface acidity to the well-dispersed metal species by XRD, CO titration, and TEM (Figs. 7.24, 7.25, and 7.26). Besides, they performed XAS characterization to determine the chemical environments of metals in the catalyst and indicated that Zn mainly maintained the form of ZnO, while CuO was partially reduced to Cu_2O during the reaction process. The shift of characteristic adsorption energy indicated there was synergistic effect between loaded metal species and zeolite support (Fig. 7.27). Wang et al. [32] found that an appropriate

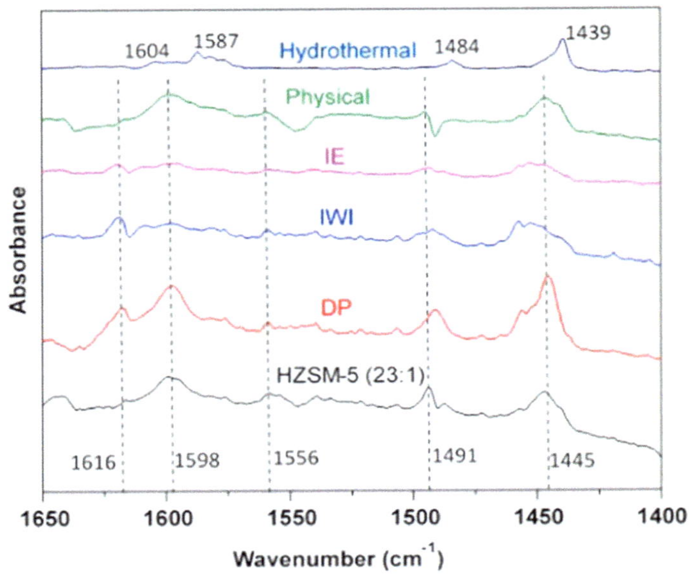

Fig. 7.22 DRIFT spectra collected over HZSM-5 and ZnCuGa/ZSM-5 catalysts prepared through deposition-precipitation (DP), incipient wetness impregnation (IWI), ion exchange (IE), physical mixing, and hydrothermal synthesis for glycerol valorization after pyridine adsorption. Reproduced with permission from Elsevier

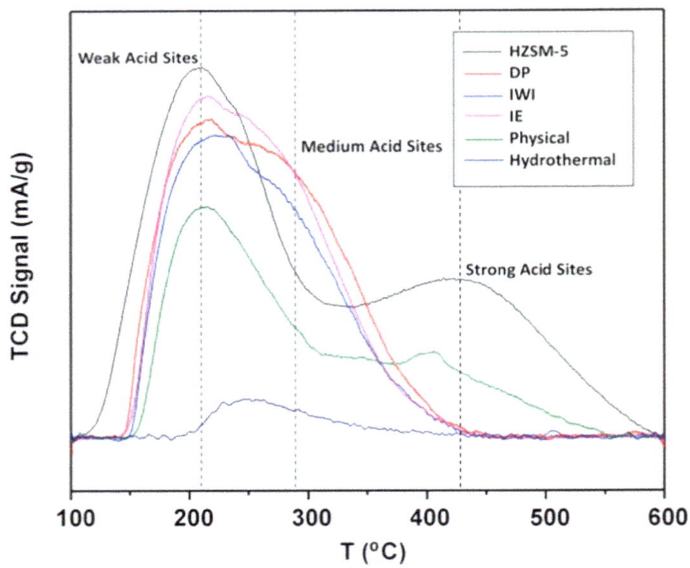

Fig. 7.23 NH_3-TPD signals collected over HZSM-5 and ZnCuGa/ZSM-5 catalysts prepared through deposition- precipitation (DP), incipient wetness impregnation (IWI), ion exchange (IE), physical mixing, and hydrothermal synthesis for glycerol valorization. Reproduced with permission from Elsevier

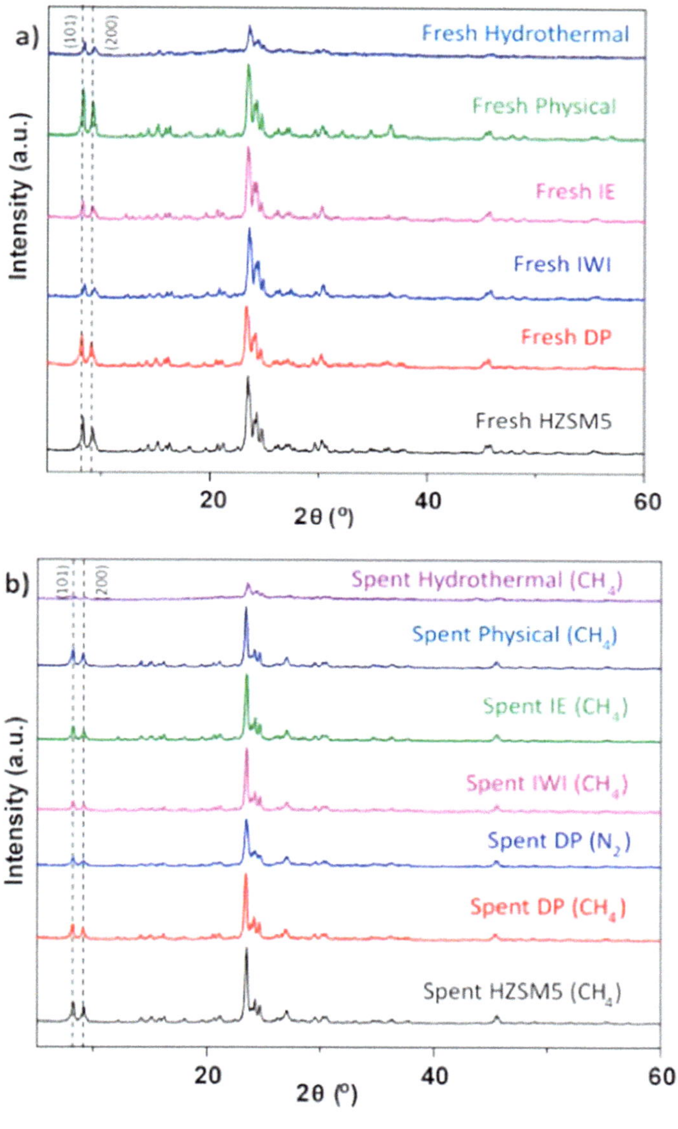

Fig. 7.24 XRD patterns collected over HZSM-5 and ZnCuGa/ZSM-5 catalysts prepared through deposition precipitation (DP), incipient wetness impregnation (IWI), ion exchange (IE), physical mixing, and hydrothermal synthesis (a) before (Fresh) and (b) after (Spent) glycerol valorization reaction under different gas environments. Reproduced with permission from Elsevier

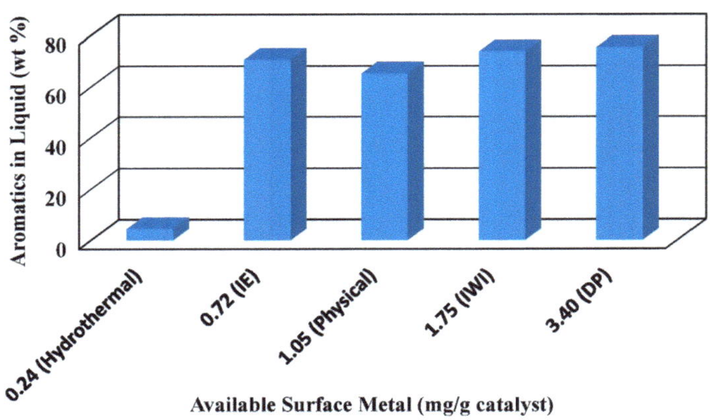

Fig. 7.25 Available surface metal determined by CO titration and corresponding to aromatic performance over HZSM-5 and ZnCuGa/ZSM-5 catalysts prepared through deposition precipitation (DP), incipient wetness impregnation (IWI), ion exchange (IE), physical mixing, and hydrothermal synthesis for glycerol valorization. Reproduced with permission from Elsevier

surface acidity of zeolite was critical for cellulose conversion under methane environment: deficient acidity led to low aromatization activity, while excessive acidity led to over-cracking. Besides, Zn addition was also crucial for the selectivity towards aromatics, especially $Zn^{\delta+}$ species (proved by XPS and XAS characterizations) introduced by chemical vapor deposition. Further evidences from DRIFTS and NH_3-TPD confirmed the better BTEX selectivity might come from enhanced medium acid sites. They also confirmed the participation of methane by isotope-labelling experiments. $^{13}CH_4$ experiments indicated methane could be incorporated into not only alkyl groups but also aromatic rings in liquid products, while CD_4 experiments proved that hydrogen atoms from methane was preferably added into aromatic and benzylic groups rather than alkyl groups. They also revealed the additional function of methane other than a reactant: TEM-EDS mapping results showed that the Zn species were better dispersed in methane environment, yet they tended to agglomerate in N_2 environment, indicating the facilitation of methane for metal redispersion and active site retention. Wang et al. [34] used XRD, TEM, and EDS mapping to confirm the dispersion of zinc and gallium species in ZnGa/ZSM-5 catalysts and found that good dispersion could be maintained under methane environment, while nitrogen environment led to agglomeration. They further determined the chemical presence of Zn and Ga by XPS, which could be ZnO and GaO^+. Further DRIFTS and NH_3-TPD characterizations indicated that abundant weak acid site facilitated the reaction between furfural and methane. Moreover, isotope-labelling experiments proved the participation of methane, which favorably occurred at benzylic site or meta position of the benzene ring. Wang et al. [33] also found that a moderate acidity of support was advantageous for lignin conversion under methane

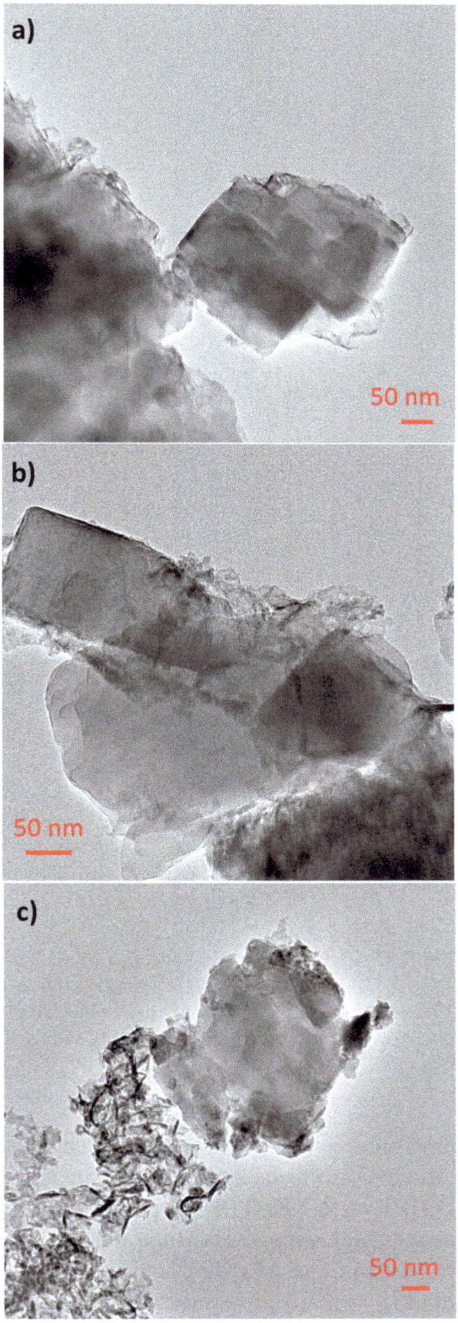

Fig. 7.26 TEM images of (**a**) fresh ZnCuGa/ZSM-5-DP, (**b**) ZnCuGa/ZSM-5-DP after reaction in CH_4 environment, and (**c**) ZnCuGa/ZSM-5-DP after reaction in N_2 environment for glycerol valorization. Reproduced with permission from Elsevier

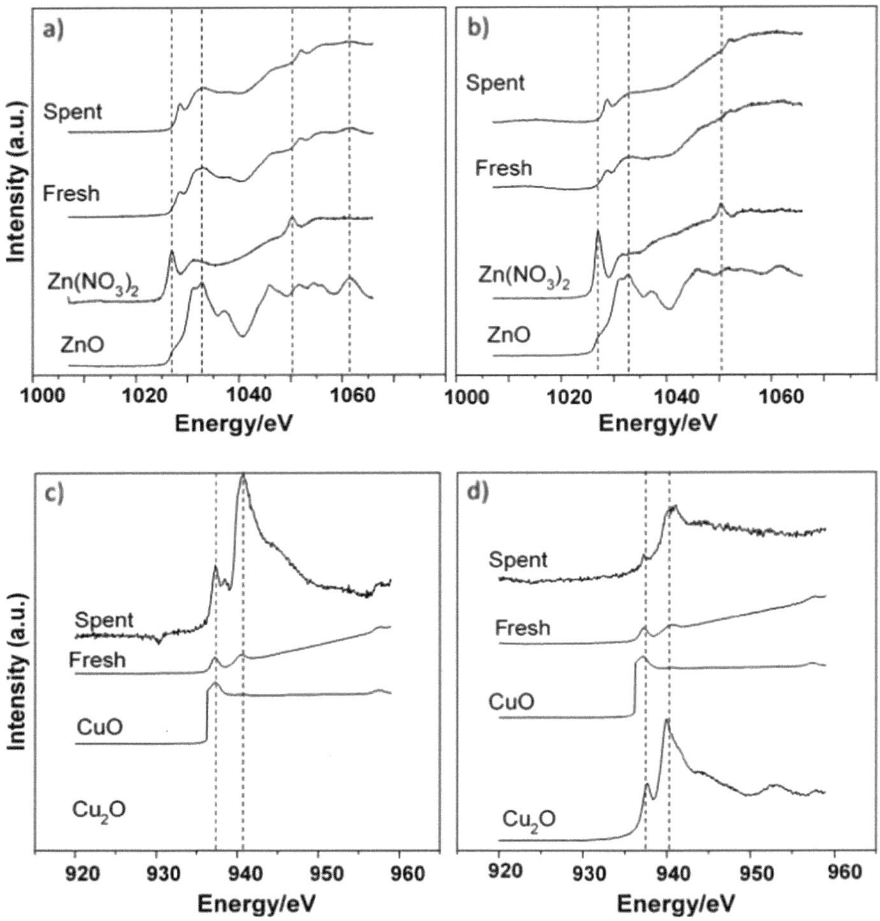

Fig. 7.27 Zn L-edge XANES spectra of ZnO, $Zn(NO_3)_2$ as reference, and fresh and spent ZnCuGa/ZSM-5-DP catalyst for glycerol valorization at the surface (**a**) and in the bulk (**b**); Cu L-edge XANES spectra of CuO, Cu_2O as reference and fresh, and spent ZnCuGa/ZSM-5-DP catalyst for glycerol valorization at the surface (**c**) and in the bulk (**d**). Reproduced with permission from Elsevier

environment due to the balance of catalyst activity (high acidity required) and inhibition of over-cracking and coking process (low acidity preferred) (Fig. 7.28). DRIFTS and NH_3-TPD results further suggested that the strength of acid sites was also a determining factor: the introduction of Zn and Ga metal components led to reduced number of strong acid sites and increased number of weak acid sites, favoring the aromatization selectivity. Besides, they proved the essential role of catalyst

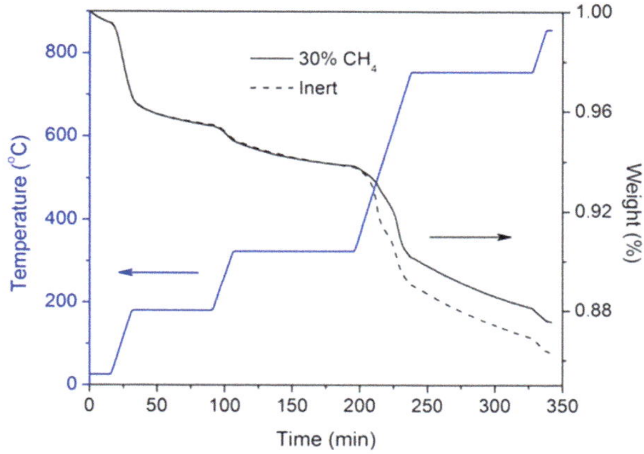

Fig. 7.28 TGA profiles of Ag/ZSM-5 for catalytic biomass conversion under methane environment. Reproduced with permission from [22]. Copyright (2014) American Chemical Society

morphology by comparing ZSM-5 zeolite and another silica-alumina material prepared through co-precipitation. Methane precipitation was again proved by isotope-labelling experiments and NMR analyses, which occurred preferably on aromatic and benzylic sites.

In 2019, Lalsare et al. [35] claimed that oxophilic metals such as Fe and Mo facilitated the conversion of oxygen in lignin-containing biomass together with methane to form CO and hydrogen. Furthermore, they suggested that the acid sites and metal species in FeMo/ZSM-5 favored water-gas shift and steam methane reformation reaction to produce high H_2/CO value in gas product. Wang et al. [36] suggested that the favorable performance of Zn/HZSM-5 for phenol deoxygenation with methane could be due to increased Lewis acidic sites, which was proved by DRIFT and NH_3-TPD characterizations. They also found that the introduction of methane facilitated the dispersion of zinc over the catalyst support by XRD, TEM, and XPS. Further XAS characterization illustrated that Zn was bonded to zeolite support and formed active sites with coordinated chemical environment differing from Zn or ZnO. Furthermore, they confirmed the methane participation by ^{13}C, ^{1}H, and ^{2}H isotope-labelling experiments and subsequent GC-MS and NMR analyses. Finally, they proposed the reaction mechanism by quenching the reaction process at various reaction times and analyze the corresponding intermediates: firstly, phenol got adsorbed and converted into aryl and phenoxy radicals; then unoccupied active sites led to the activation of methane and its interaction with phenol to form cresol and xylenol; next, the phenoxy and alkylphenol species were converted into aromatics as the final product; meanwhile, oxygen atoms came into gas phase in the form of CO through decarbonylation and got involved in secondary reactions such as water-gas shift and Boudouard reaction; and finally, the produced intermediates in the previous step might undergo complex recombination, aromatization, and Diels-Alder processes to form light hydrocarbons, BTX, and naphthalene.

7.4.2 Coke Formation

In 2011, Mendiara et al. [41] investigated a series of catalysts for biomass tar reforming and found that both methane and biomass could form carbon deposition on catalyst surface, leading to deactivation. However, the carbon deposition from methane was sensitive to water content in the gas phase, while the one from biomass did not conform the tendency. Besides, it was concluded that the presence of H_2 and CO_2 both mitigate carbon deposition processes. In 2014, He et al. [22] found that the introduction of methane could reduce graphitic coke formation due to the compensation of hydrogen atoms from methane by TGA weight loss assignment (Fig. 7.28).

7.5 Challenges for Industrialization

Despite the aforementioned achievements, it is apparent that there is still a considerable gap between laboratory work and final industrialization for methane utilization in biomass conversion processes. It should be noted that stepwise scaling up and troubleshooting practices, in terms of equipment, materials, operating procedures, etc., are inevitable to facilitate these processes. However, the scale-up of a chemical engineering process could not be simply completed by magnifying all design parameters by a fixed factor. More practical problems need to be taken into consideration. These problems are essentially caused by enormous differences between laboratory and industrial reaction processes, including but not limited to:

7.5.1 Differences on Operating Conditions

In order to better realize the evaluation and screening of catalysts, laboratories often choose more extreme experimental conditions, such as high space velocity, high temperature, high pressure, and deliberate excess of some reactants. However, many practical issues such as production efficiency, economy, and safety need to be considered in the industrial production process. The difference in concerns has led to a great difference in the operating conditions of laboratory devices and industrial devices so that the performance of catalysts in laboratory evaluation could not fully represent its industrialization performance. In order to handle the problem, kinetic studies are essential for predicting the operation performance of industrial reactors through limited laboratory experimental data. Through the study of reaction kinetics, the influence of changes in operating conditions such as temperature, pressure, and concentration on the reaction rate can be obtained, so as to predict the reaction performance of the catalyst under different operating conditions.

7.5.2 Differences in On-Stream Time

Due to the limitations of material resources and working time in the laboratory, the evaluation time of catalysts is often short, which is generally limited to a few hours to tens of hours. The difference is that in order to ensure maximum economic benefit in the continuous industrial production process, the catalyst is often expected to run for thousands or even tens of thousands of hours. Therefore, the short-term evaluation results of the catalyst laboratory are not necessarily representative of its long-term operating performance. To get better control of the long-term performance of catalysts, the deactivation process should be fully investigated, and corresponding deactivation kinetics could help in solving the problem. Through the study of deactivation mechanics, the law of catalytic performance over time can be obtained, so as to predict the lifetime of the catalyst and provide a basis for economic evaluation of the catalytic process.

7.5.3 Differences in Characteristic Parameters

In the process of catalyst scaling up, assuming that the production scale is enlarged by K times, the amount of employed catalysts is generally also expanded by K times. However, the characteristic parameters of many reactors cannot be changed in the same proportion (e.g., the size of the reactor will be magnified by $K^{1/3}$, and the cross-sectional area will be enlarged by $K^{2/3}$, while the heat transfer coefficient will not change significantly, etc.). This has resulted in the reaction characteristics and laws of the small-scale catalyst to be greatly disturbed during the scale-up process and even caused the reactor to completely deviate from the normal operating state. The abovementioned differences in many aspects will lead to the deterioration of the reaction status of the catalyst in the actual industrial scale-up process, the shortening of the catalyst life, the reduction of product conversion rate and selectivity, etc., which directly or indirectly increase the cost of use and even cause safety accidents in severe cases. Therefore, the research on the catalyst engineering process is very important for the excellent performance of the catalyst and the safety and efficiency of production. However, from public literature, there is still a lack of research on the engineering problems of biomass conversion catalysts, which is not conducive to the catalyst from the laboratory to the process of industrialization (Fig. 7.29) [48]. Reactor design and simulation are exigently needed for better control of the reaction process. Through the investigation of reactor simulation, the change of important operating parameters in the reactor can be obtained with different characteristic parameters and conditions, thus guiding the operation process of the industrial device after the scale-up.

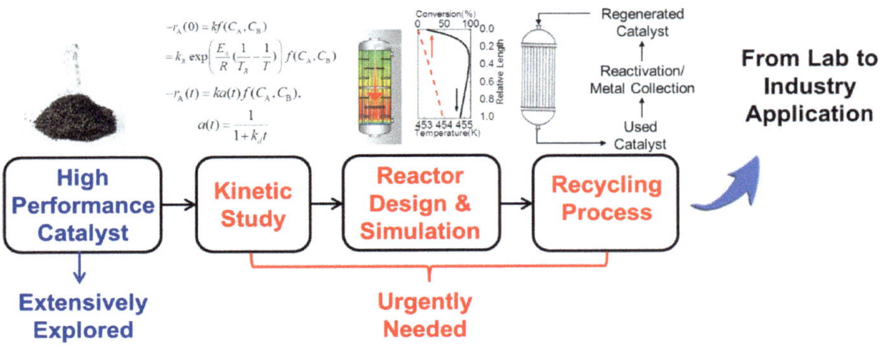

Fig. 7.29 Key steps and challenges for catalyst industrialization. Reproduced with permission from Elsevier

7.5.4 Recycling Process

As shown in the discussion above, biomass upgrading processes under methane environment are often accompanied by inevitable coking process, which is the result of balanced hydrogen-to-carbon ratio. Therefore, it can be anticipated that the catalyst will get deactivated in a considerably rapid manner. If the deactivated catalyst is not appropriately treated and recycled, the operational cost would become unfavorable. Besides, the used catalyst itself, especially the toxic and nonbiodegradable components such as heavy metals could cause difficulties on waste control and treatment and further environmental pollution. Therefore, the reuse of catalyst is of great importance and necessity. Feasible recycling processes should be fully designed and investigated to accelerate the path towards real application.

In general, in order to realize final industrialization of biomass upgrading process under methane environment, several main problems still remain unsolved, becoming the “bottle neck” of further development of corresponding techniques. The key points include but are not limited to (1) the imbalance between abundant catalyst and process investigations and scanty kinetic studies; (2) models for describing catalyst deactivation processes; (3) industrial reactor design and simulation; and (4) recycling process. The corresponding work should be greatly appreciated and highly valued.

7.6 Summary

With the developments in recent decades, biomass upgrading under methane environment has become a promising pathway for green and sustainable energy utilization. Different catalytic systems for various biomass feed have been proposed, and the reaction performance has been carefully evaluated. Furthermore, plenty of characterization methods have been employed to dig deeper into the active sites of catalysts and reaction mechanisms, especially the activation of methane in these processes. There is no doubt that creative catalyst design and the involvement of high-quality characterization methodologies should still be encouraged, while DFT calculations could also be a powerful tool for understanding the reason behind good catalytic performance.

Among all types of catalysts, metal loaded on zeolite support is most extensively studied and understood and comes closest to real application.

While the state-of-the-art catalysts are only a small step away from large-scale industrialization, important studies for the scaling up process including kinetic studies, reactor design and simulation, and recycling process are in extreme shortage and limiting the industrialization process. Efforts concerning these aspects should be emphasized and encouraged since more engineering problems can only be found and solved by practical attempts. With these efforts, the realization of economically and environmentally friendly catalytic biomass valorization under methane environment can be optimistically anticipated.

References

1. M. Ni, D.Y.C. Leung, M.K.H. Leung, K. Sumathy, An overview of hydrogen production from biomass. Fuel Process. Technol. **87**(5), 461–472 (2006)
2. M. Lee, Y.-L. Lin, P.-T. Chiueh, W. Den, Environmental and energy assessment of biomass residues to biochar as fuel: A brief review with recommendations for future bioenergy systems. J. Clean. Prod. **251** (2020)
3. L.J.R. Nunes, T.P. Causer, D. Ciolkosz, Biomass for energy: A review on supply chain management models. Renew. Sust. Energ. Rev. **120** (2020)
4. A. Wang, D. Austin, H. Song, Investigations of thermochemical upgrading of biomass and its model compounds: Opportunities for methane utilization. Fuel **246**, 443–453 (2019)
5. L.R. Matthews, A.M. Niziolek, O. Onel, N. Pinnaduwage, C.A. Floudas, Biomass to liquid transportation fuels via biological and thermochemical conversion: Process synthesis and global optimization strategies. Ind. Eng. Chem. Res. **55**(12), 3203–3225 (2015)
6. A. Molino, S. Chianese, D. Musmarra, Biomass gasification technology: The state of the art overview. J. Energy Chem. **25**(1), 10–25 (2016)
7. F. Behrendt, Y. Neubauer, M. Oevermann, B. Wilmes, N. Zobel, Direct liquefaction of biomass. Chem. Eng. Technol. **31**(5), 667–677 (2008)
8. A.V. Bridgwater, Review of fast pyrolysis of biomass and product upgrading. Biomass Bioenergy **38**, 68–94 (2012)
9. M.S. Mettler, D.G. Vlachos, P.J. Dauenhauer, Top ten fundamental challenges of biomass pyrolysis for biofuels. Energy Environ. Sci., 5(7) (2012)
10. W. Luo, W. Cao, P.C.A. Bruijnincx, L. Lin, A. Wang, T. Zhang, Zeolite-supported metal catalysts for selective hydrodeoxygenation of biomass-derived platform molecules. Green Chem. **21**(14), 3744–3768 (2019)
11. W. Jin, L. Pastor-Pérez, D. Shen, A. Sepúlveda-Escribano, S. Gu, R.T. Ramirez, Catalytic upgrading of biomass model compounds: Novel approaches and lessons learnt from traditional hydrodeoxygenation - a review. ChemCatChem **11**(3), 924–960 (2019)
12. D.C. Elliott, Historical developments in hydroprocessing bio-oils. Energy Fuel **21**, 1792–1815 (2007)

13. A. Losz, J. Elkind, The Role of Natural Gas in the Energy Transition. https://www.energypolicy.columbia.edu/research/commentary/role-natural-gas-energy-transition (2019)
14. BP Statistical Review of World Energy. BP Energy Economics, https://www.bp.com/en/global/corporate/energy-economics/statistical-review-of-world-energy.html (2019)
15. C. Davis, BP Bearish on Natural Gas Prices, LNG Exports to 2021. Natural Gas Intelligence, https://www.naturalgasintel.com/bp-bearish-on-natural-gas-prices-lng-exports-to-2021/ (2019)
16. D. Bradley, EIA Trims 2020 Natural Gas Price Forecast to $2.33; Relief Seen in 2021. (2020)
17. V.R. Choudhary, A.K. Kinage, T.V. Choudhary, Low-temperature nonoxidative activation of methane over H-galloaluminosilicate (MFI) zeolite. Science **275**, 1286–1288 (1997)
18. O. Anunziata, Improvement of methane activation using n-hexane as co-reactant over Zn/HZSM-11 zeolite. Catal. Commun. **5**(8), 401–405 (2004)
19. M.V. Luzgin, V.A. Rogov, S.S. Arzumanov, A.V. Toktarev, A.G. Stepanov, V.N. Parmon, Methane aromatization on Zn-modified zeolite in the presence of a co-reactant higher alkane: How does it occur? Catal. Today **144**(3–4), 265–272 (2009)
20. V.R. Choudhary, K.C. Mondal, S.A. Mulla, Simultaneous conversion of methane and methanol into gasoline over bifunctional Ga-, Zn-, in-, and/or Mo-modified ZSM-5 zeolites. Angew. Chem. **44**(28), 4381–4385 (2005)
21. O.A. Anunziata, G.V.G. Mercado, L.B. Pierella, Catalytic activation of methane using n-pentane as co-reactant over Zn/H-ZSM-11 zeolite. Catal. Lett. **87**, 167–171 (2003)
22. P. He, H. Song, Catalytic conversion of biomass by natural gas for oil quality upgrading. Ind. Eng. Chem. Res. **53**(41), 15862–15870 (2014)
23. P. He, W. Shan, Y. Xiao, H. Song, Performance of Zn/ZSM-5 for in situ catalytic upgrading of pyrolysis bio-oil by methane. Top. Catal. **59**(1), 86–93 (2015)
24. Z. Yang, A. Kumar, A. Apblett, Integration of biomass catalytic pyrolysis and methane aromatization over Mo/HZSM-5 catalysts. J. Anal. Appl. Pyrolysis **120**, 484–492 (2016)
25. A. Wang, P. He, M. Yung, H. Zeng, H. Qian, H. Song, Catalytic co-aromatization of ethanol and methane. Appl. Catal. B Environ. **198**, 480–492 (2016)
26. A. Wang, D. Austin, A. Karmakar, G.M. Bernard, V.K. Michaelis, M.M. Yung, et al., Methane upgrading of acetic acid as a model compound for a biomass-derived liquid over a modified zeolite catalyst. ACS Catal. **7**(5), 3681–3692 (2017)
27. Z. Yang, A. Kumar, A.W. Apblett, A.M. Moneeb, Co-pyrolysis of torrefied biomass and methane over molybdenum modified bimetallic HZSM-5 catalyst for hydrocarbons production. Green Chem. **19**(3), 757–768 (2017)
28. D.A. Gunawardena, S.D. Fernando, Catalytic conversion of glucose micropyrolysis vapors in methane-using isotope labeling to reveal reaction pathways. Energ. Technol. **5**(5), 708–714 (2017)
29. D. Austin, A. Wang, J.H. Harrhy, X. Mao, H. Zeng, H. Song, Catalytic aromatization of acetone as a model compound for biomass-derived oil under a methane environment. Cat. Sci. Technol. **8**(19), 5104–5114 (2018)
30. H. Shafaghat, J. Jae, S.-C. Jung, J.-K. Jeon, C.H. Ko, Y.-K. Park, Effect of methane co-feeding on product selectivity of catalytic pyrolysis of biomass. Catal. Today **303**, 200–206 (2018)
31. D. Austin, A. Wang, P. He, H. Qian, H. Zeng, H. Song, Catalytic valorization of biomass derived glycerol under methane: Effect of catalyst synthesis method. Fuel **216**, 218–226 (2018)
32. A. Wang, D. Austin, P. He, X. Mao, H. Zeng, H. Song, Direct catalytic co-conversion of cellulose and methane to renewable petrochemicals. Cat. Sci. Technol. **8**(21), 5632–5645 (2018)
33. A. Wang, H. Song, Maximizing the production of aromatic hydrocarbons from lignin conversion by coupling methane activation. Bioresour. Technol. **268**, 505–513 (2018)
34. A. Wang, D. Austin, H. Qian, H. Zeng, H. Song, Catalytic valorization of furfural under methane environment. ACS Sustain. Chem. Eng. **6**(7), 8891–8903 (2018)
35. A. Lalsare, Y. Wang, Q. Li, A. Sivri, R.J. Vukmanovich, C.E. Dumitrescu, et al., Hydrogen-rich syngas production through synergistic methane-activated catalytic biomass gasification. ACS Sustain. Chem. Eng. **7**(19), 16060–16071 (2019)

36. A. Wang, D. Austin, P. He, M. Ha, V.K. Michaelis, L. Liu, et al., Mechanistic investigation on catalytic deoxygenation of phenol as a model compound of biocrude under methane. ACS Sustain. Chem. Eng. **7**(1), 1512–1523 (2018)
37. V. Sadykov, N. Mezentseva, G. Alikina, R. Bunina, V. Pelipenko, A. Lukashevich, et al., Nanocomposite catalysts for internal steam reforming of methane and biofuels in solid oxide fuel cells: Design and performance. Catal. Today **146**(1–2), 132–140 (2009)
38. Y. Xiao, A. Varma, Catalytic deoxygenation of guaiacol using methane. ACS Sustain. Chem. Eng. **3**(11), 2606–2610 (2015)
39. C. Huang, C. Xu, B. Wang, X. Hu, J. Li, J. Liu, et al., High production of syngas from catalytic steam reforming of biomass glycerol in the presence of methane. Biomass Bioenergy **119**, 173–178 (2018)
40. Y. Shen, Y. Liu, H. Yu, Enhancement of the quality of syngas from catalytic steam gasification of biomass by the addition of methane/model biogas. Int. J. Hydrog. Energy **43**(45), 20428–20437 (2018)
41. T. Mendiara, J.M. Johansen, R. Utrilla, A.D. Jensen, P. Glarborg, Evaluation of different oxygen carriers for biomass tar reforming (II): Carbon deposition in experiments with methane and other gases. Fuel **90**(4), 1370–1382 (2011)
42. S.A. Chattanathan, S. Adhikari, S. Taylor, Conversion of carbon dioxide and methane in biomass synthesis gas for liquid fuels production. Int. J. Hydrog. Energy **37**(23), 18031–18039 (2012)
43. R. Horn, R. Schlögl, Methane activation by heterogeneous catalysis. Catal. Lett. **145**(1), 23–39 (2014)
44. P. Tang, Q. Zhu, Z. Wu, D. Ma, Methane activation: The past and future. Energy Environ. Sci. **7**(8), 2580–2591 (2014)
45. Y. Xu, S. Liu, L. Wang, M. Xie, X. Guo, Methane activation without using oxidants over Mo / HZSM-5 zeolite catalysts. Catal. Lett. **30**, 135–149 (1995)
46. A.A. Gabrienko, S.S. Arzumanov, M.V. Luzgin, A.G. Stepanov, V.N. Parmon, Methane activation on Zn2+ -exchanged ZSM-5 zeolites. The effect of molecular oxygen addition. J. Phys. Chem. C **119**(44), 24910–24918 (2015)
47. J. Xu, A. Zheng, X. Wang, G. Qi, J. Su, J. Du, et al., Room temperature activation of methane over Zn modified H-ZSM-5 zeolites: Insight from solid-state NMR and theoretical calculations. Chem. Sci., 3(10) (2012)
48. H. Xu, G. Luo, Green production of PVC from laboratory to industrialization: State-of-the-art review of heterogeneous non-mercury catalysts for acetylene hydrochlorination. J. Ind. Eng. Chem. **65**, 13–25 (2018)

Chapter 8
Mechanism Studies on Biofuel Conversion Under Methane Environment

8.1 Introduction

Biofuel is a generic terminology referring to the fuels derived from biomass, such as plant, algae, or biowaste. In contrast to traditional fossil fuels which are generated through long-term geological processes (which might take centuries), biofuel is regarded as a renewable energy resource since the feedstock can be regenerated easily in a limited period of time. However, like other natural resources such as coal, petroleum, and natural gas, the composition of biofuels is also complicated and generally varies from each other according to different sources. The complexity of biofuel constituents causes challenges and difficulties for better understanding of the reaction mechanism, since the evolved chemical reactions are of huge numbers and most of which are highly convoluted. Therefore, it is necessary to select several typical model compounds which could roughly represent the main components of biofuel, so that better understanding of the reaction pathways and overall control of the reaction processes can be realized. For example, an estimation of several crucial properties of the product oil including the gas/liquid/coke yield, the aromatic content, the oxygen content, the total acidic number, the viscosity, the density, the average molecular weight, and the yields of typical distillates can be achieved and used as a reference to control the upgrading process of real biofuel feedstocks. This chapter mainly focuses on the studies regarding mechanisms of biofuel conversion under methane environment, which is mainly embodied by the evolution of model compounds.

In spite of the complexity and diversity, the main components in biofuel and related products can be roughly classified into five groups, including (1) carboxylic acids such as formic, acetic, and propionic acids, which contain at least one carboxyl group (-COOH); (2) alcohols and phenols such as methanol, ethanol, ethylene glycol, and methyl phenol, which contain at least one hydroxyl group (-OH); (3) esters such as methyl formate, ethyl acetate, and butyrolactone, which contain at least one ester group (-COO-); (4) aldehydes such as formaldehyde, acetaldehyde, and ethanedial,

H. Song et al., *Methane Activation and Utilization in the Petrochemical and Biofuel Industries*, https://doi.org/10.1007/978-3-030-88424-6_8

which contain at least one aldehyde group (-CHO); and (5) ketones such as acetone, acetophenone, and diacetyl, which contain at least one carbonyl group (-CO-). However, it should be noted that multiple aforementioned groups can be present in one molecule, constructing more complex compounds such as furans (furan and its derivatives such as furfuryl alcohol and furfural), sugars (such as 1,6-anhydroglucose and acetol), and guaiacols (such as isoeugenol, eugenol and 4-methyl propyl syringol) [1]. According to the classification, it is clear that the presence of oxygen is one of the most important characteristics of biofuel, which deserves extra attention in relevant studies. Although all typical biofuel constituents are mainly composed of the same elements (C, H, and O), the reactivity during the upgrading process can be notably different. For this reason, a further identification of the controlling reactions for typical biofuel upgrading processes can be critical, which enables a general understanding and overall control of the real upgrading practices. Besides, mechanistic studies provide further guidance for the development of catalysts with better performances, such as higher conversion and selectivity toward desired products, as well as longer lifetime.

Based on the classification of model compounds in biofuel, the upgrading process could contain a bulk of reactions, which could be roughly derived into these steps: (1) the cracking of large molecules and deoxygenation as the first step; (2) some intermediates generated from the first step might undergo a series of reactions including secondary cracking to form even smaller radicals and molecules, oligomerization reaction towards polymers, and dehydrogenation reactions to form olefins and cycloalkanes; and (3) termination reactions to form relatively stable final products, such as alkylation to add side chain alkyl groups to a molecule, isomerization to reconstruct the chain structure, disproportionation to form highly unsaturated compounds (such as dienes), and saturated compounds (such as alkanes) through hydrogen transfer and condensation reactions to form addition products (such as esters) and small molecules (such as water) [2]. A large quantity of model compounds has been used to mimic the real biofuel conversion process so that the reaction mechanisms can be better understood for the modeling of practical reaction processes. Several typical compounds which are widely found in biofuel include but not limited to alcohols (such as ethanol), acids (such as acetic acid), furan derivatives (such as furfural), and the "basic units" of botanical biomass such as glucose, cellulose, and lignin. The corresponding investigation of these model compounds is summarized to illustrate their essential roles in biomass conversion process by proposing corresponding reaction pathways as follows.

8.2 Catalytic Conversion of Alcohols Under Methane Environment

Alcohols are relatively simple oxygen-containing molecules which are commonly reported as model compounds for biofuel upgrading process. The typical model compounds in this category consist of methanol, ethanol, and glycerol. Catalysts used for corresponding studies are mainly focused on catalyst with acid sites such as zeolite [3, 4]. The investigations of the simplest alcohol methanol over HZSM-5 catalyst suggested that olefins including ethylene and propene are generated through different pathways, and the selectivity towards ethylene and propene can be controlled by catalyst design [5]. It should be noted that the reaction of methanol is crucial for coal-based chemistry, since the feedstock methanol can be directly synthesized through syngas from coal. Therefore, corresponding studies such as MTO (methanol to olefins), MTG (methanol to gasoline), and MTA (methanol to aromatics) have been extensively explored. The most common methodologies for tuning the upgrading process can be the modification of the amount and strength of acid sites as well as the micropore and mesopore structures on the catalyst [5]. For ethanol, it is reported that the dehydration process is not mandatory for the formation of hydrocarbons with low carbon numbers. The influences of physical and chemical properties on the upgrading performance of biomass-derived ethanol over zeolite-based catalysts are comprehensively studied in previous work. The key points that control the catalytic activity and selectivity include the topology and surface structure of zeolite support, the impregnation of active metal components, the amount of weak and strong acid sites, and the reaction conditions [6, 7].

Among the solid fundamental work concerning alcohol conversion to useful hydrocarbons, the co-aromatization of ethanol and methane deserves extra attention. It has been proved that the addition of methane in ethanol upgrading process over Ag/ZSM-5 catalyst leads to better aromatization performance evidenced by higher aromatic yield [8]. During the process, methane can be directly incorporated into the final aromatic products. It is found that the well-dispersed silver nanoclusters on ZSM-5 support and the abundance of Bronsted and Lewis acid sites work together for the realization of the desired catalytic performances. This newly reported process builds up a new route for the co-utilization of ethanol derived from biomass and the natural gas resources, which are both economically favorable. Glycerol is another common alcoholic substance derived from biomass, and a further valorization process is preferred. It is found that the reaction of glycerol under methane environment leads to high yields of aromatics such as benzene, toluene, and xylenes [9]. Compared with the catalytic upgrading process under inert gas atmosphere, it is proved that the addition of methane not only results in the incorporation of methyl group into the final product but also leads to better dispersion of active metal components (i.e., Ag nanoparticles). The interesting behavior of methane on facilitating the dispersion of active metal sites deserves extra attention. Besides, it is proved that high crystallinity of zeolite support, sufficient amount of medium acid sites, and good metal dispersion are all crucial for the favorable catalytic performances. A synergy between methane and active metal species is also proposed, since the active metal sites are responsible for methane activation and the latter in turn results in higher activity and stability of metal nanoparticles.

The possible mechanism of alcohol conversion under methane environment is illustrated in Fig. 8.1. As described in Sect. 8.1, the first steps generally involve cracking and deoxygenation. For small alcoholic molecules such as methanol and ethanol, the deoxygenation reaction is generally realized through dehydration process, during which olefins with 2–4 carbon atoms and H_2O can be produced. Then, the formed olefin species undergo further carbon chain growth and hydrogen transfer process, during which more unsaturated species with higher carbon numbers (typically 6–8 C atoms) can be generated. In this step, methane participation can be

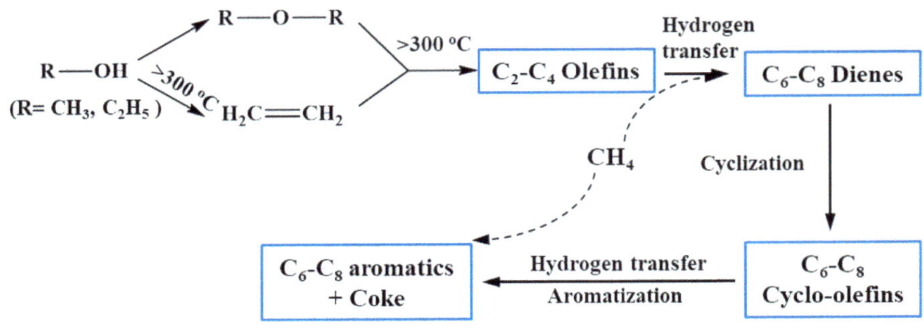

Fig. 8.1 Possible reaction pathway for ethanol transformation into higher hydrocarbons under methane. Reproduced with the permission from Elsevier

anticipated upon activation, which leads to more unsaturated hydrocarbons with higher carbon atoms. Once a considerable number of unsaturated hydrocarbons with suitable carbon chain length such as C_6–C_8 dienes are present, the cyclization process could proceed with the assistance of zeolite structure, leading to the formation of corresponding cycloolefins. Finally, these cycloolefin species could be converted into aromatic species easily on the acid sites of zeolite support through aromatization process, during which hydrogen transfer from acid site to the unsaturated cycloolefin molecule and subsequent dehydrogenation is realized. Once the aromatic species is formed, methane could also react with them through alkylation process, leading to side chain methyl substituents. From the proposed mechanism, it can be clearly seen than the presence of methane promotes the formation of C_6–C_8 aromatic species in different ways. These C_6–C_8 aromatics are also commonly called "BTEX families," which are of great value in petrochemical industry. Therefore, the addition of methane during the upgrading process could be an effective way for enhancing the added value of alcohols. However, despite the large quantity of theoretical studies on alcohol conversion under methane environment, there is still a gap between laboratory research and industrial application, which might be due to the lack of detailed process optimization and the resulting low profitability. In order to accelerate the industrialization process, careful optimization of reaction conditions, appropriate reactor design, and long-term catalyst stability test can be critical.

The positive effect of methane addition in ethanol upgrading process is validated in Fig. 8.2. Compared with a control experiment under N_2 environment, it is evident that the selectivity of aromatics increases dramatically from 13.5 to 86.8 wt.%, while the formation of undesired ether products deceases correspondingly. Besides, the ethanol conversion is also increased significantly. The excellent catalytic performances can also be deduced from the notably increase of water content from 2.8 wt.% to 74.9 wt.% in the final liquid product, indicating more oxygen atoms are successfully removed from the hydrocarbon chain with the presence of methane.

Besides, it is found that a reduction of Ag/ZSM-5 catalyst prior to the upgrading reaction leads to better catalytic performances, as shown in Fig. 8.3. It is observed that the ethanol conversion can be slightly increased, while the selectivity towards aromatic products can be further improved evidently, along with decreasing yield of liquid oxygenates under methane environment. These favorable results are even

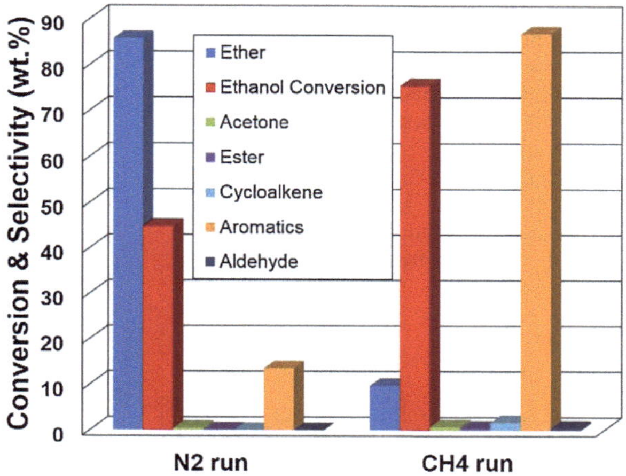

Fig. 8.2 Ethanol conversion and liquid product selectivity from the runs over 1%Ag/ZSM-5 at 400 °C under N_2 and CH_4 environments. Reproduced with the permission from Elsevier

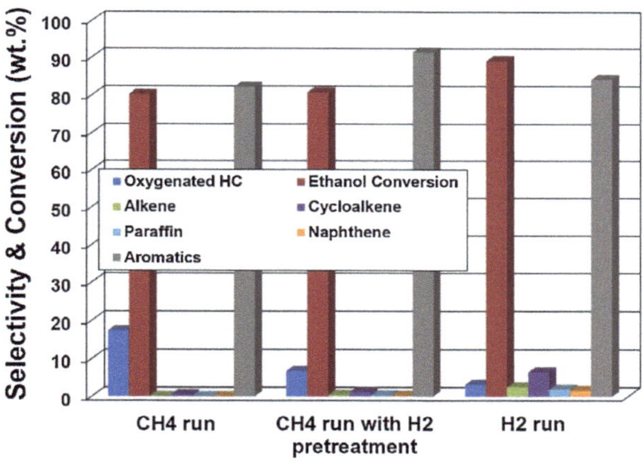

Fig. 8.3 Ethanol conversion and liquid product selectivity from the runs over 1%Ag/ZSM-5 with and without H_2 pretreatment at 400 °C under CH_4 and H_2 environments. Reproduced with the permission from Elsevier

comparable with those under H_2 environment, especially in terms of aromatic selectivity. Moreover, the yield of organic-phase product (oil) is also enhanced from 6.6 wt.% to 11.2 wt.% after the catalyst is treated by hydrogen in advance. These phenomena all indicate that the silver metal components are more effective at lower valances (i.e., metal form rather than oxide form).

In order to further confirm the participation of methane in ethanol upgrading process, the carbon number distribution of aromatic products in the liquid phase is analyzed and shown in Fig. 8.4. It can be seen that with the presence of methane, the carbon number distribution clearly shifts to the heavier end (more products with higher carbon number are formed), especially using the catalyst without hydrogen treatment. In fact, the increased carbon number can be mainly due to the higher number of side chain-substituted methyl groups according to the detailed compositional analysis. Therefore, it is clear that methane can participate in the reaction through alkylation reactions as side chain methyl substituent.

More direct evidences are derived from ^{13}C isotope-labelling experiments, as shown in Fig. 8.5, where the nuclear magnetic resonance (NMR) spectra of liquid

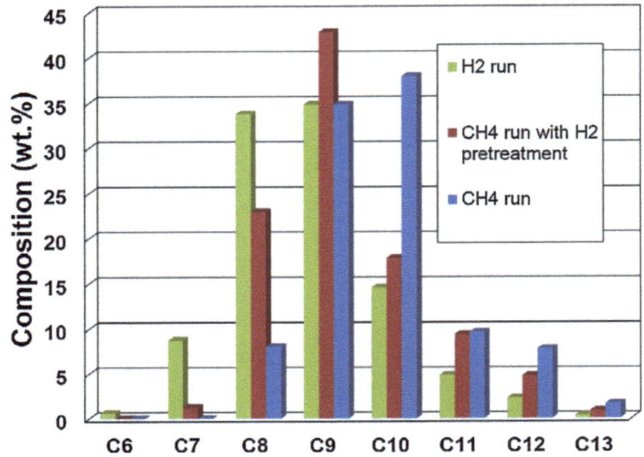

Fig. 8.4 The carbon number distribution in the aromatics portion of the liquid product collected over 1%Ag/ZSM-5 with and without H_2 pretreatment at 400°C under CH_4 and H_2 environments. Reproduced with the permission from Elsevier

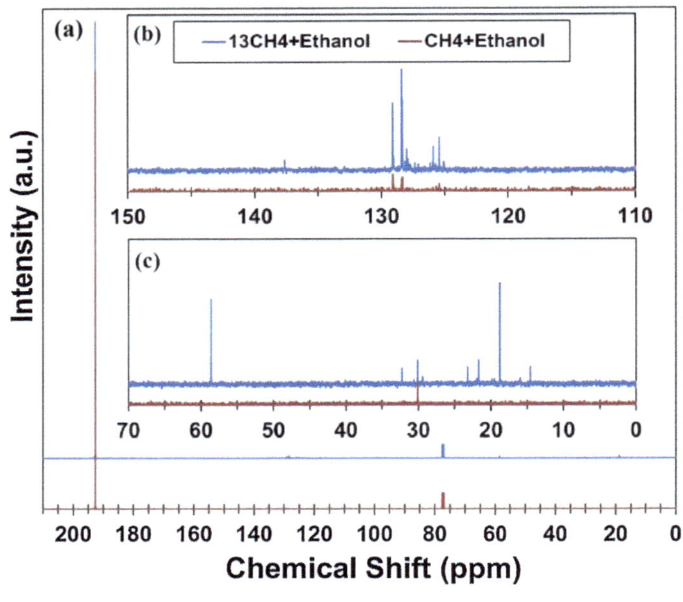

Fig. 8.5 (**a**) ^{13}C NMR spectra and its zoom-in views at the chemical shift region of (**b**) 110–150 ppm and (**c**) 0–70 ppm of the liquid products collected from ethanol conversion under $^{13}CH_4$ and CH_4 environments, respectively, at 400 °C and 1 atm. Reproduced with the permission from Elsevier

products using methane with concentrated ^{13}C and naturally distributed carbon isotopes are compared. The peak at 192.6 ppm can be attributed to carbon disulfide solvent used for extraction, and the peak at 77.2 ppm can be due to the addition of deuterochloroform used for calibrating chemical shifts. After excluding these two peaks, the detailed comparison of ^{13}C NMR peaks from products are shown in the inlets. It can be seen that the ^{13}C signal is increased evidently for both aliphatic (chemical shift at 15–30 ppm) and aromatic (chemical shift at 125–130 ppm) carbons. More specifically, the signals at 129.2 and 127.9 are designated to the carbon atom in the benzene ring on the ortho/meta positions with respect to the substituted alkyl group, and the peak at 125.4 ppm corresponds to the carbon atom in the aromatic ring at para site. Another peak at 30.5 ppm belongs to the benzylic carbon, and the signal at 58.6 can be attributed to the carbon bonded to an oxygen atom. Several other peaks at 30.5, 23.2, 21.7, 18.8, and 14.5 can be attributed to alkyl groups, and the different chemical shifts resulted from the atom or moiety they are linked to. The differences of intensity in all these observed peaks provide a strong evidence for the methane incorporation, which further explains the shifted carbon number distribution in Fig. 8.4.

To further confirm whether there is a synergy between the reaction of ethanol and methane, thermalgravimetric analysis (TGA) and differential scanning calorimetry (DSC) results of spent catalysts using different feedstocks are also collected and illustrated in Fig. 8.6. In a general process, the catalyst is firstly loaded in the aluminum crucible, and the mass is recorded. Then, the feedstock is fed, and the increased mass due to adsorption of reactants and the heat behavior are monitored. It is observed that the mass change and heat behavior of the catalyst with different feedstocks are easily differentiable. The main mass gain under nitrogen environment can be attributed to the adsorption of ethanol over the catalyst. After the addition of methane, the mass gain is increased notably, along with the drastically intensified heat flow, indicating more species are adsorbed (exothermic) and reacted (endothermic). It should be noted that the gained mass in entry ethanol + methane is much higher than the sum of values from the separate cases (i.e., ethanol and methane alone), indicating more species can be adsorbed with the presence of both two reactants. Besides, the enhanced endothermic heat behavior is also supported by thermodynamic calculations, since the co-aromatization of methane and ethanol has the reaction heat of 244.8 kJ mol^{-1}, which is much higher than that of the pure ethanol aromatization process (80.0 kJ mol^{-1}).

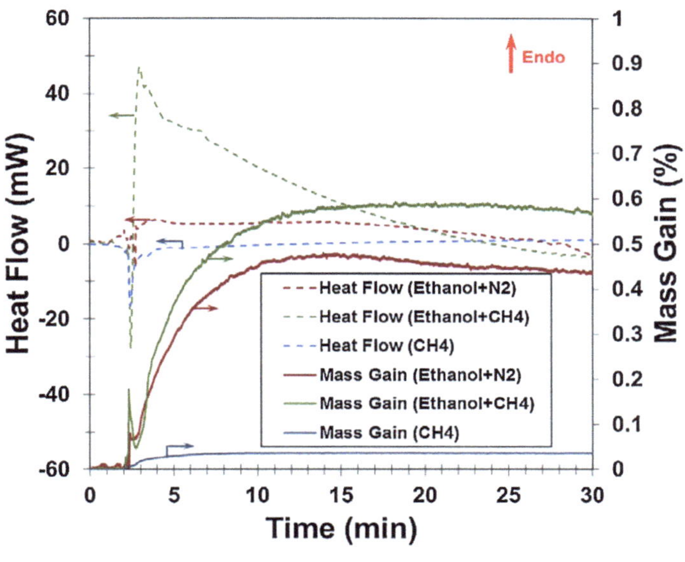

Fig. 8.6 TGA-DSC profiles of spent 1%Ag/ZSM-5 catalysts after the upgrading process with various feedstocks at 400 °C and 1 atm. Reproduced with the permission from Elsevier

Therefore, methane participation in the aromatization process can be confirmed. This synergistic effect between ethanol and methane is important not only for ethanol upgrading but also for methane utilization. It can be seen from the TGA curve of the pure methane entry that the mass gain is extremely limited, indicating methane cannot be effectively utilized alone under given reaction conditions, which can be due to the intrinsic inertness. However, after ethanol is co-fed, the amount of adsorbed methane can be evidently enhanced, suggesting the presence of other species might also facilitate methane utilization considerably. This concept is not only applicable in the case of ethanol upgrading but also crucial for the valorization of other feedstocks under methane environment, which deserves extra attention.

The characterizations of catalysts employed in this study are also performed, which provides valuable understanding of the active sites in this catalytic process. The in situ diffuse reflectance infrared Fourier transform spectroscopy (DRIFTS) results in Fig. 8.7 demonstrate that only peaks at 1300–1400 cm^{-1} are detectable. These peaks can be attributed to C-O stretching and C-H bending of adsorbed ethanol molecules. However, no characteristic aromatic peaks at 1500–1700 cm^{-1} can be detected, which might be explained by the formed aromatic products that are mainly in the inner pores of zeolite structure and are thus difficult to be detected on the surface by DRIFTS technique.

X-ray diffraction (XRD) patterns of several catalysts dedicated for ethanol upgrading process are shown in Fig. 8.8. No peaks other than the characteristic ones from the zeolitic structures can be observed, suggesting the loaded metal components are well dispersed into fine particles, whose crystalline peaks cannot be detected by XRD technique. Moreover, the position and intensity of all peaks derived from

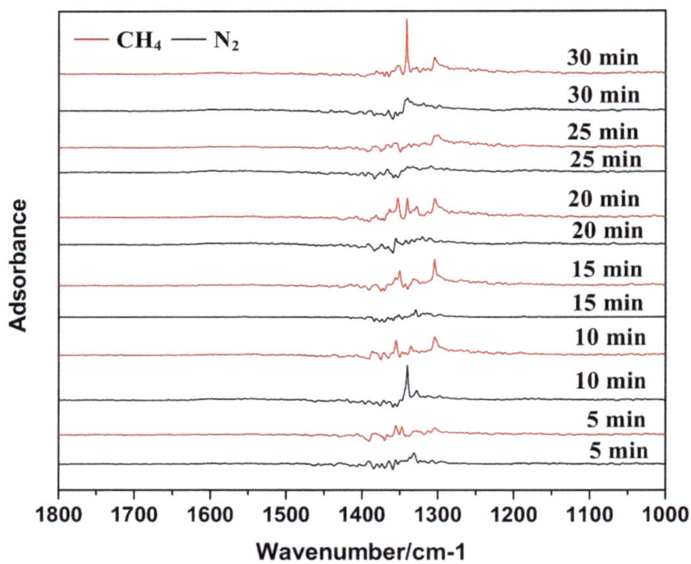

Fig. 8.7 DRIFTS spectra collected at different time over 1%Ag/ZSM-5 during ethanol conversion under various environments at 400 °C and 1 atm. Reproduced with the permission from Elsevier

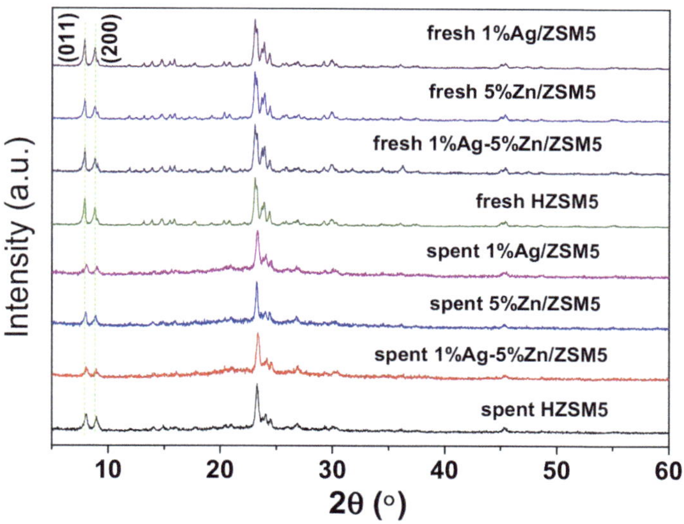

Fig. 8.8 XRD patterns of fresh and spent HZSM-5, Zn/ZSM-5, Ag-Zn/ZSM-5, and Ag/ZSM-5 catalysts. Reproduced with the permission from Elsevier

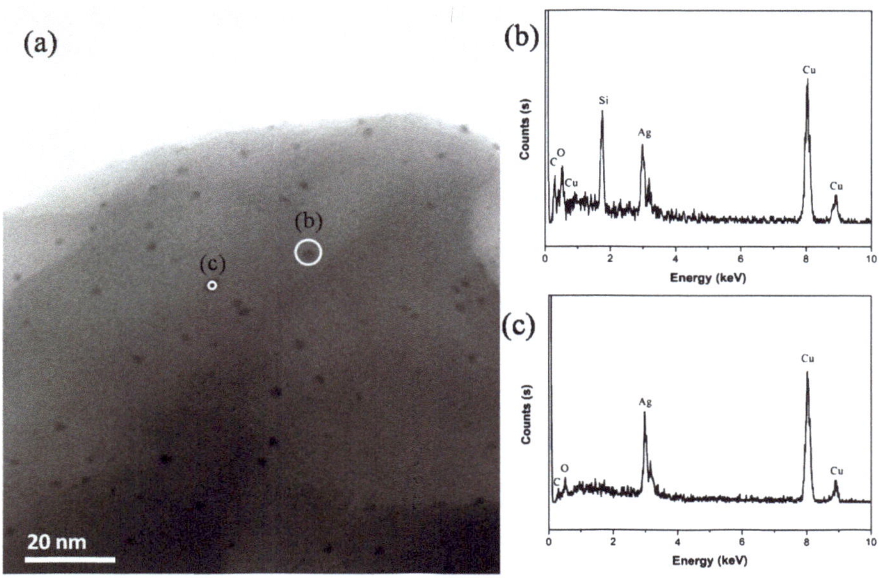

Fig. 8.9 TEM image of fresh Ag/ZSM-5 (a) catalyst and (b, c) the corresponding EDX spectra of the areas circled in (a). Reproduced with the permission from Elsevier

zeolite framework are similar in all fresh and spent catalysts, indicating the favorable stability of the employed catalytic structures in the reaction process.

The conclusion regarding metal dispersion is further substantiated by the transmission electron microscope (TEM) images, as shown in Fig. 8.9. It is observed that the silver metal components (the elemental composition is confirmed by energy dispersive X-ray spectroscopy (EDX) spectra) loaded on the zeolite support are small particles with the average diameter of ~2 nm. However, the agglomeration of Ag particles can be observed as illustrated in Fig. 8.10, which could be unfavorable for the stability of Ag/ZSM-5 catalyst in long-term operation.

The acidity of several typical catalysts is further evaluated by DRIFTS through pyridine adsorption, as shown in Fig. 8.11. It can be seen that HZSM-5 and Ag/ZSM-5 have abundant Lewis and Bronsted acid sites, while the amount is decreased after Zn addition. The surface acidity is believed to be closely related to the catalytic performances. As summarized in Table 8.1, the catalyst with higher acidity leads to higher ethanol conversion and aromatic selectivity, accompanied with the disadvantages including low liquid yield and high coke formation rate. Therefore, a balanced acidity of catalyst can be crucial for the catalytic co-aromatization of ethanol and methane.

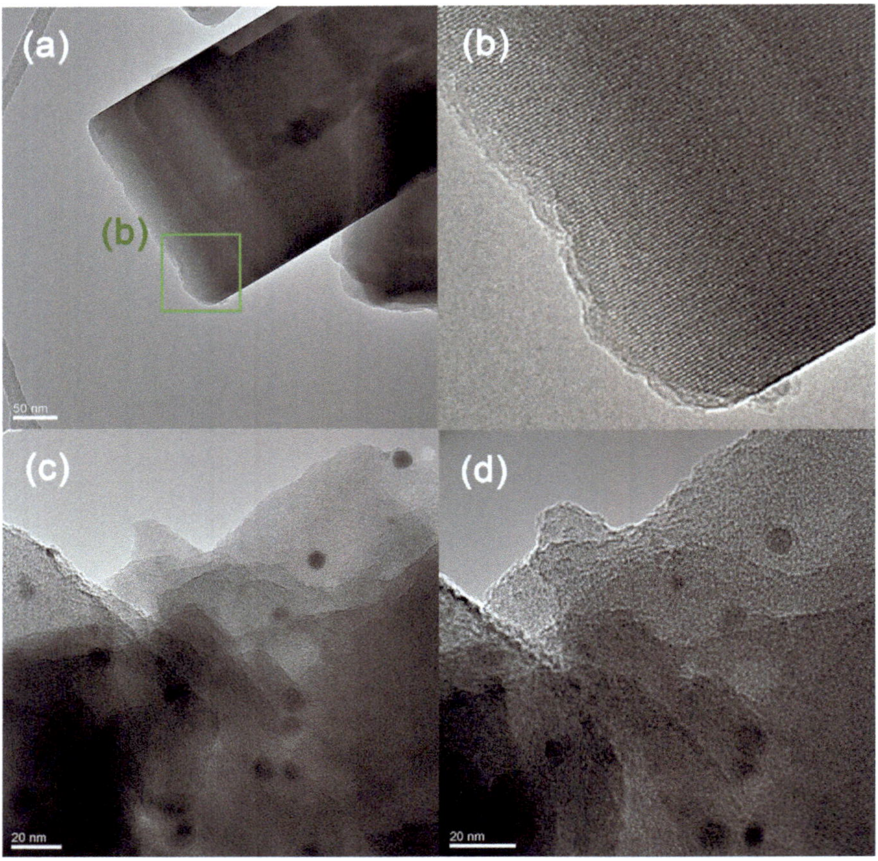

Fig. 8.10 TEM images of spent Ag/ZSM-5 catalyst (a) with ZSM-5 support focused, (b) zoom-in view of the squared area in (a), (c) and (d) with active metals focused at different resolutions. Reproduced with the permission from Elsevier

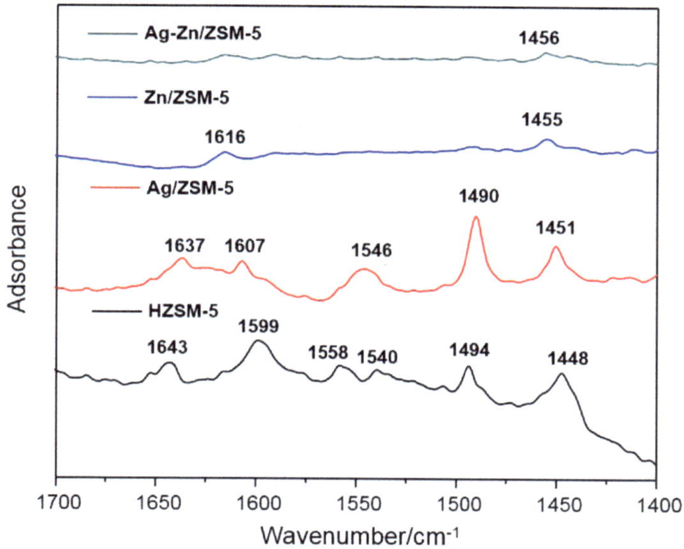

Fig. 8.11 DRIFTS spectra of several typical catalysts after pyridine adsorption for ethanol upgrading process. Reproduced with the permission from Elsevier

Table. 8.1 Catalytic performances and corresponding coke formation rate of several catalysts for ethanol upgrading under methane environment

Catalyst	Ethanol conversion (%)	Liquid yield (wt.%)	Aromatic selectivity (wt.%)	Coke formation rate (10^{-5} s^{-1})
HZSM-5	46.19	33.52	75.38	1.51
Ag/ZSM-5	75.43	35.08	86.85	3.41
Zn/ZSM-5	6.56	41.08	0.94	1.09
AgZn/ ZSM-5	44.20	69.12	0.86	1.12

8.3 Catalytic Conversion of Oxygenates with Carbonyl Group Under Methane Environment

Carbonyl (-C=O) is another type of moiety which is widely present in biomass. Due to the high polarity of carbonyl group, especially carboxylic acids, their existence in biofuel generally leads to undesirable physical and chemical properties, including, but not limited to, high viscosity, corrosiveness (high total acid number and low pH), low energy density, poor stability, and incompatibility with other petroleum fractions [10]. Therefore, the removal of species with these carbonyl groups could be greatly helpful for improving the quality of biofuel products. The simplest molecules with carbonyl group consist of acetic acid, formaldehyde, and acetone, which are used as model compound for corresponding studies.

Although the conversion of acetic acid into alcohols or alkanes through a hydrodeoxygenation process is already relatively mature, the reaction is still energy extensive with low profitability [11, 12]. It is also reported that acetic acid can be converted into hydrogen through steam reforming reaction, while the stringent reaction conditions are still unfavorable [13]. If acetic acid can be directly deoxygenated and useful hydrocarbons can be produced under milder conditions, the economic and environmental friendliness can be greatly improved. It has been reported that the co-feeding of methanol and acetic acid over ZSM-5 zeolite leads to desired high value-added product such as higher hydrocarbons, indicating the synergy between the two feedstock and providing a new route for the effective utilization of biomass feedstocks [14]. Meanwhile, the upgrading of acetic acid under methane environment also demonstrates promising results [15]. It is found that methane can be effectively activated by the loaded silver-zinc active metal components, interacting with acetic acid and producing desirable aromatic products. Liquid NMR results suggest that methane could be preferentially incorporated as side chain-substituted aliphatic groups. Further mechanistic investigation indicates that the acetic acid can be firstly cracked into free hydrocarbon radicals such as methyl ($CH_3\cdot$), ethyl ($CH_3CH_2\cdot$), and methylene ($\cdot CH_2\cdot$) groups through decarboxylation and decarbonylation process, during which the carbonyl group is removed from acetic acid as carbon oxides [16]. These produced hydrocarbon radicals are highly active, and molecules with higher carbon numbers can be generated through chain extension caused by their self-combination. After a carbon chain with suitable carbon atoms (~6–8) is formed, the ZSM-5 zeolite structure is responsible for a subsequent aromatization process, which is similar to that in the case of alcohol upgrading. Then, these aromatic products can further react with hydrocarbon or hydroxyl free radicals to form substituted aromatic species such as alkylbenzene and phenol derivatives.

The catalytic valorization of acetone has also been investigated under methane environment. Similarly, it is found that methane can be incorporated into the final liquid products, enhancing liquid yield, especially the yields of aromatics, which is evidenced by isotope-labelling experiments and NMR spectroscopy [17]. It is suggested that the main factors for the excellent selectivity of desired aromatic species (such as BTEX) could be the well-dispersed active metal nanoparticles in the inner pores of zeolite structure, the suitable amount of weak, medium and strong acid sites, and the maintained chemical environment of active metal components. It is again proved that methane is not only functioning as a reactant to be incorporated into final products but also demonstrating synergy with loaded metal components in terms of facilitation of metal dispersion and stabilization of formed active structures. Besides, the structure of zeolite is also investigated, and it is believed that the hierarchical porosity established by the micropores of zeolite and additional mesopore and macropore structures caused by the morphology of the support is essential for the high selectivity of favorable aromatic species (i.e., BTEX species) and correspondingly low selectivity of aromatic molecules with higher (>9) carbon atoms. This phenomenon is further explained by the relative activity of different pathways derived from the modulated amount and strength of acid sites and shape selectivity caused by the control of zeolite structure and morphology [18]. More specifically, the zeolite structure with high Bronsted acidity (which directly donates protons) leads to the high conversion of acetone, while the selectivity is effectively tuned by the modification of zeolite structure through ion exchange process or addition of extra metal components [19]. A reaction mechanism is also proposed as shown in Fig. 8.12. Since acetone is detected as an intermediate in the upgrading process of acetic acid, it is proposed that the first step for the conversion of acetic acid could be a ketonization process, during which two acetic acid molecules are combined to form one acetone molecule and the remaining carbon and oxygen atoms form carbon oxides (CO_2 or CO). This pathway is supported by the presence of carbon oxide in the composition analysis of product gas. Then, starting from acetone, two different pathways are considered. The first route is a direct condensation and dehydration process, forming short-chain olefins (such as isobutene). These species can be further converted into long-chain olefins (such as C_8 olefin) through self-combination. The unsaturated species with suitable carbon numbers could undergo a cyclization process after diffusing into the inner pores of zeolite structure, forming corresponding naphthenes (cycloalkanes). In this step, activated methane could play a role for the connection of short carbon chains and compensation of appropriate carbon number for the formation of C_6–C_8 cycloalkanes. These cycloalkane species could be further converted into aromatics through hydrogen transfer, during which the dehydrogenation of naphthenes towards cycloolefin could occur and hydrogen (H_2) could be produced. The cycloolefin could be further dehydrogenated to form aromatics, and the produced hydrogen molecules could be used to saturate other present olefin species (alkenes) to produce saturated species (alkanes). The second route is also a well-known mechanism with the isophorone intermediate [20]. In this route, two acetone molecules can be combined together through an aldol condensation and subsequent β-scission, generating one mesityl oxide molecule and a water molecule. Then, mesityl oxide could interact with another acetone molecule to form isophorone, which is also an important chemical product used as solvent and polymer precursor. With the presence of zeolite support, isophorone can be further converted into phenol derivatives through dehydrogenation. In this step, activated

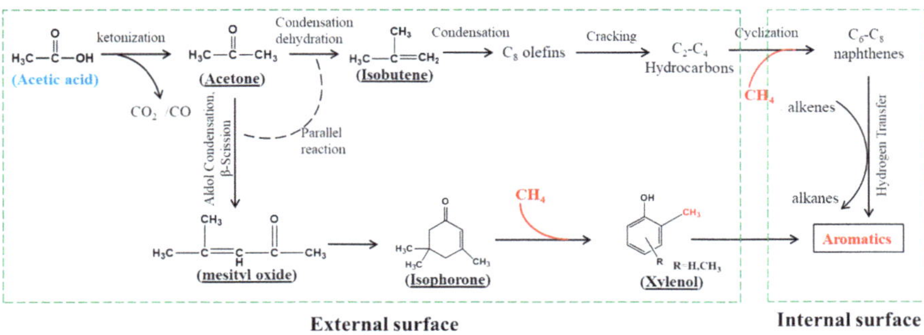

Fig. 8.12 Possible reaction pathways for upgrading acetic acid/acetone under methane environment. Reproduced with the permission from Elsevier

methane could react with the phenol ring through alkylation, adding a side chain methyl substituent to the molecule. Besides, phenol derivatives could also be dehydrated into aromatics with the presence of free hydrogen radicals. From the mechanism, it can be deduced that the deprivation of carbonyl group is essential for the formation of hydrocarbon species. Therefore, the timely removal of produced oxygen containing molecules such as carbon oxides and water could be useful for high conversion and selectivity towards favorable products.

The reaction performances of several catalysts for upgrading acetic acid under methane environment is shown in Fig. 8.13. Firstly, it is apparent that the employed catalysts lead to significantly higher conversion of acetic acid, liquid yield, and methane conversion compared with the no catalyst entry, indicating the important role of catalysts. Among a series of catalysts, the function of each metal component can be concluded as follows: (1) silver results in higher liquid yield; (2) the coke yield is lower over zinc-based catalysts; and (3) methane conversion can be further enhanced, and the coke formation can be inhibited to a certain extent after the introduction of gallium. The advantageous effects of metal components greatly improve the feasibility of the valorization process, which deserve further investigation.

The detailed gas and liquid product distributions of the abovementioned catalysts are illustrated in Figs. 8.14 and 8.15, respectively. In general, over 85 wt.% high aromatic selectivity in the liquid phase is observed over all charged catalysts, which can be easily separated and further used in petrochemical processes. The main gas products are carbon dioxide and C_3 species, suggesting the reaction should proceed through decarboxylation and ketonization-dehydration. During the ketonization process, a small amount of acetone could be produced, which is also detected in the liquid phase. Besides, a considerable yield of phenol derivatives can be explained

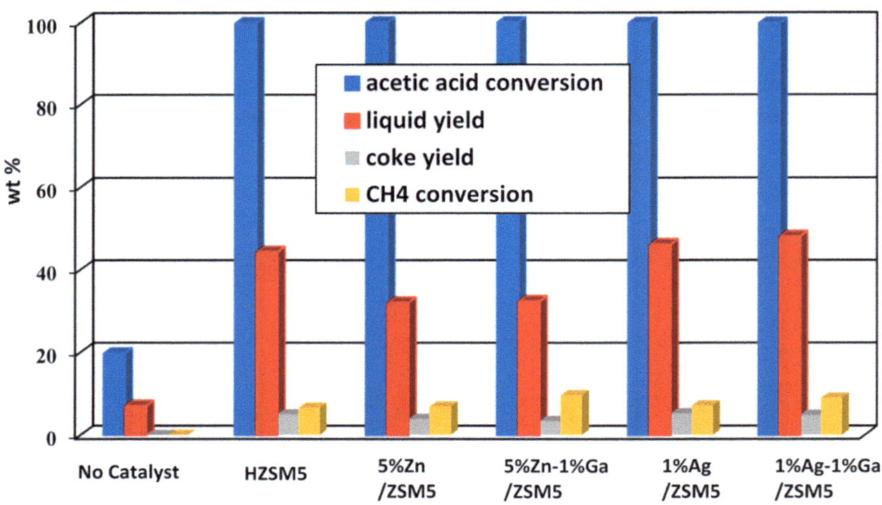

Fig. 8.13 Acetic acid conversion, liquid yield, coke yield, and methane conversion obtained over various catalysts or without a catalyst in acetic acid upgrading process under methane environment. Reproduced with permission from Reference [15]. Copyright (2017) American Chemical Society

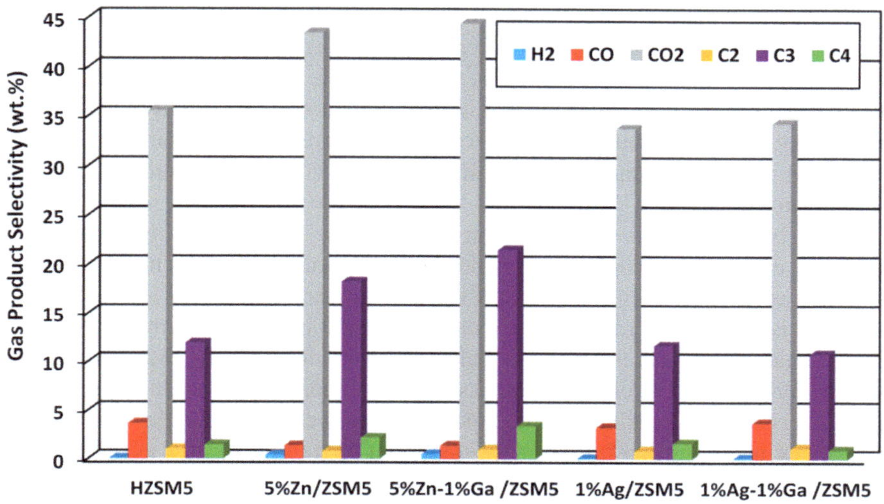

Fig. 8.14 Selectivity of gas products collected over various catalysts in acetic acid upgrading process under methane environment. Reproduced with permission from Reference [15]. Copyright (2017) American Chemical Society

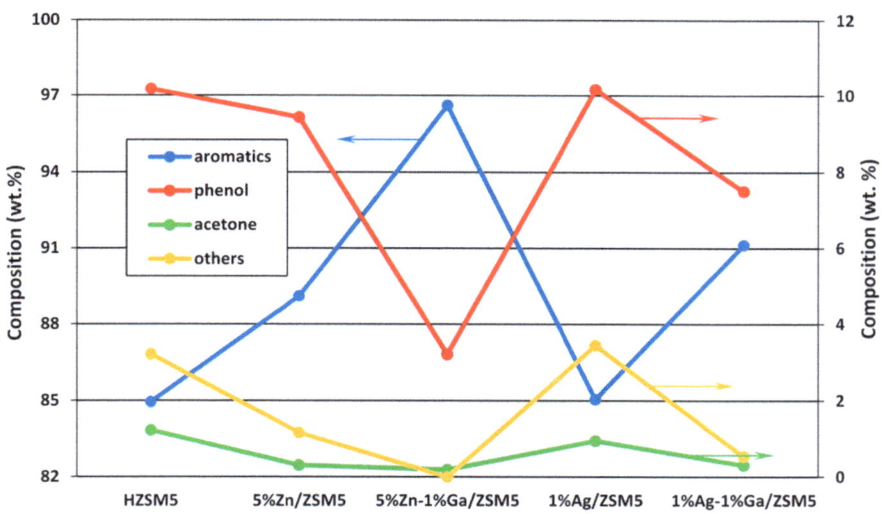

Fig. 8.15 The composition of liquid products collected over various catalysts in acetic acid upgrading process under methane environment. Reproduced with permission from Reference [15]. Copyright (2017) American Chemical Society

by the formation and evolution of isophorone, which is derived from acetone through a well-known mechanism [20]. Other products only account for a small fraction in the final liquid product, further demonstrating the promising selectivity of the upgrading process.

The carbon number distribution of liquid products is compared in Fig. 8.16. After a comprehensive evaluation of catalytic performances shown in Figs. 8.13, 8.15, and 8.16, 5%Zn-1%Ga/ZSM-5 is selected as the preferred catalyst, due to the formation of more aromatics with 6–8 carbon numbers (BTEX family), less phenol selectivity, low coke production, higher methane conversion, and compromised liquid yield, which demonstrates highest potential for industrialization among the investigated catalysts.

The advantages of methane addition become more evident after the comparison between experiments carried out under methane and unreactive nitrogen atmospheres, as shown in Figs. 8.17 and 8.18. The aromatic selectivity is higher when methane is present regardless of the charged catalyst, indicating methane facilitates the aromatization process towards valuable aromatic products. Meanwhile, the selectivity of phenol is reduced in all cases after methane introduction. Since phenol is another oxygen-containing compound with low energy density, the inhibition of phenol formation further improves the quality of upgrading product as biofuel.

To obtain direct evidences for methane participation, isotope-labelling experiments are performed, and the NMR spectra of liquid products under $^{13}CH_4$ and natural CH_4 are shown in Fig. 8.19. An apparent enhancement of ^{13}C signal intensity is observed in the case under $^{13}CH4$ and eloquently proves the incorporation of methane in the final liquid products. A zoomed in comparison in Fig. 8.19b regarding the peak of phenyl and benzyl carbon atoms (chemical shift at 130.12, 129.80, 129.17,

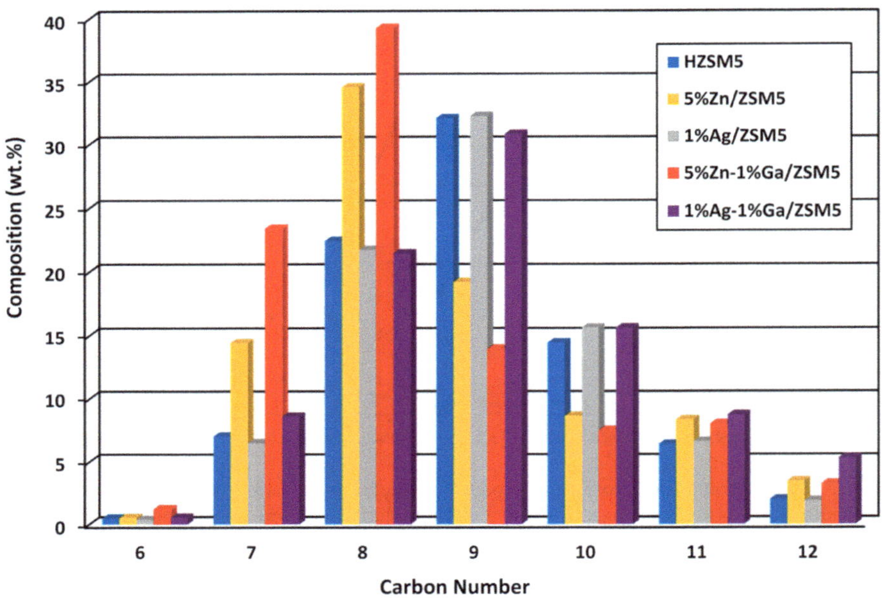

Fig. 8.16 Carbon number distribution in the aromatic portion of the liquid products collected over various catalysts in acetic acid upgrading process under methane environment. Reproduced with permission from Reference [15]. Copyright (2017) American Chemical Society

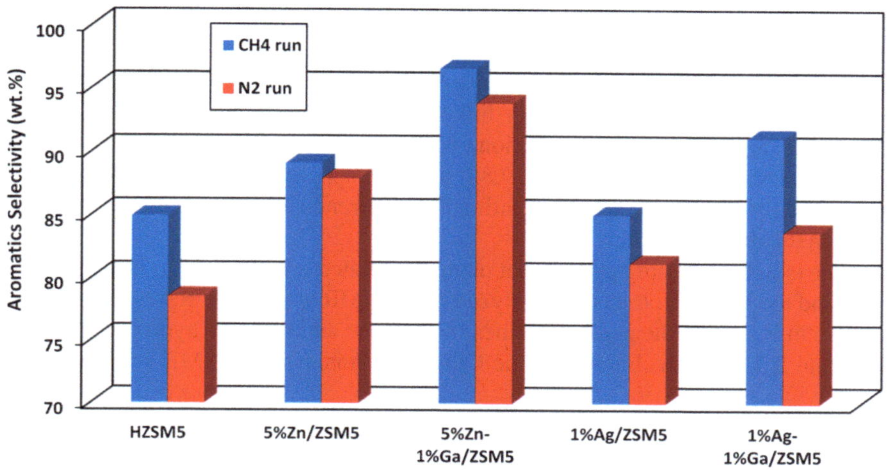

Fig. 8.17 Aromatics concentration in liquid collected from the runs over various catalysts in acetic acid upgrading process under methane and nitrogen environment. Reproduced with permission from Reference [15]. Copyright (2017) American Chemical Society

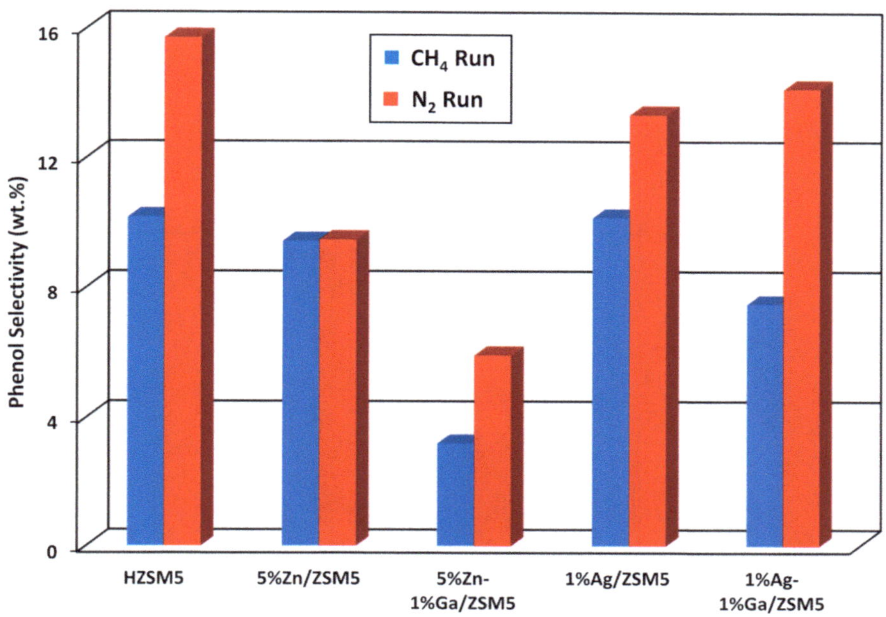

Fig. 8.18 Phenol selectivity in liquid collected from the runs over various catalysts in acetic acid upgrading process under methane and nitrogen environment. Reproduced with permission from Reference [15]. Copyright (2017) American Chemical Society

128.48, 126.26, and 125.56 ppm) suggests that the carbon from methane could end up directly in the aromatic ring of the liquid aromatic species. However, the intensities of peaks at 37.55 and 134.30 ppm are relatively lower, indicating the carbon in the aromatic ring which is directly linked by a side chain alkyl group that is not favorable for methane incorporation. Moreover, the peaks of aliphatic groups (chemical shift at 29.45, 21.50, 19.85, and 15.94 ppm) are increased even more significantly in the $^{13}CH_4$ run, indicating the preference of methane incorporation as side chain aliphatic groups.

The isotope-labelling experiment using deuterated methane (CD_4) is also carried out, and the liquid products are analyzed by ^1H NMR and ^2H NMR, respectively. As shown in Fig. 8.20, the peak intensities of aromatic and benzylic hydrogen (7.6, 7.1, 2.8, and 2.4 ppm) decline under the CD_4 environment in ^1H NMR while increase

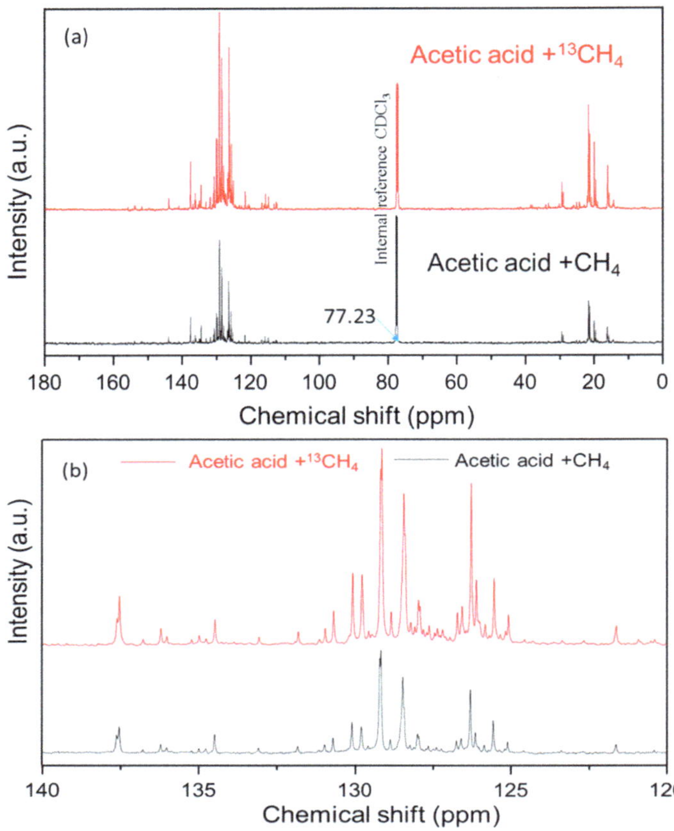

Fig. 8.19 ^{13}C liquid NMR spectra (**a**) and enlarged views in the 120–140 ppm chemical shift region (**b**) of the liquid products collected from the reaction between acetic acid and $^{13}CH_4/CH_4$. The peaks are calibrated at 77.23 ppm as the chemical shift of internal reference deuterochloroform. Reproduced with permission from Reference [15]. Copyright (2017) American Chemical Society

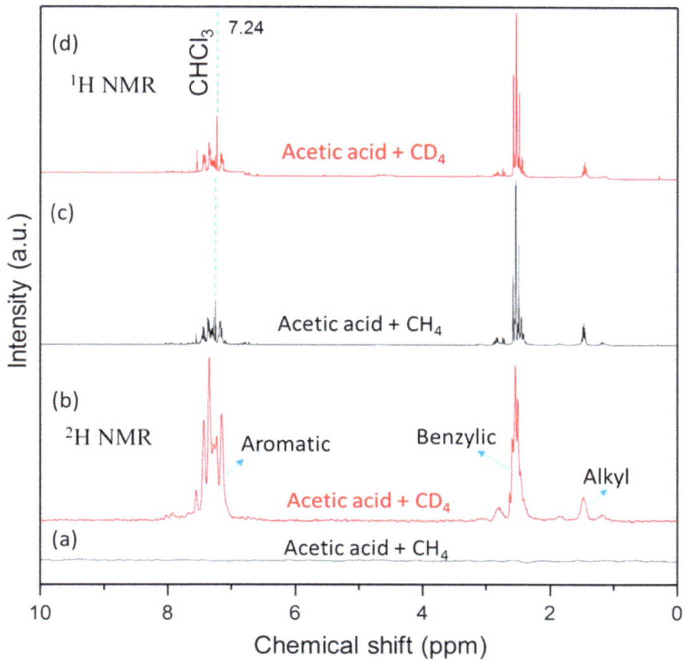

Fig. 8.20 ^{2}H NMR spectra (**a**, **b**) and ^{1}H NMR spectra (**c**, **d**) of the liquid products collected from the reaction between acetic acid and CH_4/CD_4. The peaks are calibrated at 7.24 ppm as the chemical shift of internal reference deuterochloroform. Reproduced with permission from Reference [15]. Copyright (2017) American Chemical Society

drastically in ^{2}H NMR, suggesting that a large portion of these hydrogen atoms come from methane. However, the change of alkyl groups which are not directly linked to aromatic ring (<2 ppm) are relatively insignificant, indicating these hydrogen atoms more likely come from acetic acid.

More insights can be acquired in the experiments using two types of isotope-labelled acetic acids, namely, $^{13}CH_3COOH$ and $CH_3{}^{13}COOH$. It is clear that the signal of ^{13}C in liquid NMR increases drastically after natural acetic acid is replaced by an isotope-labelled counterpart, indicating the involvement of acetic acid in the final liquid products. More specifically, if $CH_3{}^{13}COOH$ is used as the feedstock, the signals of phenyl carbon with direct linked side chain substituent (chemical shift at 137.55 and 134.30 ppm) and benzyl carbon (the carbon atom directly linked to aromatic ring, chemical shift at 39.23, 21.5, and 19.85 ppm) are intensified the most, suggesting the preference of carboxyl group towards the formation of these groups. Meanwhile, the methyl group in acetic acid does not demonstrate this tendency, which is better incorporated in the aromatic ring without connecting a side chain group (chemical shift at 130.12, 129.8, 129.17, 128.38, 126.26, and 125.56 ppm). Besides, comparing the signal enhancement after adding isotope-labelled methane (~2 times higher) and acetic acid (~10 times higher), it can be concluded that the carbon in the final products mainly comes from acetic acid and the carbon in methane contributes a small portion (Fig. 8.21).

Further characterizations of the optimized catalyst are carried out to confirm the active sites. The XPS results are shown in Fig. 8.22. The 2p spectra of Ga indicates that it is mainly preserved in oxide form (Ga_2O_3) throughout the reaction with high stability. According to Zn 2p spectra, it is also confirmed that Zn is present in the form of ZnO. However, the peak intensity in spent catalyst is reduced under methane environment while enhanced under nitrogen environment. This phenomenon suggests that the zinc oxide in fresh catalyst is not immutable. It is subject to agglomeration under inert atmosphere, while is better dispersed with the presence of methane. The possible effect of methane for facilitating the dispersion of active metal components deserves extra attention, which could be another advantage of using methane as the feed gas beyond a reactant. This conclusion is further substantiated by O 1s spectra, in which the peak intensity is decreased under nitrogen atmosphere while increased under methane. The binding energy 532.55 eV corresponds to the oxygen atom in SiO_2 from zeolite structure, and the increase of peak intensity can be explained by the better exposure of skeleton O atom to the external surface which resulted from a better metal dispersion, consisting with Zn 2p results. An

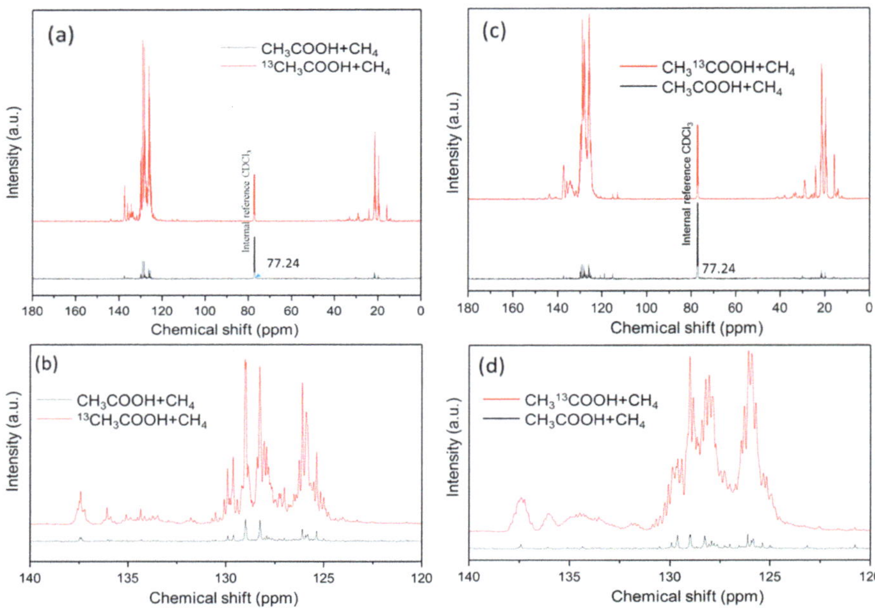

Fig. 8.21 ^{13}C liquid NMR spectra (**a, c**) and an enlarged view of the 120–140 ppm chemical shift region (**b, d**) of the liquid products collected from the reactions between $^{13}CH_3COOH/CH_3{}^{13}COOH$ and CH_4. Reproduced with permission from Reference [15]. Copyright (2017) American Chemical Society

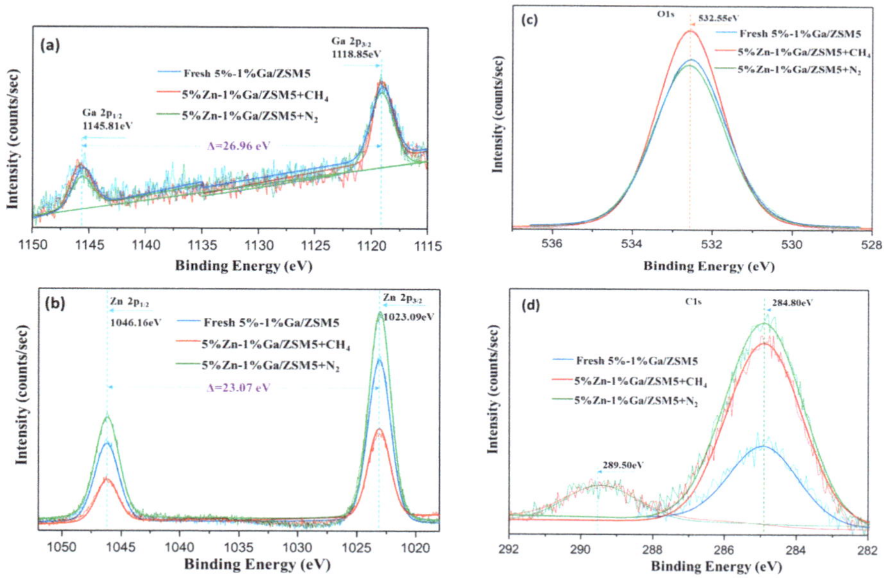

Fig. 8.22 XPS spectra of 5%Zn-1%Ga/ZSM-5 before and after reactions under different environments in the (**a**) Ga 2p, (**b**) Zn 2p, (**c**) O 1s, and (**d**) C 1s regions. Reproduced with permission from Reference [15]. Copyright (2017) American Chemical Society

evident intensified carbon peak is observed in spent catalysts in C 1s spectra. Nevertheless, the peak height is lower after the catalyst is used in methane than that in nitrogen, indicating less coke is formed when methane is present. This tendency is also beneficial for the stability of catalyst in long-term operations.

The mechanism of the reaction is further investigated through a series of reactions with controlled reaction times, which is realized through quenching the reactor in water after a certain period of time. The results are shown in Figs. 8.23 and 8.24. At the beginning (<6 min), a rapid generation of acetone, C_4 (mainly contains isobutene), and CO_2 is observed, which could be the primary intermediates derived from ketonization, condensation, and dehydration, as illustrated in Fig. 8.12. Along with the reaction process, the concentration of aromatics increases monotonously, while the amounts of C_4 in gas phase and acetone in liquid phase are decreasing. Besides, the concentration of phenol peaks at 3 min and declines rapidly afterwards. To sum up these evidences, it can be concluded that acetone, isobutene, and phenolics are all possible intermediates for the formation of final aromatic species. However, the mechanism behind could be different due to the varied structures and chemical properties. Besides CO_2, the final gas product mainly contains C_3 species, which might come from the cleavage reaction of carbon chain during the co-aromatization process as by-products.

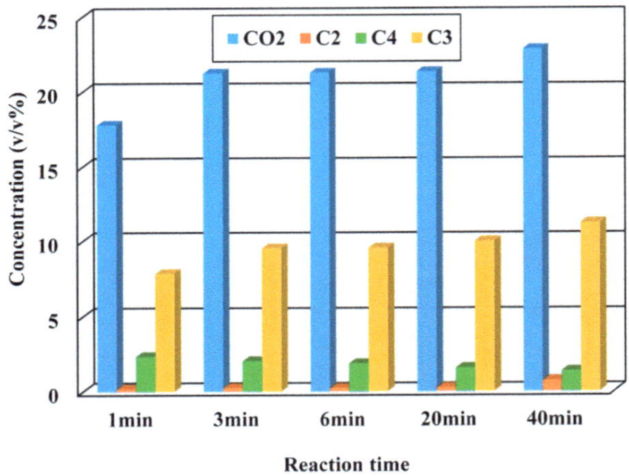

Fig. 8.23 The concentration of CO_2 and hydrocarbons in the gas phase collected after various reaction times in acetic acid upgrading process under methane environment. Reproduced with permission from Reference [15]. Copyright (2017) American Chemical Society

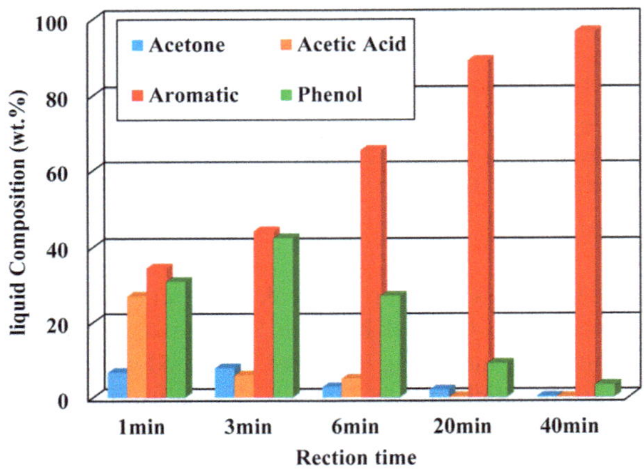

Fig. 8.24 The composition of liquid products collected after different reaction times in acetic acid upgrading process under methane environment. Reproduced with permission from Reference [15]. Copyright (2017) American Chemical Society

8.4 Catalytic Conversion of Phenolics and Lignin Under Methane Environment

Phenolic compounds such as phenol, cresol, anisole, and guaiacol are commonly found as the pyrolysis products derived from lignin [18, 21–29]. Correspondingly, lignin can be regarded as an intricated polymer of phenolics. It is widely observed in vascular plants and some microbes as the structure supporting material. It can be deduced from the function of lignin that it is a kind of rigid substances which cannot be decomposed easily, making its utilization difficult. Several techniques have been reported for effective lignin conversion, such as hydrodeoxygenation under hydrogen atmosphere [30–34] . However, due to the high cost of hydrogen, this route is regarded as economically unfavorable, and the replacement of H_2 by other reductant (hydrogen donor) is preferred. There have been successful examples of these hydrogen donors such as propanol, tetralin, decalin, and naphthalene [35, 36]. Methane, as an inexpensive hydrogen donor, could be an ideal choice for the upgrading of lignin. It has been proved that methane could demonstrate comparable deoxygenation performance as hydrogen for guaiacol valorization [29, 37]. This study greatly boosts the confidence of methane utilization on lignin upgrading process. During the process, the effective ways for lignin decomposition could be critical for its effective conversion. Therefore, how to weaken the C-O bond becomes the focus of research, which is generally realized by adding electron-withdrawing group or saturating the aromatic ring, so that the C-O bond dissociation energy can be decreased. In the case of lignin conversion under methane environment, it is reported that the presence of methane leads to several advantages including higher lignin conversion, better yield of aromatics, and higher selectivity towards BTEX species [38]. The employed catalyst mainly contains zeolite structure impregnated by zinc-silver active metals. Similar to the upgrading process of aforementioned alcohol and acetone, the function of catalyst can be roughly divided into two parts. On the one hand, active metal species are used for methane activation, which can be further involved in deoxygenation of phenolic species as hydrogen donor. On the other hand, zeolitic structure is responsible for the formation of aromatics after the suitable hydrocarbon chain structure is generated. The participation of methane in the reaction process is also proved by the comparison of NMR data using natural and ^{13}C isotope-labelled methane as feed gas. It is found that methane can be incorporated not only in side chain alkyl groups but also in the aromatic ring, indicating it is interacting with the decomposition products of lignin in many ways, including the alkylation process, to form substituent methyl group on the aromatic ring and the co-aromatization process to produce more aromatic products. These effects positively influence the lignin upgrading process compared to thermal pyrolysis, which significantly improve the economic and energy effectiveness and are thus promising for industrial applications.

The possible mechanism of lignin upgrading process under methane environment is shown in Fig. 8.25. It should be noted that lignin must be firstly degraded into phenolics, which can be further converted into other useful hydrocarbons with the assistance of catalyst and methane. Phenol is used as a representative for clarifying the reaction pathways in Fig. 8.25. In fact, several routes have been proposed for phenol deoxygenation under hydrogen environment, including saturation-dehydration, direct deoxygenation, and tautomerization [39–49]. However, saturation-dehydration is less possible under methane environment, since the direct hydrogenation of aromatic ring is a mandatory first step in this pathway, while the reaction is highly thermodynamically unspontaneous when methane is used as the hydrogen donor. Therefore, two main pathways are considered for phenol upgrading with the presence of methane. The first pathway is the direct deoxygenation of phenol and its derivatives such as cresol, leading to the formation of aryl radicals. It has been discussed repeatedly that methane is capable to participate in alkylation process when a molecule with aromatic structure is present. Therefore, it is not surprising that methane can also react with phenol to form alkyl-substituted derivatives in this step, which can be deoxygenized into aryl radicals with side chain

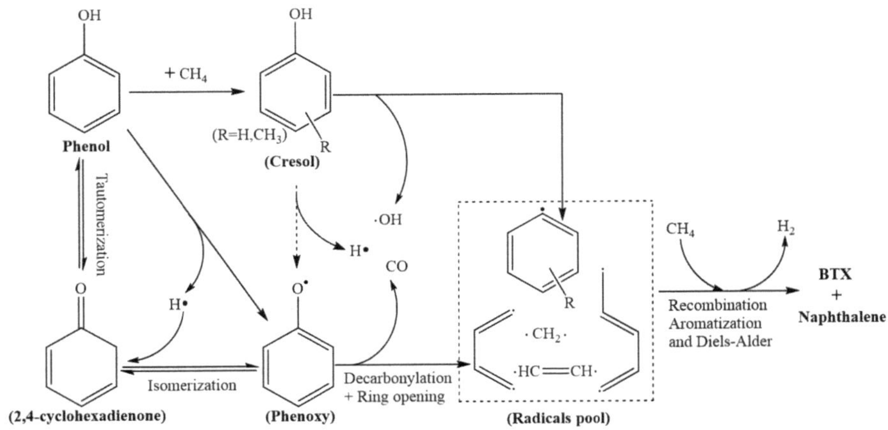

Fig. 8.25 Proposed reaction pathways for upgrading of phenolic compounds under methane environment. Reproduced with the permission from Elsevier

methyl group correspondingly. These radicals can be further stabilized by combining with other free radicals (such as Ḥ) to produce aromatic products such as benzene, toluene, and xylenes. The second pathway could be the tautomerization (isomerization reaction which is readily reversible), during which phenol is firstly tautomerized into 2,4-cyclohexadienone, which more easily accepts a proton to form a phenoxy radical through a reverse isomerization. This radical is more active than phenol, and decarbonylation is possible through breaking the ring, releasing a carbon atom and forming a carbon monoxide molecule. Then the remaining part could be cracked into small hydrocarbon radicals such as methylene, ethylene, and diene, which could be further recombined into aromatics in the inner pore of zeolite catalyst. Similarly, methane can also take part in this reaction for the carbon chain prolongation. Besides, different from the upgrading of small non-aromatic molecules such as methanol and acetone, the conjugated diene structure produced from the ring opening of phenol enables the possibility of forming bicyclic aromatics such as naphthalene through the Diels-Alder reaction. The oxygen atoms from phenol finally end up in small molecules such as carbon oxide and water, which can be easily separated from the desired hydrocarbon products. It is again observed that the participation of methane leads to more aromatic products, which is favorable for the utilization of both lignin and natural gas in a more efficient way.

It is reported that the yield of aromatic can be enhanced from 56.2 C to 85.1 C mol% if the gas atmosphere is changed from nitrogen to methane in phenol upgrading process, accompanied with an increase of BTEX selectivity from 66.7 C to 82.0 C mol%, indicating the positive impact of methane addition [50]. To obtain more direct evidence of methane incorporation, isotope-labelling experiments are carried out. According to the results in Fig. 8.26, a shift in mass spectra towards the higher end is observed in all major liquid products after $^{13}CH_4$ is used as the gas feed instead of natural CH_4, proving the existence of carbon atom from methane in these

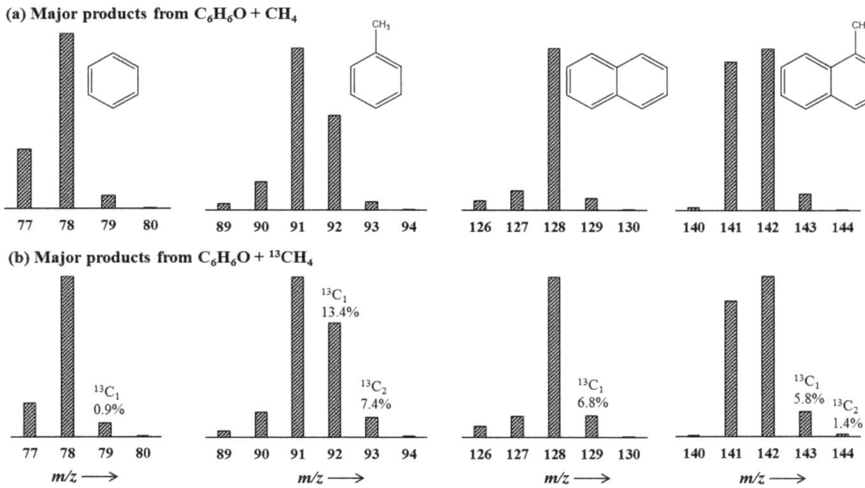

Fig. 8.26 Mass spectra of major products of benzene, toluene, naphthalene, and methylnaphthalene: (**a**) with the natural abundance of ^{13}C and (**b**) formed from ^{13}C-labeled methane and phenol at 400 °C (60 min) on 1%Zn/ZSM5 with the estimated isotopic composition (mol%). Reproduced with permission from Reference [50]. Copyright (2018) American Chemical Society

products. Consistent conclusions can be drawn from the NMR spectra shown in Fig. 8.27. The peak intensity of ^{13}C is evidently increased in the $^{13}CH_4$ run, suggesting the participation of carbon atoms from methane, which end up in the products. Interestingly, the intensity increases of different carbon groups are highly uneven. An over tenfold enhancement of benzylic carbon (chemical shift at 21.70 ppm) is observed, indicating methane is preferably incorporated as the methyl group directly linked to aromatic ring through alkylation reaction. The signals of carbon atoms in the aromatic ring (chemical shift at 137.55, 134.30, 129.05, 128.34, and 125.42) are also intensified with the presence of $^{13}CH_4$, confirming the carbon from methane could also be incorporated in these positions. However, this increase is less significant than that of benzylic carbon. This observation is also in coincidence with mass spectra, since the ^{13}C m/z signal increase of benzene (containing only aromatic C atoms) is much less dominant than that of toluene (containing both aromatic and benzylic C atoms).

Similar isotope-labelling experiment is also performed using phenol-1-^{13}C (the carbon directly linked to the hydroxyl group is ^{13}C-concentrated) to get further understanding of the reaction mechanism, as shown in Fig. 8.28. The notable higher intensity of ^{13}C signals in the isotope-labelled run indicates the incorporation of 1-C of phenol in liquid products, preferably in aromatic rings (chemical shift at 130–125 ppm). However, the peak of benzylic carbon at 21.72 ppm is also increased, indicating the possibility of a ring expansion-contraction pathway for the exchange of carbons in the aromatic ring and on the side chain.

The exploration of reaction intermediates is performed through reaction evaluation with controlled reactions in short (1 s) and long (1 h) period of times. The

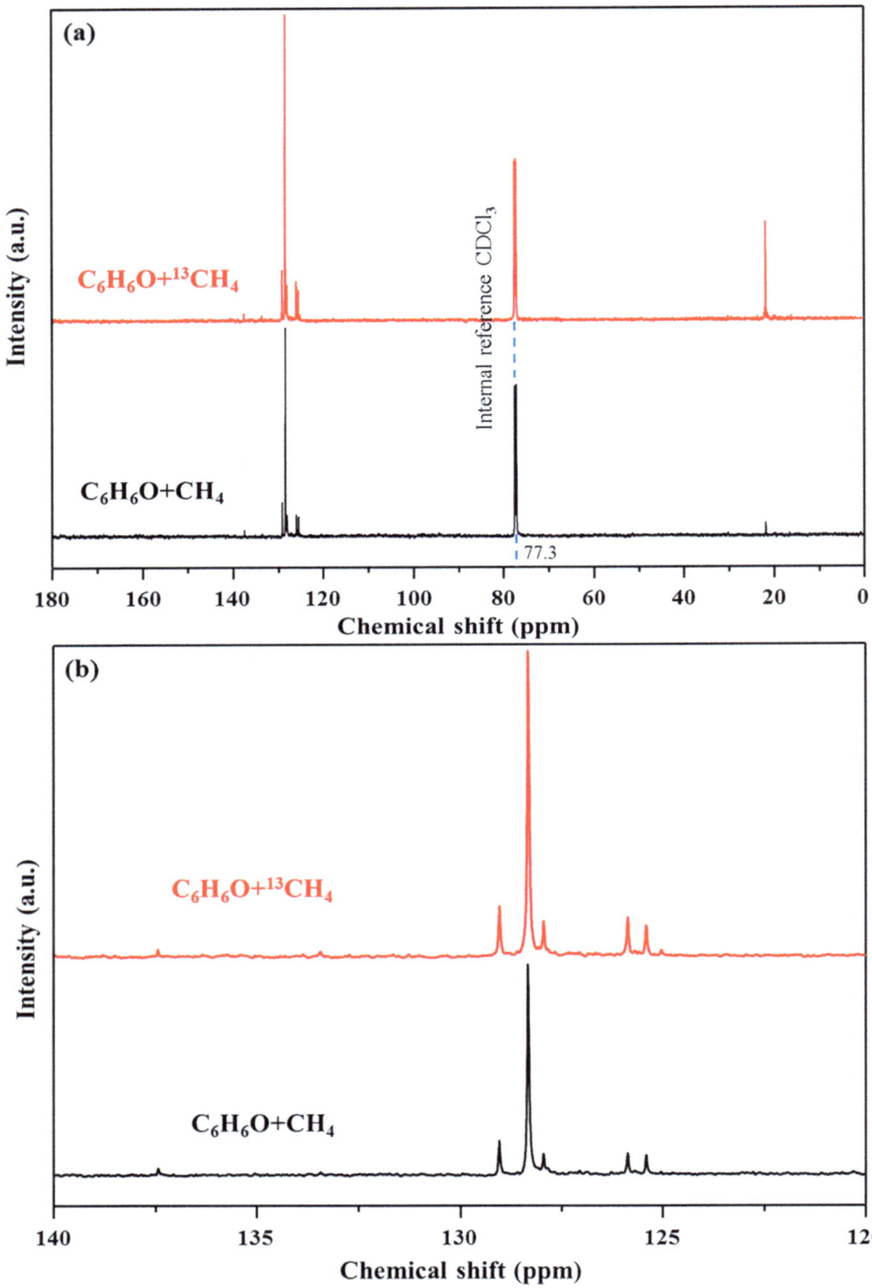

Fig. 8.27 ^{13}C liquid NMR spectra (a) and the zoom-in views (b) in the 120–140 ppm chemical shift region of the liquid products collected from the reaction between phenol and $^{13}CH_4/CH_4$. The triplet peak centered at 77.3 ppm belongs to $CDCl_3$ which serves as the internal reference for more accurate comparison. Reproduced with permission from Reference [50]. Copyright (2018) American Chemical Society

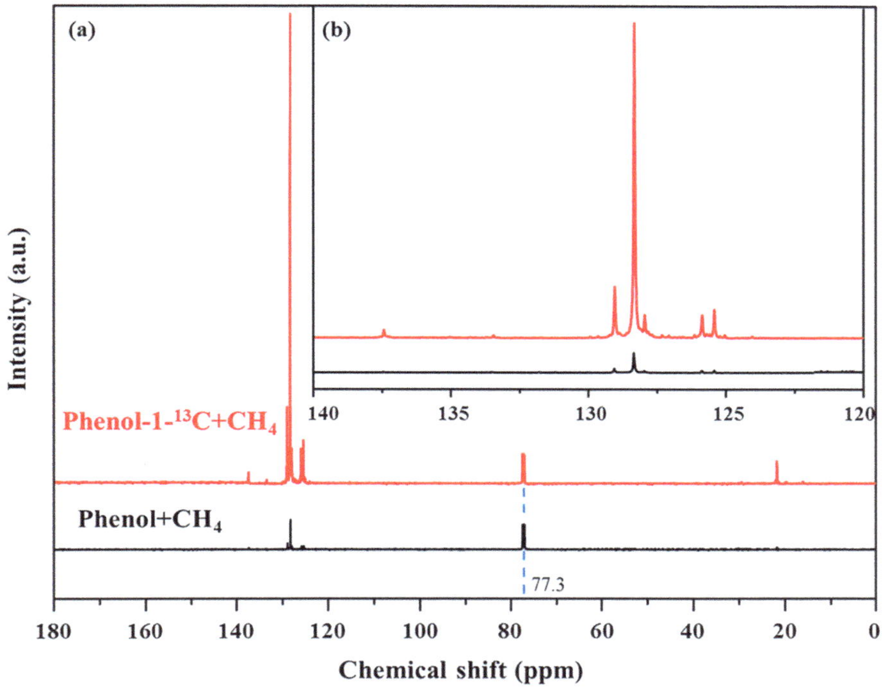

Fig. 8.28 (**a**) ^{13}C liquid NMR spectra and (**b**) an expansion of the 120–140 ppm chemical shift region of the liquid products collected from the reactions between phenol/phenol-1-^{13}C and CH_4. The triplet peak centered at 77.3 ppm belongs to $CDCl_3$ which serves as the internal reference for more accurate comparison. Reproduced with permission from Reference [50]. Copyright (2018) American Chemical Society

reactor is quenched in cooling water, and the catalyst is collected for solid-state NMR characterizations, which is shown in Fig. 8.29. In the case where $^{13}CH_4$ is used as the feed gas, a sharp increase of benzylic carbon at 19.6 ppm is observed in a short period of time (1 s), indicating the activation of methane, and subsequent alkylation can be the first step of the reaction. In this rapid process, almost no carbon from methane is interchanged into the aromatic ring, as evidenced by the similar intensity of aromatic carbon at 129.1 ppm. However, after a relatively long time period (1 h), the signal of aromatic carbon also increases notably, indicating the exchange of carbon between the methyl group and aromatic ring. When 1-^{13}C phenol is used as the feedstock, it can be seen that the signal of aromatic carbon goes up rapidly in a short period, which could be attributed to the large amount of unreacted phenol. The peak at 115.6 ppm deserves extra attention, which can be designated as the carbon in π-conjugated complexes (such as CH=CḤ and CH=CH-CH=CḤ). These species might be the transient reaction intermediates derived from phenol decomposition, which are reconstructed into aromatic products afterwards, as proved by the disappearance of peak at 115.6 ppm and the dominant peak at 129.1 ppm after 1 h. This observation further favors the direct deoxygenation reaction pathway since the carbon from phenol is well preserved on the

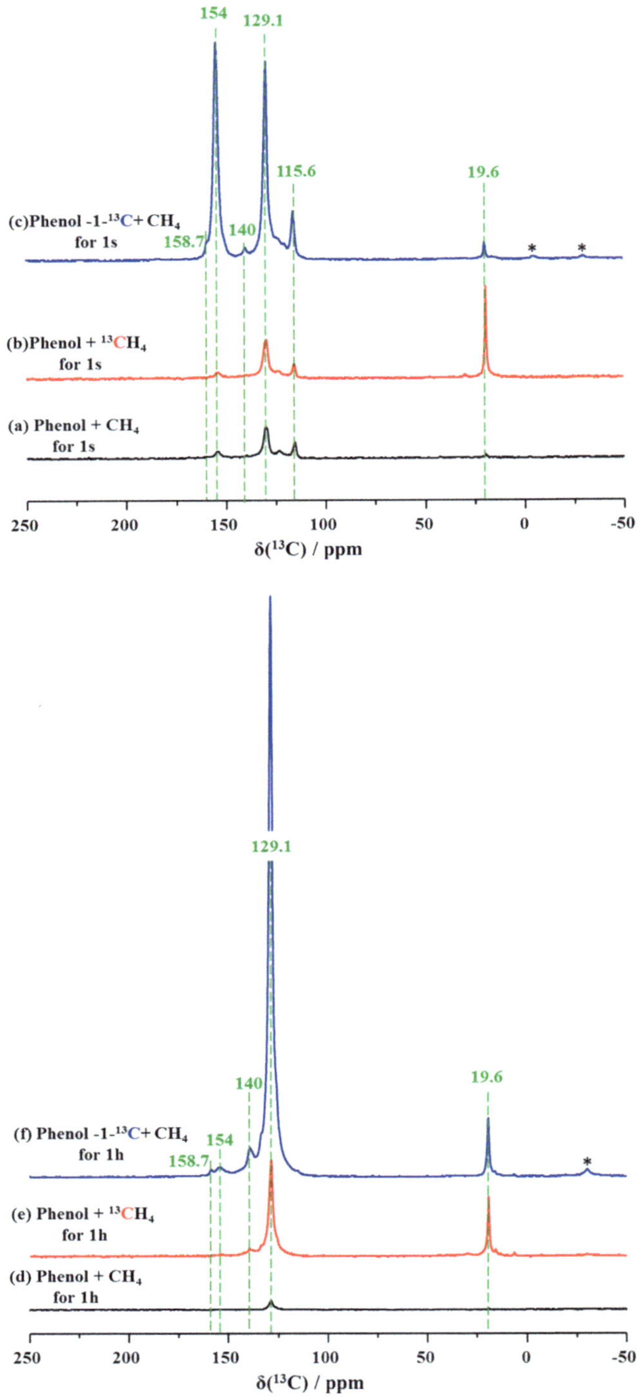

Fig. 8.29 Solid-state ^{13}C CP/MAS NMR investigation into the reaction of methane and phenol over Zn/ZSM-5 zeolite. The ^{13}C CP/MAS NMR spectra were recorded upon the reaction of phenol and CH_4 for (**a**) 1 s and (**d**) 1 h, phenol and $^{13}CH_4$ for (**b**) 1 s and (**e**) 1 h, and phenol-1-^{13}C and CH_4 for (**c**) 1 s and (**f**) 1 h at 400 °C. *Denotes the spinning side bands. Reproduced with permission from Reference [50]. Copyright (2018) American Chemical Society

surface of catalyst after ring opening of phenol rather than fully released as carbon oxide into the gas phase.

Besides the detailed investigation of reaction mechanism, the active sites over the employed catalyst is also characterized. After Zn/ZSM-5 is selected as the optimized catalyst for phenol upgrading process, the results of Zn L_3-edge X-ray adsorption spectroscopy (XAS) is carried out and shown in Fig. 8.30. Comparing the spectra of fresh catalyst and spent catalyst under methane and nitrogen environment, it is interestingly found that the catalyst used with the presence of methane leads to higher peak intensity at 1028 eV, which can be attributed to unoccupied Zn s and d orbital states caused by oxygen vacancies. This unique chemical state of Zn is different from metal zinc and zinc oxide as illustrated in Fig. 8.30, which could be the active sites on zeolite support for phenol upgrading process. This observation suggests that methane is not only acting as a reactant but also as an activator for the facilitation of active site formation in the process. The beneficial effect of methane beyond reactant also corroborate the importance and appropriateness of methane utilization in biofuel upgrading processes.

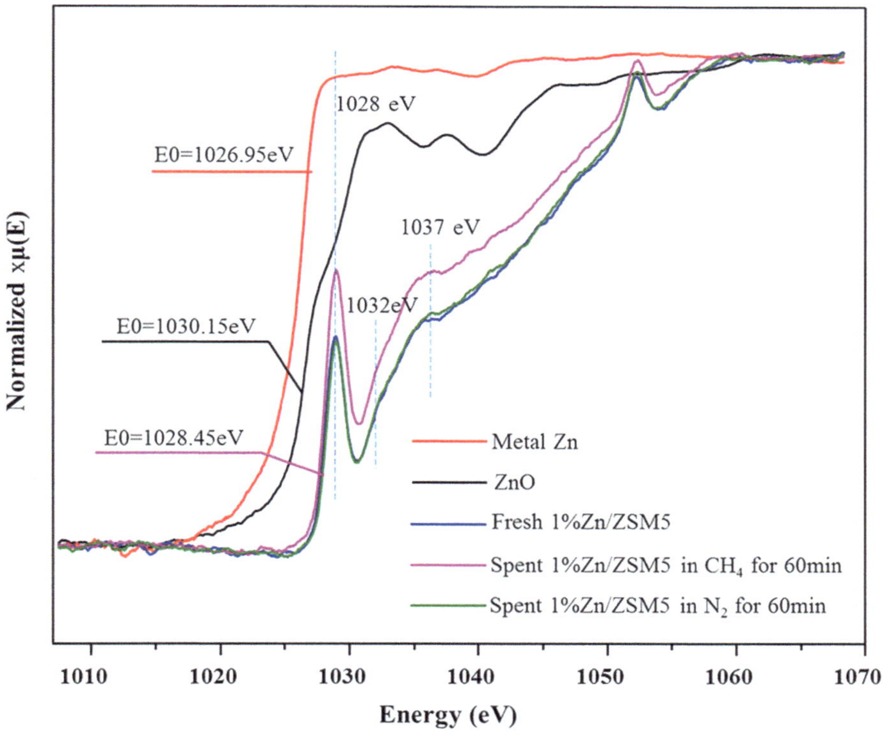

Fig. 8.30 Zinc L_3-edge XAS spectra collected in fluorescence mode from metal Zn, ZnO references, fresh catalyst, and spent catalysts under CH_4/N_2 environment at room temperature. The spectra are normalized by subtracting a straight line fitted to zero before L_3-edge. Reproduced with permission from Reference [50]. Copyright (2018) American Chemical Society

8.5 Catalytic Conversion of Cellulose Under Methane Environment

Cellulose is another macromolecule which is constructed by glucose units and is thus a type of polysaccharide. Cellulose is the most abundant natural polymer in the world, which is extremely widely distributed in the cell wall of green plants and microbes. However, different from sugars which can be directly digested by human beings as a main source of nutrition, the glucose units in cellulose are linked by β-1,4-glycosidic bond, resulting in the nondigestible properties for most carnivores and omnivores. Therefore, cellulose is a type of material with enormous reserve, and the realization of its better utilization is of great importance. Extensive efforts have been focused on the chemical conversion of cellulose into valuable products such as 5-hydro-xymethylfurfural (5-HMF), glycolic acid, lactic acid, and furan derivatives for its valorization [51–56]. However, due to its relatively inert physical and chemical properties, the direct conversion faces specific challenges such as severe mass transfer and kinetic limitations, especially under heterogeneous conditions, resulting in low reactivity. Several techniques have been developed to improve the situation, such as functionalization of the catalyst surface for better adsorption, addition of ionic liquids for enhanced solubility and depolymerization, applying microwave or ultraviolet light to overcome the mass transfer and/or kinetic barriers, design of a two-phase reactor system, coupling with other reactions, and introduction of new types of catalysts such as acid-based catalysts and nanofibers [57–63].

Among these pioneer investigations, one of the most promising direction could be the effective conversion of cellulose to aromatics or alkanes, which is highly valuable in petrochemical industry. Model compounds such as xylose (which can be easily converted into xylitol) and glucose (which can be readily converted into sorbitol) are selected for the exploration of relevant reaction pathways [60]. Pure thermal pyrolysis of glucose is already investigated, which indicates glucose is firstly decomposed into small oxygen-containing molecules, and a secondary molecular recombination process could proceed afterwards to give useful hydrocarbons. However, the stringent reaction conditions (high temperature and pressure) are not favorable for practical applications in a cost-effective way [64]. Then, the reaction under hydrogen atmosphere is also attempted. It is reported that the hydrogenolysis of xylitol and sorbitol in aqueous phase could be used to product *n*-alkanes with high yields, indicating the C-O bond can be selectively broken without the cleavage of C-C bond [60]. Another attempt to convert glucose/sorbitol to hydrocarbons is also realized, generating highly branched alkanes and alkenes as well as alkylated aromatics [65]. These species generally have low boiling points and are thus used as liquid transportation fuels. Isotope-labelling techniques are also applied to reveal the reaction pathways of catalytic glucose conversion [66]. It is suggested that monocyclic aromatic species are most possibly generated from a hydrocarbon pool after the oxygen atom is removed in the primary pyrolysis process [67]. Based on this theory, appropriate adjustment of the hydrocarbon pool could be essential for the better selectivity towards desired hydrocarbons such as aromatics. More in-depth investigation suggests that glucose is firstly decomposed in the rapid primary pyrolysis process to form small oxygen-containing molecules through retro-aldol condensation at relatively low temperatures [68]. Then, aromatics can be generated in the inner pores of employed zeolite catalysts. However, at high temperature, other intermediates such as anhydrosugars and furans can be produced, which compete with aromatization to form coke through polymerization. Therefore, the selection of a suitable reaction temperature is critical for high selectivity towards desired aromatics. Based on these studies, the upgrading of glucose under methane environment is also performed. It is found that methane can be successfully utilized as a hydrogen donor for the deoxygenation of small oxygen-containing molecules produced in the fast pyrolysis process through a reaction pathway similar to methane steam reforming. Besides, methane could also participate in the co-aromatization process as discussed in previous sections, as well as in a complicated cross-coupling reaction. Therefore, the aromatic yield can be dramatically increased, which further demonstrates the advantages of methane addition in glucose upgrading process [69].

The possible reaction mechanism for cellulose conversion under methane environment is shown in Fig. 8.31. As proved by many studies, the first step for the process is mainly thermal pyrolysis where cellulose is decomposed into small oxygenates (such as furan and furfural) through depolymerization, dehydration, and decarboxylation [68, 70–73]. The derived oxygenates could be further deoxygenated into even smaller hydrocarbon fractions, and oxygen atom can be released in the form of carbon oxide or water. After the hydrocarbon pool is formed, the following steps are similar to the case of lignin upgrading processes. Namely, methane could participate in the reaction in at least two ways upon activation. The first way is direct interaction with the hydrocarbon pool to form aromatics through co-aromatization and Diels-Alder reactions. Another possibility could be the alkylation reaction after the aromatics are already generated, resulting in methyl-substituted

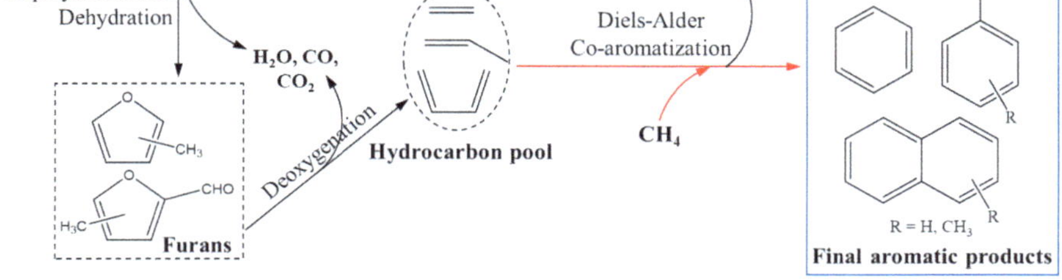

Fig. 8.31 Possible reaction pathway involved in the co-conversion of cellulose and methane to aromatics. Reproduced with the permission from Elsevier

aromatic products. Besides, bicyclic aromatic species could also be generated, most probably through Diels-Alder reactions between monocyclic aromatics and the conjugated diene intermediates in the hydrocarbon pool. It can be deduced that methane again favors the formation of desirable products including BTEX species, which could improve the profitability of the valorization process.

Silica-to-alumina (STA) ratio is a determining factor for the acidity of zeolite support, which is essential for catalytic processes under methane environment. The influence of STA value of the zeolite support on cellulose upgrading process is studied as shown in Fig. 8.32. It is observed that zeolite support with an STA value of 30 demonstrates optimal catalytic performances in terms of high liquid yield and BTEX yield, high oxygen removal efficiency, and low carbon residue. This phenomenon can be explained by the fact that balanced surface acidity is necessary for cellulose upgrading. If the acidity is excessively high (in the case of STA = 23), the adsorption and subsequent reaction of decomposed cellulose will be too severe, so that undesired side reactions including over-cracking and coking might be more significant, as evidenced in Fig. 8.32. On the other side, if too less acidity is present over zeolite catalysts (in the case of STA = 50, 80, and 280), the deoxygenation capability is not sufficient enough for removing oxygen atoms from cellulose and decomposition products. This statement is supported by the trend of lower oxygen removal ability, reduced liquid yield, and higher carbon residue along with

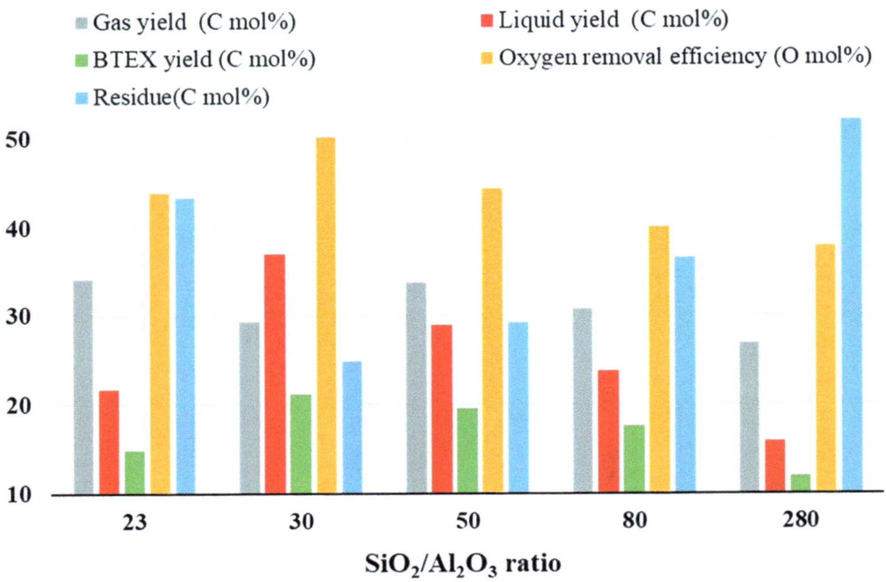

Fig. 8.32 Overall catalytic performance of zeolite supports with different SiO_2/Al_2O_3 ratios on cellulose conversion. Reaction conditions: 0.10–0.12 g of cellulose with 1.0 g catalyst, 2.5 MPa of CH_4 at 400 °C, 60 min. BTEX: benzene and toluene, with small proportion of ethylbenzene and xylene. Reproduced with the permission from the Royal Society of Chemistry

increasing STA value. This understanding might also be useful for other catalytic upgrading processes relating to oxygen-containing biofuel molecules.

The participation of methane is again confirmed by isotope-labelling experiments, and the results are exhibited in Figs. 8.33 and 8.34. The liquid ^{13}C NMR spectra in Fig. 8.33 clearly illustrates the increase of peak intensity after the feed gas is changed from natural CH_4 to isotope-labelled $^{13}CH_4$, indicating the incorporation of C atoms from methane into final liquid products. Interestingly, not only the peaks of alkyl groups (chemical shift at 21.72 ppm) but also those of aromatic groups (chemical shift at 120–140 ppm) are intensified, suggesting methane can be incorporated both as side chain aliphatic groups (mainly methyl group according to mass spectra) and in aromatic ring. The side chain methyl group can be attributed to the alkylation reaction, while the carbon in the aromatic ring can be formed in at least two ways. Firstly, methane might directly participate in the co-aromatization reaction with existing hydrocarbon intermediates derived from the ring opening of the

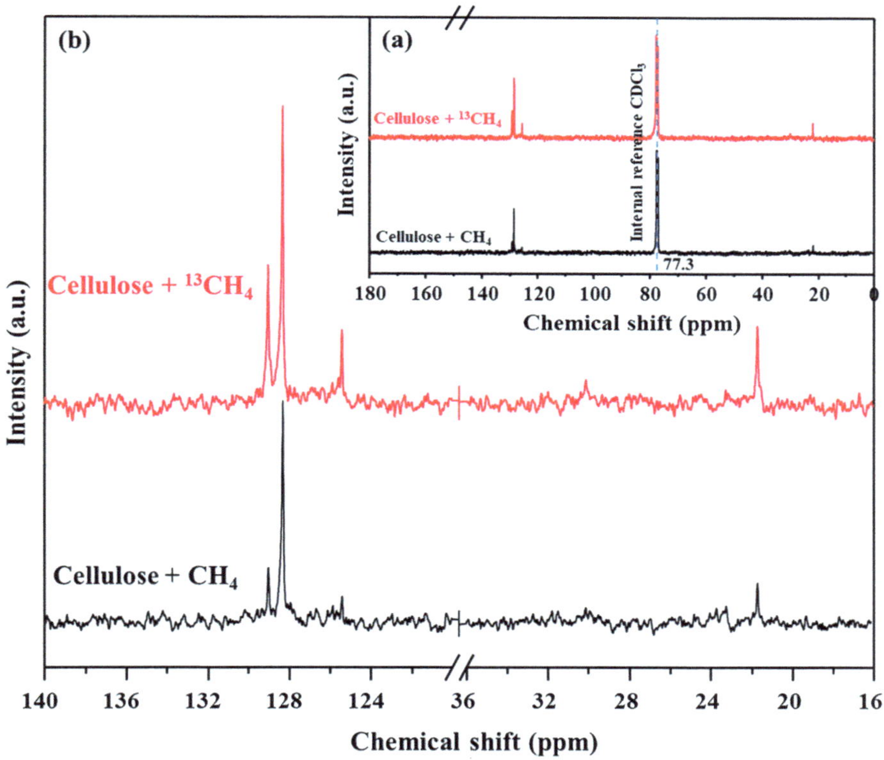

Fig. 8.33 ^{13}C liquid NMR spectra (**a**) and its zoom-in views in the 120–140 ppm and 16–36 ppm chemical shift regions (**b**) of the liquid products collected from the reaction between cellulose and $^{13}CH_4/CH_4$. The triplet peak centered at 77.3 ppm belongs to $CDCl_3$ which serves as the internal reference for a more accurate comparison. Reproduced with the permission from the Royal Society of Chemistry

cellulose. Secondly, the side chain methyl group might also be interchanged into the aromatic ring through a ring expansion-contraction process.

In addition to ^{13}C isotope-labelling experiments and NMR spectra, corresponding comparison experiments using CD_4 (deuterium-concentrated methane) are also performed. As shown in Fig. 8.34, the peak intensities in ^{1}H NMR spectra decreases dramatically after CH_4 is replaced by CD_4, especially those for aliphatic hydrogen atoms (chemical shift at 2.4 ppm). Meanwhile, the corresponding peaks come out in ^{2}H NMR when CD_4 is used as the feed gas. These evidences further prove the presence of atoms from methane in the final products, particularly in the side chain methyl (-CH_3) groups.

A series of Zn-loaded zeolite catalysts are used for cellulose upgrading under methane environment, among which 3%Zn(II)-$Zn^{\delta+}$/ZSM-5 prepared by a combination of wet impregnation (WI) and chemical vapor deposition (CVD) demonstrates optimized catalytic performances. The corresponding characterization results are shown in Figs. 8.35, 8.36, and 8.37.

Pyridine adsorption diffuse reflectance infrared Fourier transform spectroscopy (py-DRIFTS) is used to analyze the amount and type of surface acidity over the prepared catalysts. As shown in Fig. 8.35, both signals of Bronsted acid sites (1492,

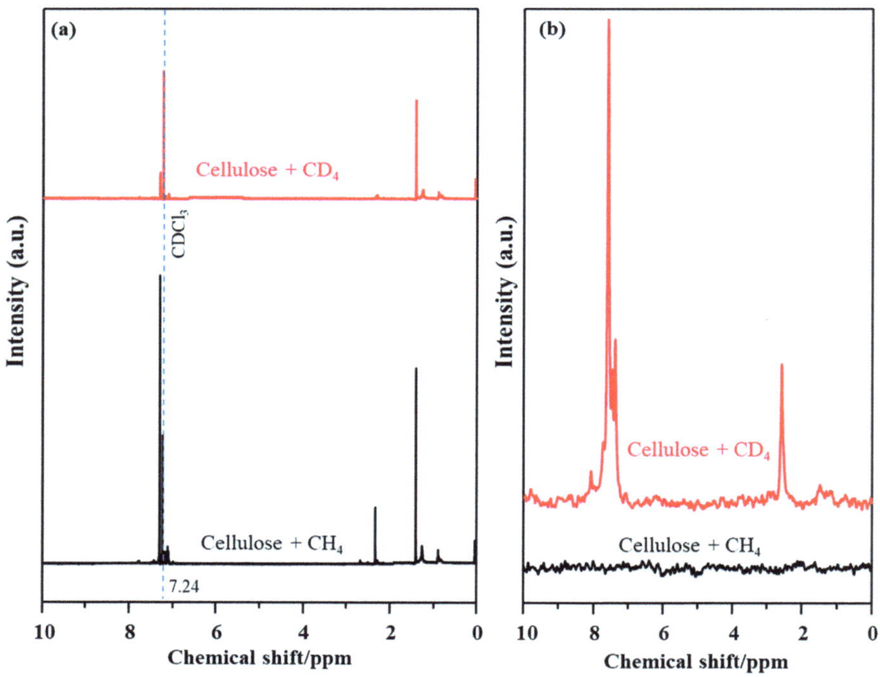

Fig. 8.34 ^{1}H NMR spectra (**a**) and ^{2}H NMR spectra (**b**) of the liquid products collected from the reactions between cellulose and CH_4/CD_4. The peak at 7.24 ppm belongs to $CDCl_3$ which serves as the internal reference. Reproduced with the permission from the Royal Society of Chemistry

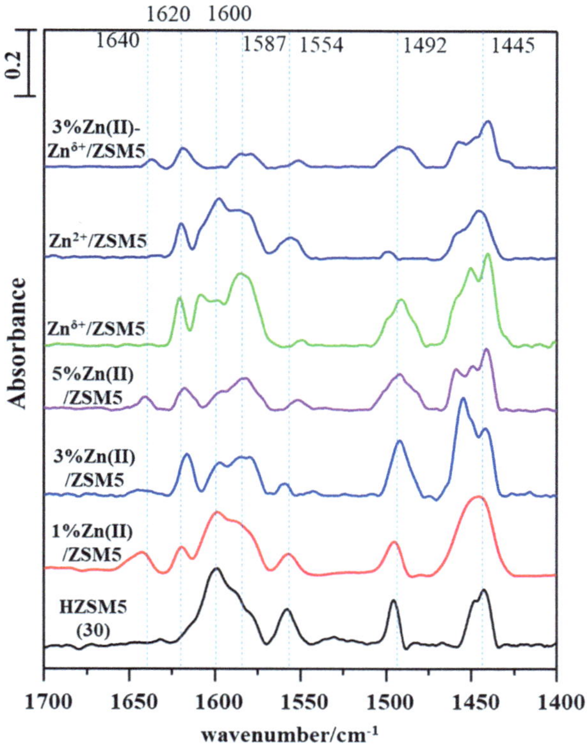

Fig. 8.35 DRIFT spectra collected over various fresh catalysts upon pyridine adsorption. Reproduced with the permission from the Royal Society of Chemistry

1554, and 1600 cm^{-1}) and Lewis acid sites (1445 and 1492 cm^{-1}) can be clearly observed. The 3%Zn(II)-Zn$^{\delta+}$/ZSM-5 shows balanced distribution of Bronsted and Lewis acid sites, which could be responsible for the improved catalytic performances. This unique surface acid properties can be attributed to the combined synthesis strategy, during which wet impregnation introduces Zn(II) and chemical vapor deposition introduces Zn$^{\delta+}$.

X-ray photoelectron spectroscopy (XPS) provides additional evidences for the chemical environments of Zn in the catalysts, as shown in Fig. 8.36. It can be observed that the peak of Zn^{2+} (originally at 1023.3 eV) clearly shifts to the lower binding energy in Zn$^{\delta+}$ catalyst (1022.6 eV). However, this peak is also different from metallic Zn(0) which is located at 1021.7 eV. Therefore, this peak belongs to Zn with a valence between 0 and +2, as labelled by δ+, which can be attributed to the presence of oxygen vacancies. In addition, the peak positions remain unchanged after reaction under either CH_4 or N_2 environment, proving the good stability of Zn active species during the reaction process. Interestingly, the peak of Zn$^{\delta+}$ over 3%Zn$^{\delta+}$/ZSM-5 is slightly decreased under methane environment while increased under nitrogen, indicating there might be a migration of Zn$^{\delta+}$ from the surface to inner pores of zeolitic structure with the assistance of methane. This deduction is also supported by the intensified metal oxide peak (532.5 eV) under methane environment

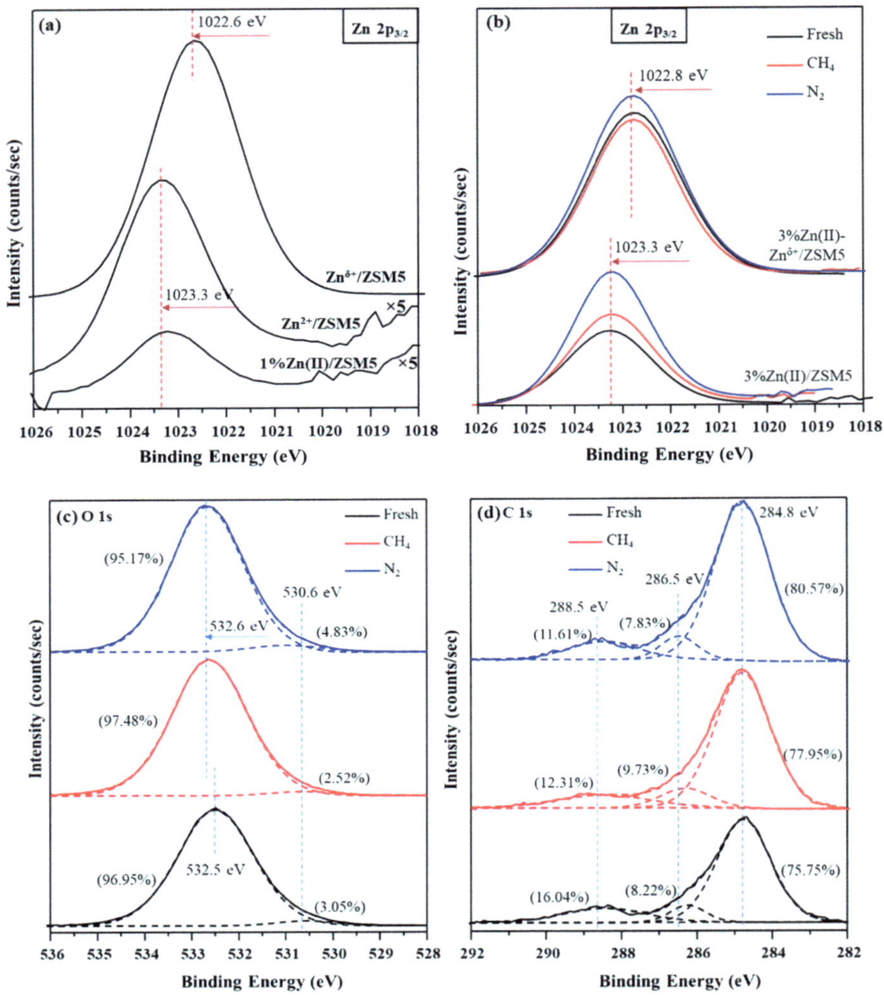

Fig. 8.36 X-ray photoelectron spectroscopy of fresh and spent catalysts. (**a**) Zn 2p3/2 XPS spectra region of fresh 1%Zn(II)/ZSM5, Zn^{2+}/ZSM5, and $Zn^{\delta+}$/ZSM5 catalysts. (**b**) Zn 2p3/2 XPS spectra region of fresh (black line) and spent catalysts (3%Zn(II)/ZSM5, 3%Zn(II)-$Zn^{\delta+}$/ZSM5) collected from CH_4 (red line) and N_2 (blue line) runs. O1s (**c**) and C1s (**d**) XPS spectra of fresh (black line) and spent 3%Zn(II)-$Zn^{\delta+}$/ZSM5 catalyst collected from CH_4 (red line) and N_2 (blue line) runs. XPS spectra are fitted with Lorentzian-Gaussian peaks of 30:70 ratio. The dotted line represents the fitting curve, and the solid line represents the experimental data. The fitted peak positions and concentrations are indicated on the spectra. Reproduced with the permission from the Royal Society of Chemistry

and its declined intensity under nitrogen in O 1s spectra. This unique effect brought by methane can be favorable for better catalyst performances. In C1s spectra, the increase of peak at 284.8 eV can be attributed to coke formation. It is observed that the increase rate under methane environment is lower than that under nitrogen

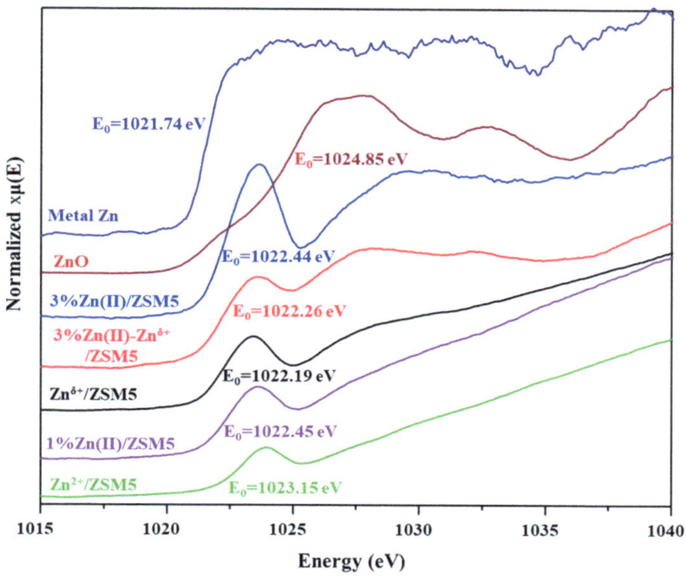

Fig. 8.37 Zinc L_3-edge XAS spectra collected in fluorescence mode from metal Zn, ZnO references, and various fresh supported Zn catalysts at room temperature. The spectra are normalized by subtracting a straight line fitted to zero before the L_3-edge

environment, suggesting coking is partially inhibited with the presence of methane, which is another advantage of adopting methane as the gas atmosphere.

The abovementioned conclusions are further corroborated by X-ray adsorption spectroscopy (XAS) in Fig. 8.37. The energies of L_3-edge show the sequence Zn (1021.74 eV) < $Zn^{\delta+}$/ZSM-5 (1022.19 eV) < 3%Zn(II)-$Zn^{\delta+}$/ZSM-5 (1022.26 eV) < 1% Zn(II)/ZSM-5 (1022.45 eV) ≈ 3%Zn(II)/ZSM-5 (1022.44 eV) < Zn^{2+}/ZSM-5 (1023.15 eV) < ZnO (1024.85 eV), which is highly consistent with XPS data. It is observed that $Zn^{\delta+}$ species leads to intermediate sharpness of Zn L_3-edge, which corresponds to the balanced amount of unoccupied Zn 4s- and 4d-derived states and might be beneficial for the catalytic cellulose upgrading process.

8.6 Catalytic Conversion of Furans Under Methane Environment

Furans is commonly used to represent furan derivatives, namely, compounds containing a furan ring, such as furfural, furfuryl alcohol, and methyl substituted furans [69, 74]. These substances are widely accepted as intermediate in biomass pyrolysis processes. Owing to its characteristic chemical structure, distinct reaction behaviors can be observed compared to other oxygen-containing species. It has been reported that furfural can be converted to 1,5-pentanediol with high selectivity, suggesting that the ring opening via C-O bond and subsequent hydrogenation could occur without breaking the carbon chain structure [75]. Another example of highly selective furfural conversion gives the product cyclopentanone, while a more typical product could be 2-methylfuran, which is generated through hydrogenation of the aldehyde group and hydrogenolysis of C-O bond afterwards [76]. In addition, the formation of aromatics through the Diels-Alder reaction is characteristic for the upgrading of furans, which can be attributed to its unique structure consisting of a five-member ring with one oxygen heteroatom and a conjugated diene structure [70]. This process leads to distinctive products such as benzofuran, which is generally taking place in the channels of zeolite catalyst. It is proved that the selectivity towards aromatics in the conversion of furans can be manipulated by the concentration of other species such as hydrogen, olefin, and methanol [77, 78]. This characteristic is also proved to be closely related to the synergistic effects between furans and other short-chain species in the Diels-Alder reaction. Another case is the selective conversion of dimethylfuran and ethylene towards *p*-xylene, which is the most useful xylene species among the isomers. It is suggested that Bronsted acid sites are responsible for the dehydration of the additive product in the Diels-Alder reaction, producing *p*-xylene and H_2O. Kinetic studies demonstrate that the catalytic dehydration can proceed rapidly, and the Diels-Alder cycloaddition thus becomes the rate-controlling step.

The catalytic conversion of bioderived furans under methane environment also ends up with promising results [79]. High selectivity towards monocyclic aromatics can be achieved, and methane can still be used in the reaction process in many ways. The proposed mechanisms are shown in Fig. 8.38. The reaction pathways are basically similar to those in the case of cellulose upgrading after the primary thermal pyrolysis step as shown in Fig. 8.31, while more details are given for further clarification. Besides the typical aromatic products derived from the aromatization of the radicals from the hydrocarbon pool and methane, the conjugated structure in furan also enables the direct Diels-Alder reaction to form benzofuran, which can be regarded as a starting point for the production of additional bicyclic aromatic products. Methane can also participate in the process, facilitating the deoxygenation of benzofuran (and its derivatives) into naphthalene. The understanding of the reaction pathways can be regarded as the basis for effective catalytic upgrading of furans under methane environment, which is expected to accelerate the real applications for the utilization of inexpensive methane and biomass resources simultaneously.

Zn-Ga/ZSM-5 catalyst is employed for catalytic furfural valorization, and the reaction evaluation results of five consecutive runs without catalyst regeneration is shown in Fig. 8.39. Constant catalytic performances in terms of complete furfural conversion, aromatic selectivity, gas yield, and methane conversion are observed, proving the excellent activity and stability of the prepared catalyst, which could be used readily in long-term operations.

The participation of methane is confirmed again by isotope-labelling experiments. It is observed that the mass spectra of main liquid products shift towards higher m/z values after natural methane is replaced by $^{13}CH_4$, indicating the

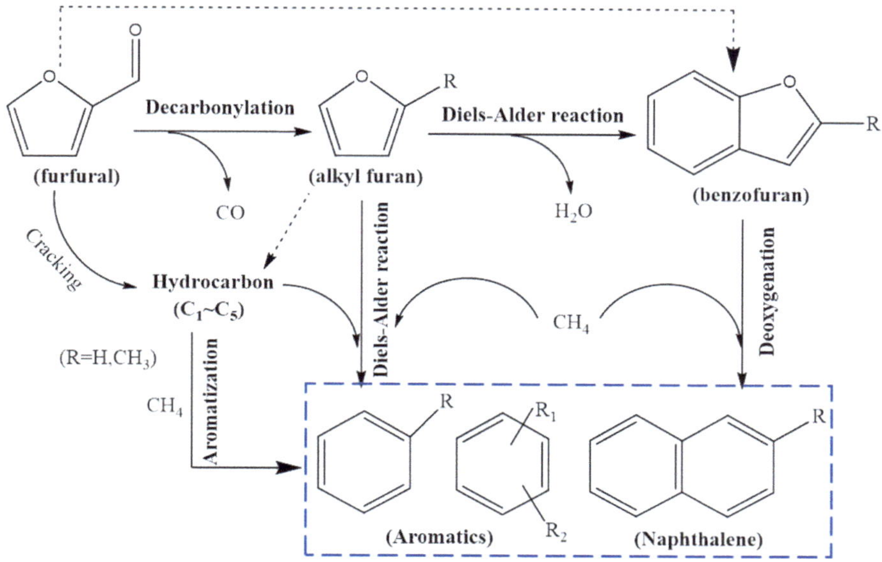

Fig. 8.38 Possible reaction pathway involved in the co-conversion of cellulose and methane to aromatics. Reproduced with the permission from Elsevier

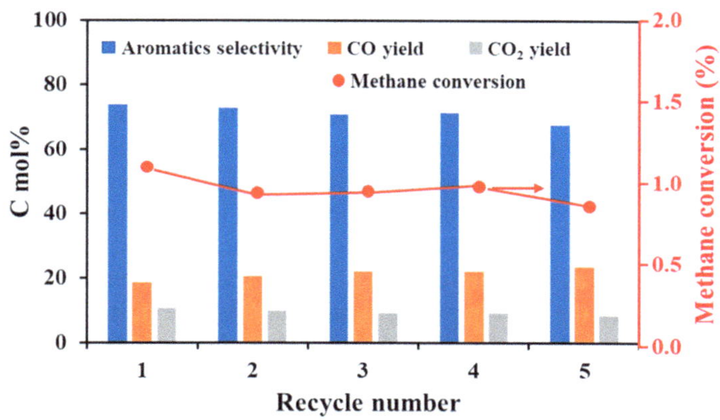

Fig. 8.39 Catalytic performances of Zn-Ga/ZSM5 for furfural valorization under methane environment in five consecutive runs. Reaction conditions: mass ratio of furfural to catalyst of 0.11–0.12, methane pressure at 2.5 MPa, and temperature at 400 °C. The errors of each conversion are estimated to be within ±3%. Reproduced with the permission from [79]. Copyright (2018) American Chemical Society

incorporation of ^{13}C from methane. However, the ^{13}C signal increase rate in aromatics without side chain groups (such as benzene and naphthalene) is not as significant as that in aromatics with side chain methyl groups (such as toluene and xylene), indicating the carbon from methane is preferably located at side chains (Fig. 8.40).

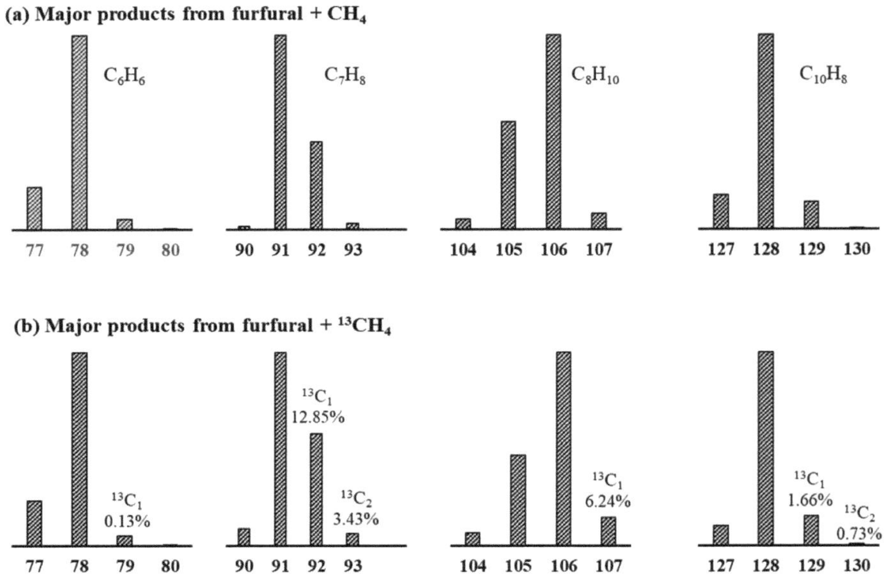

Fig. 8.40 Mass spectra of major products (benzene C_6H_6, toluene C_7H_8, xylene C_8H_{10}, and naphthalene $C_{10}H_8$): (**a**) using natural CH_4 as the gas feed and (**b**) using $^{13}CH_4$ as the gas feed for furfural upgrading over Zn-Ga/ZSM5 catalyst. The values indicate the changes of relative intensities. Reproduced with the permission from [79]. Copyright (2018) American Chemical Society

More cogent evidences can be obtained through liquid NMR spectra, as shown in Figs. 8.41 and 8.42. An evident ^{13}C signal intensification can be detected when methane is replaced by the ^{13}C isotope-concentrated counterpart, strongly proving the carbon atoms from methane is involved in final liquid products. Comparing aliphatic C (30.15 and 21.72 ppm) and aromatic C (120–140 ppm), it can also be concluded that carbon from methane tends to be incorporated easily in aliphatic positions, especially as benzylic carbon (the side chain carbon directly linked to aromatic ring), since the corresponding signals increase the most.

Similar conclusions can be drawn from ^1H NMR and ^2H NMR spectra in Fig. 8.42. It is observed that the ^1H signals decrease apparently and ^2H signals increase correspondingly when CD_4 is used as the feed gas instead of CH_4. The variation is much more significant for benzylic hydrogen (chemical shift at 2.3 ppm) than hydrogen in aromatic rings (7.0–8.0 ppm) and other aliphatic hydrogens (0.0–2.0 ppm), indicating activated methane is preferably linked to the aromatic ring as a side chain methyl substituent.

The reaction mechanism is further studied by control experiments operated at different temperatures and reaction times as illustrated in Figs. 8.43 and 8.44, which is realized through quenching the reaction in cold water. It is observed that furfural is consumed rapidly at the beginning of the reaction process, and alkyl furans are main products at lower temperatures. These species can be the intermediates for the formation of benzofuran through the Diels-Alder reaction, which is proved by the

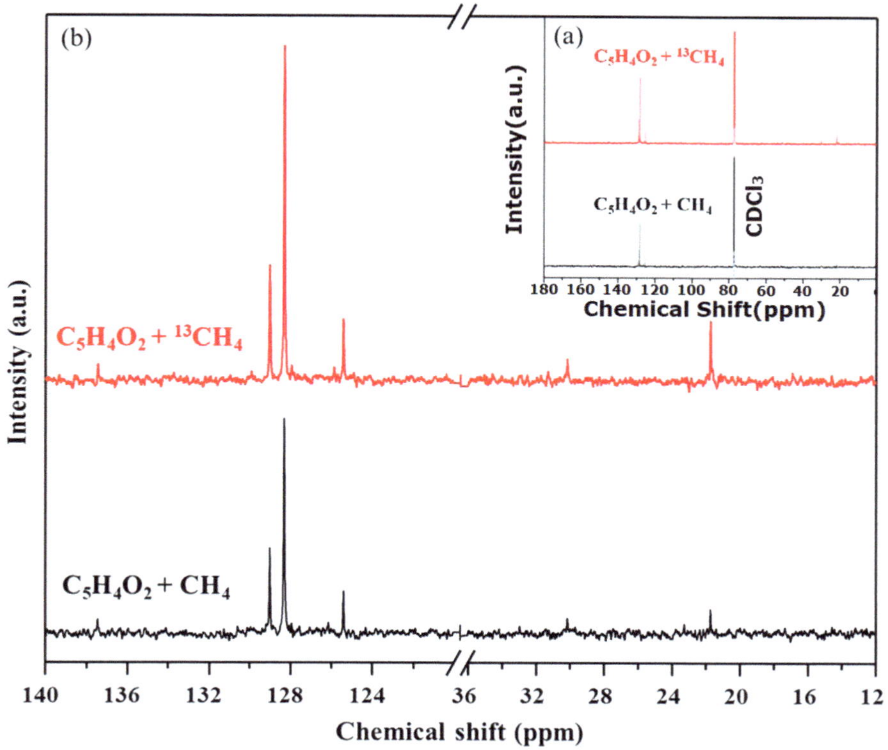

Fig. 8.41 ^{13}C liquids NMR spectra (**a**) and its zoom-in views in 120–140 ppm and 12–36 ppm chemical shift regions (**b**) of the liquid products collected from the reactions between furfural and $^{13}CH_4/CH_4$. The peak at 77.3 ppm belongs to the internal reference $CDCl_3$. Reproduced with the permission from [79]. Copyright (2018) American Chemical Society

increase of benzofuran concentration corresponding to decreasing methyl furan concentration. This reaction is unique for catalytic upgrading of furans due to their characteristic molecular structures. Meanwhile, as reaction continue to proceed, the concentration of benzofuran also starts to decline, along with the higher concentration of aromatic (monocyclic) and naphthalene species, indicating benzofuran could undergo a further deoxygenation reaction to produce aromatics.

As for gas phase, C_1–C_5 hydrocarbons are observed at the initial stage, which can be ring-opening products from furfural and are converted into aromatics afterwards through co-aromatization. After the primary steps, carbon oxides (CO and CO_2) become main gas products, which can be derived from decarbonylation of furfural. However, the concentration of CO_2 is increasing monotonously, while the CO concentration starts to decrease after 30 min at 400 °C. This different trend can be explained by secondary gas-phase reactions such as the Boudouard reaction ($2CO = C + CO_2$) and water-gas shift reaction ($H_2O + CO = H_2 + CO_2$). The latter reaction is also evidenced by the presence of a small amount of H_2. Besides, H_2 can

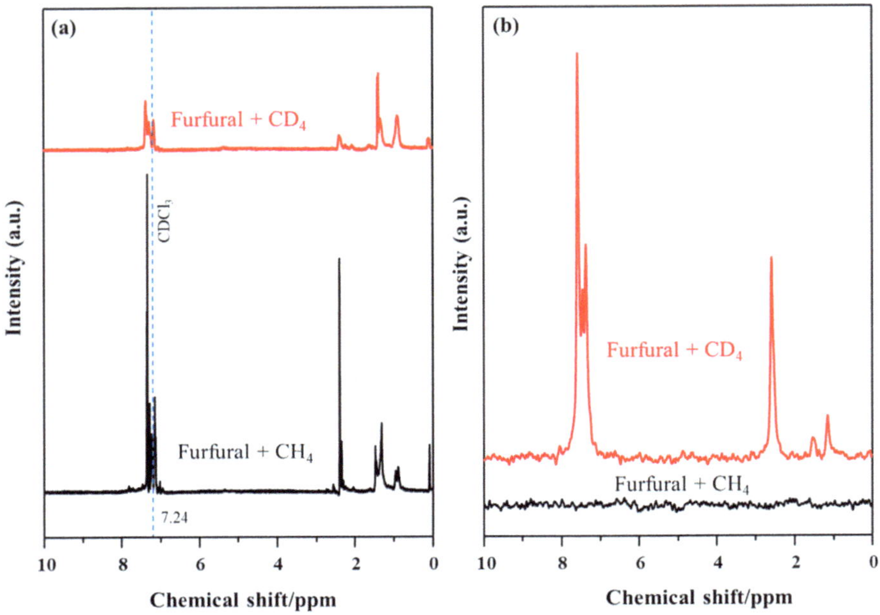

Fig. 8.42 ^{1}H NMR spectra (**a**) and ^{2}H NMR spectra (**b**) of the liquid products collected from the reaction between furfural and CH_4/CD_4. The peak at 7.24 ppm was attributed to the internal reference $CDCl_3$. Reproduced with the permission from [79]. Copyright (2018) American Chemical Society

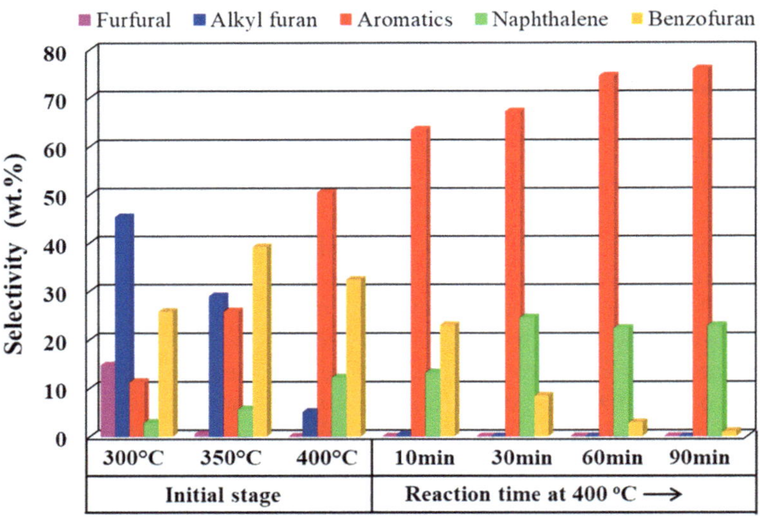

Fig. 8.43 Selectivity of liquid products collected in furfural conversion into aromatics under CH_4 environment at different temperature and reaction times. Alkyl furan: furan and alkyl furan; aromatics: C_6–C_{10} monoaromatics and a small amount of indene derivatives; naphthalene: naphthalene and alkyl naphthalene; benzofuran: benzofuran and alkyl benzofuran. Reproduced with the permission from [79]. Copyright (2018) American Chemical Society

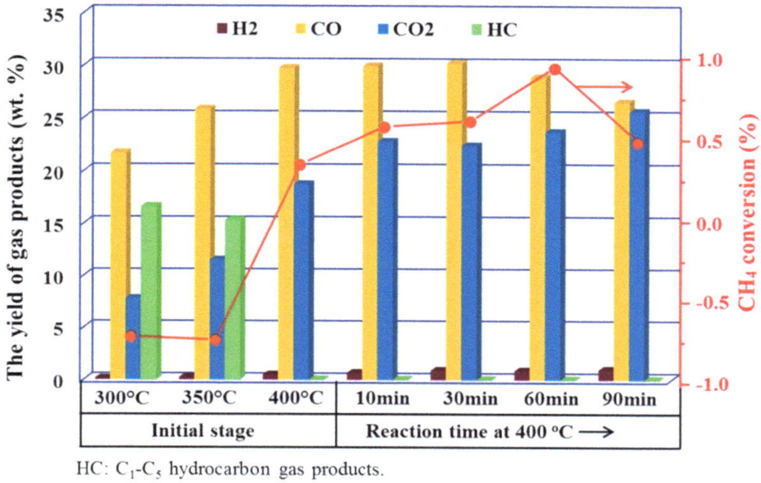

Fig. 8.44 Yields of gas products and methane conversion collected in furfural conversion into aromatics under CH_4 environment at different temperature and reaction times. Reproduced with the permission from [79]. Copyright (2018) American Chemical Society

also be produced during the aromatization reaction over zeolite catalyst, which is also widely observed in other upgrading processes.

Characterizations of catalysts are performed to reveal the active sites for furfural upgrading under methane environment. X-ray diffraction (XRD) patterns of several typical catalysts are shown in Fig. 8.45. Although the crystalline peaks of ZSM-5 zeolite structure are observed in all catalysts, the metal-related peaks are hardly detected expect for Pt, which demonstrates two characteristic peaks at $2\theta = 40°$ and 46°. It is interesting to have a detailed view of fresh and spent Zn-Ga/ZSM-5 catalysts under different gas atmosphere. It seems that the peaks for the zeolite crystal planes, particularly (101) and (020), are intensified after using under methane atmosphere while weakened under nitrogen. This might indicate that methane facilitates the better dispersion of metal species and leads to better exposure of zeolitic structure while an inert gas like nitrogen results in the opposite effect of metal agglomeration.

To dig deeper into the property-performance relationship, surface acidities of several catalysts are evaluated by both diffraction reflectance infrared Fourier transform spectroscopy (DRIFTS) and ammonia temperature-programmed desorption (NH_3-TPD), as shown in Figs. 8.46 and 8.47. The peaks derived from DRIFTS in Fig. 8.46 can be divided into Bronsted acid sites (1485, 1588, and 1604 cm^{-1}) and Lewis acid sites (1440, 1485, and 1619 cm^{-1}). Two types of acid sites are found in all employed catalysts, while the distribution can be different. From the results in Fig. 8.47, the acid sites can be divided into weak, medium, and strong ones, according to the ammonia desorption temperature. Among all the results, the acid properties of zinc-containing catalyst could be distinctive, since a much higher peak of weak acid sites can be observed. Considering the preferred catalytic performances of Zn-Ga/ZSM-5, it can be hypothesized that the abundance of weak acid sites might be favorable for selective furfural conversion towards aromatics.

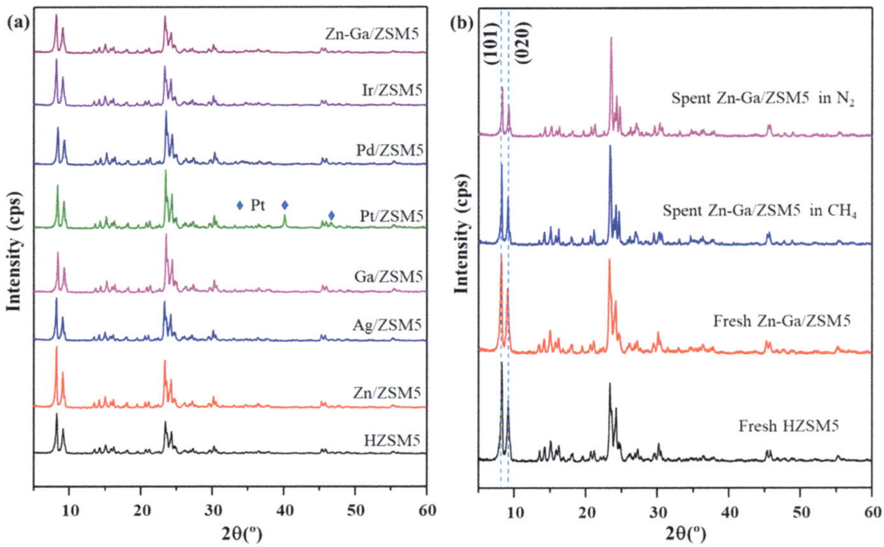

Fig. 8.45 Powder XRD patterns of (**a**) various fresh catalysts synthesized by incipient wetness impregnation and (**b**) fresh HZSM-5 and Zn-Ga/ZSM-5, as well as spent Zn-Ga/ZSM-5 catalysts collected under CH_4/N_2 environments. Reproduced with the permission from [79]. Copyright (2018) American Chemical Society

Transmission electron microscope (TEM) is used to observe microscopic structures and distribution of active metal components over the catalysts; the images are shown in Fig. 8.48. It is found that Zn and Ga particles are well dispersed in all randomly selected areas, which is confirmed by energy dispersive spectroscopy (EDS) mapping results. It is worth noting that the average sizes of metal particles in fresh catalyst, spent catalyst under methane environment, and spent catalyst under nitrogen environment are estimated to be 5.7, 3.6, and 8.5 nm, respectively. These results indicate that methane could facilitate the dispersion of active metals, while an inert atmosphere like nitrogen leads to agglomeration. This unique effect resulted from methane provides additional evidences for the advantages of introducing methane into the furfural upgrading process beyond a reactant, which could be favorable for better catalyst activity and stability.

X-ray photoelectron spectroscopy (XPS) is further applied to investigate the chemical environment of active species in the catalysts as shown in Fig. 8.49. Zinc is observed to be present as ZnO on all catalyst before and after reaction. However, the peak intensity of ZnO increases dramatically under nitrogen environment, while it is slightly reduced under methane environment after reaction. This suggests that more surface ZnO species with larger size can be found after the catalyst is used in an inert atmosphere, while better dispersion can be found under methane. This observation is highly consistent with those derived from XRD patterns and TEM

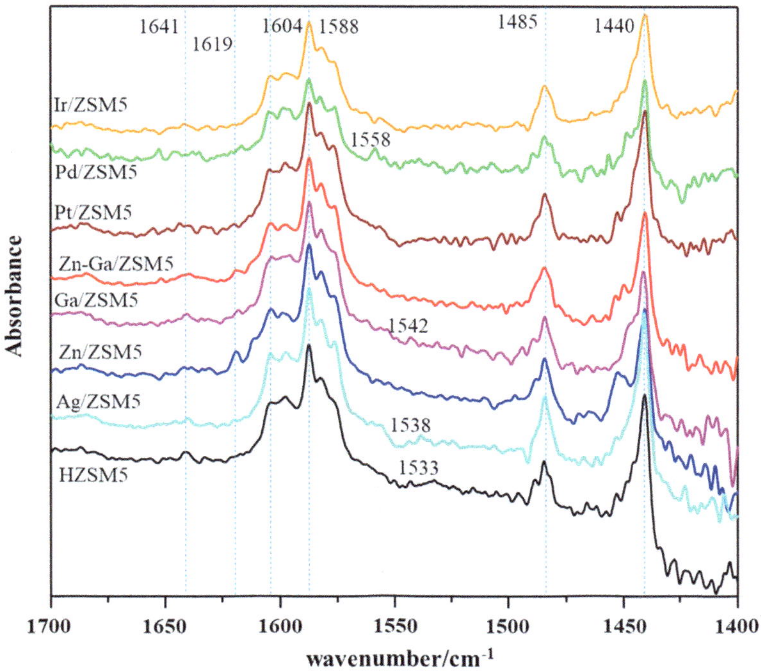

Fig. 8.46 DRIFT spectra of various fresh catalysts for furfural valorization under methane environment upon pyridine adsorption. Reproduced with the permission from [79]. Copyright (2018) American Chemical Society

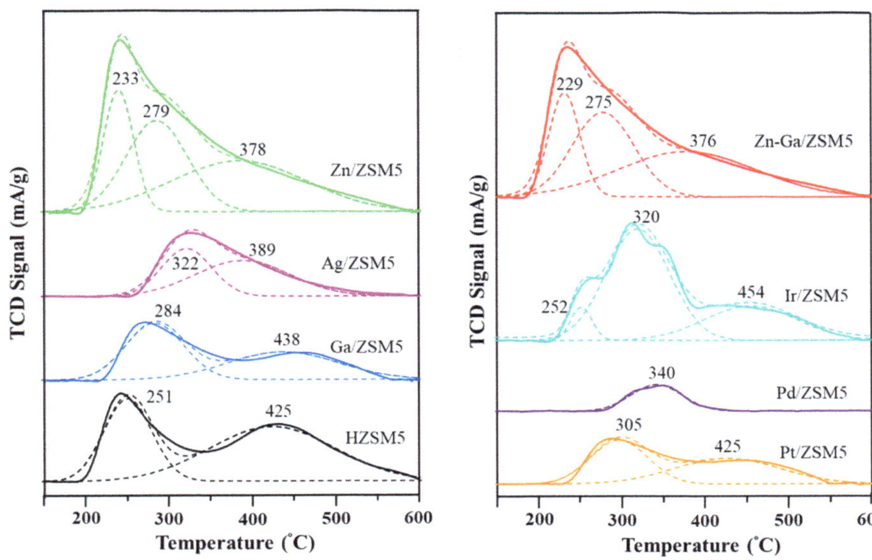

Fig. 8.47 NH_3-TPD profiles and Gaussian-deconvoluted peaks of various fresh catalysts for furfural valorization under methane environment. Reproduced with the permission from [79]. Copyright (2018) American Chemical Society

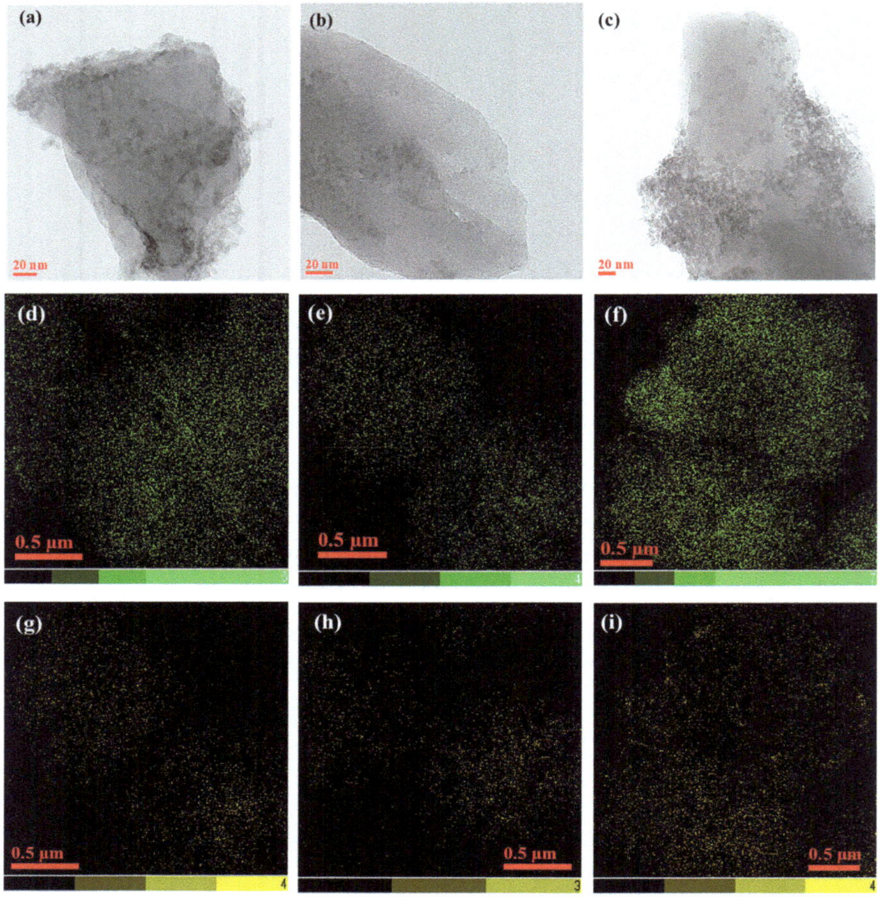

Fig. 8.48 TEM images of fresh Zn-Ga/ZSM5 catalyst (**a**) and spent Zn-Ga/ZSM5 catalyst collected under environments of CH_4 (**b**) and N_2 (**c**). EDS mapping analysis of Zn for fresh Zn-Ga/ZSM5 catalyst (**d**) and spent Zn-Ga/ZSM5 catalyst collected under environments of CH_4 (**e**) and N_2 (**f**). EDS mapping analysis of Ga for fresh Zn-Ga/ZSM5 catalyst (**g**) and spent Zn-Ga/ZSM5 catalyst collected under environments of CH_4 (**h**) and N_2 (**i**). Reproduced with the permission from [79]. Copyright (2018) American Chemical Society

images in Fig. 8.48. Similar conclusions can also be drawn for the other active metal Ga. Moreover, two main peaks can be observed in O 1s spectra after deconvolution. The major peak can be attributed to the O atom bonded to Si in zeolite structure, while the small peak can be attributed to metal oxides. This peak is decreased after the catalyst is used under methane environment while increased if nitrogen is the gas atmosphere, again substantiating the conclusion regarding metal oxide dispersion as observed in Zn and Ga spectra. Finally, three main peaks can be obtained in C 1 s spectra. The biggest peak at 284.8 can be attributed to C-C bonds, which is intensified after reaction due to the coking effect. However, it can be seen that the

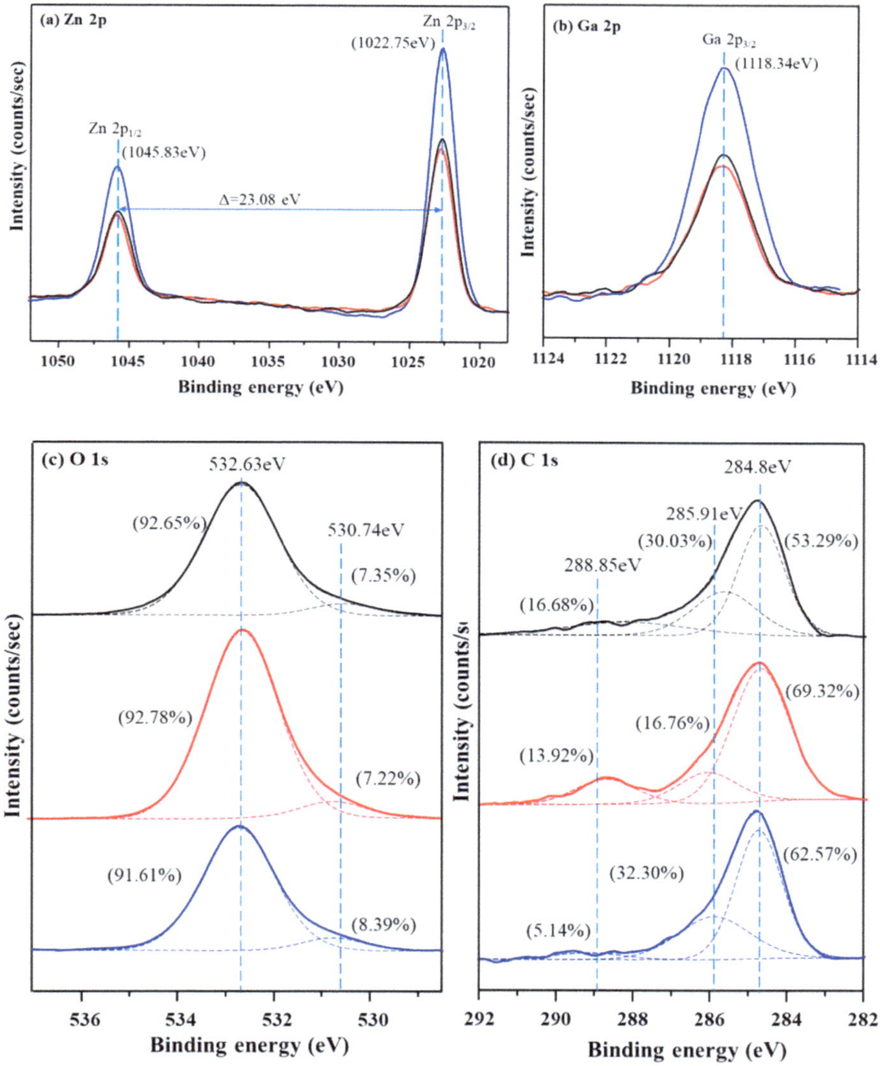

Fig. 8.49 Zn 2p (**a**), Ga 2p (**b**), O 1 s (**c**), and C 1 s (**d**) XPS spectra of fresh Zn-Ga/ZSM5 (black line) and spent Zn-Ga/ZSM5 collected from CH_4 (red line) and N_2 (blue line) runs. C 1 s and O 1 s XPS spectra are fitted with Lorentzian-Gaussian peaks of 30:70 ratio. Dotted line represents the fitting curve; solid line represents the experimental data. The fitted peak positions and concentration are indicated on the spectra. Reproduced with the permission from [79]. Copyright (2018) American Chemical Society

growth rate is higher under nitrogen environment than methane, indicating less coke is formed with the presence of methane. After comprehensive analyses of products and catalysts, the beneficial effects of methane become more definite, which strongly supports the reasonability of its utilization in biofuel upgrading processes.

8.7 Summary and Perspectives

Great achievements have been made for the model compound studies regarding biofuel conversion under methane environment. A series of catalysts have been synthesized and evaluated, indicating promising feasibility and profitability of the process. Comprehensive mechanistic studies have also been carried out to elucidate the reaction pathways and demonstrate the function of methane in the process. Several typical characteristics are listed as follows:

1. In spite of the varied structures of biofuel compounds, the presence of oxygen atom can be one of the commonalities in biofuel-related molecules. Oxygen is without dispute a crucial element for the construction and metabolism of biological species. However, it brings a series of difficulties for effective utilization of bio-derived feedstocks as energy-generating resources, including, but not limited to, low energy density since existing oxygen atoms cannot be further oxidized during the exothermic combustion reactions; high corrosiveness due to the presence of carboxyl groups and the corresponding low pH; immiscibility with other fuel fractions due to the high polarity of C-O, O-H, and C=O bonds; and high viscosity caused by the strong intermolecular interactions. Therefore, deoxygenation, namely, the removal of oxygen atoms, is generally of paramount importance for biofuel upgrading processes.
2. In order to realize deoxygenation, the presence of hydrogen donor is necessary. Methane is the hydrocarbon species with the highest hydrogen-to-carbon ratio (4:1). The abundant reserve, inexpensive cost and ample hydrogen atoms make it one of the most appropriate reductants (hydrogen donor) for biomass treatment. However, due to the intrinsic inertness of methane evidenced by high C-H bond energy, it can only be effectively utilized upon activation. Therefore, an appropriate design of catalyst is crucial for a successful co-conversion of biofuel and methane. The reported catalysts for methane activation mainly focus on metal-loaded zeolite catalysts. Silver, zinc, gallium are commonly used active metal components and an MFI zeolite structure with abundant acid sites is widely adopted as the support and catalyst for aromatization reactions.
3. Due to the complexity of naturally derived biofuel feedstocks, the selection of model compounds is necessary and effective for better understanding of the upgrading process, which in turn provides precious experiences for the control of real biofuel valorization practices. A large number of model compounds have been selected, which can be roughly derived into these groups: alcohols, carboxyl acids, aldehydes and ketones, phenolics, cellulose and sugars, and furans.

It should be noted that there are still miscellaneous oxygenates, such as glycolaldehyde and acetol, which contains two or more characteristic functional groups as listed above. However, the chosen model compounds are still representative for illustrating the mechanism since the same characteristic functional group could demonstrate similar reaction behaviors in different molecules, and the evolution of these miscellaneous oxygenates can be regarded as the combination of corresponding reactions of each functional group.

4. The participation of methane in the upgrading process can be demonstrated in at least two ways. On the one hand, once a hydrocarbon pool of short-chain hydrocarbon radicals is generated, the activated methane can be involved in the co-aromatization reaction, leading to the formation of more aromatic products as a carbon contributor. This route usually takes place in the inner pore of catalytic zeolite structures, where abundant acid sites facilitate the protonation and dehydrogenation of hydrocarbon chain with a suitable length. The pore structures with appropriate pore size also provide shape selectivity (space confinement) towards aromatics rather than products with longer carbon chains. On the other hand, after the aromatic products are formed, the abundant activated methane could also be added as side chain methyl substituents through alkylation reaction, leading to higher yield of aromatics due to the increase molecular weight of aromatic products. It should be emphasized that the effect of methane cannot be limited to the two aspects. Although simple model compounds are selected, the reaction network could still be considerably intricated, and many reactions could proceed simultaneously. For example, methane might also take part in the deoxygenation process as a hydrogen doner. It is also confirmed that the side chain alkyl groups of generated aromatic products are interchangeable. Besides, there is also a possibility that the carbon from the side chain could be transferred into the aromatic ring through a ring expansion-contraction mechanism.

It should be pointed out that although the biofuel upgrading process under methane environment is technically feasible, there are still a series of practical challenges for the real application. Some of the obstacles have been discussed in Chap. 7, and a brief summary and extra complements are listed as follows. (1) The oligomerization reactions are still widely observed in all relevant reactions, resulting in the formation of polyaromatic hydrocarbons (PAHs) and coke. These side reactions lower the utilization efficiency of biomass feedstocks and is also detrimental for the stability of catalysts, adversely influencing long-term operations. (2) Although the addition of methane leads to significantly improved upgrading performances, the absolute conversion of methane is still extremely low. It is not surprising since the inert chemical properties limits the effective activation of methane, especially at low temperatures. The low methane conversion might cause difficult engineering problems such as the separation of gas products and the recycling of methane. (3) It has been reported that the liquid yield is improved after the introduction of methane, but the value is still not satisfactory due to the inevitable formation of carbon oxides and small hydrocarbon molecules in the gas phase, again leading to low carbon utilization efficiency. (4) Due to the complexity and variability of biofuel feedstock (the composition and physical and chemical properties of biofuels from different sources can be completely different), no uniform criteria are established for quantitative evaluation of catalyst performance. The current investigations are generally based on a case-by-case mode, so that the mutual comparison and reference are of low efficiency. It is also not in favor of attracting the interest of corresponding users, slowing down the speed for practical applications.

In order to overcome the abovementioned problems, efforts have been made, and several concepts are listed here as possible solutions. (1) Process optimization should be comprehensively applied to figure out the influence of each parameters for the quality of products as well as the utilization efficiencies of biofuel and methane. (2) The introduction of new techniques for methane activation, such as UV-visible light, plasma, microwave, and biocatalyst using enzymes, deserves further evaluation. (3) More in-depth investigations of catalytic active sites should still be highly encouraged, so that the extent of methane activation can be notably improved. The corresponding studies could be closely related to the introduction of the aforementioned new techniques. It is expected that the coke formation will be further inhibited and the yield of desired liquid products can be greatly augmented upon more effective methane activation, especially if methane plays a better role as a hydrogen donor.

At last, it is worth noting that, despite the difficulties and challenges, this economically and environmentally friendly route for energy supply will attract more and more attention with the foreseeable rapidly decreasing reserve of petroleum and relatively stable supply of biofuel resources, and the pertinent research is still of great significant for sustainable development of human beings.

References

1. G.W. Huber, S. Iborra, A. Corma, Synthesis of transportation fuels from biomass: Chemistry, catalysts, and engineering. Chem. Rev. **106**, 4044–4098 (2006)
2. J.D. Adjaye, N.N. Bakhsh, Catalytic conversion of a biomass-derived oil to fuels and chemicals I: Model compound studies and reaction pathways. Biomass Bioenergy **8**(3), 131–149 (1995)
3. C.D. Chang, A.J. Silvestri, The conversion of methanol and other o-compounds to hydrocarbons over zeolite catalysts. J. Catal. **47**, 249–259 (1977)
4. C.D. Chang, C.T.-W. Chu, R.F. Socha, Methanol conversion to olefins over ZSM-5 I. effect of temperature and zeolite Si02/A1203. J. Catal. **86**, 289–296 (1984)
5. M. Bjorgen, S. Svelle, F. Joensen, J. Nerlov, S. Kolboe, F. Bonino, et al., Conversion of methanol to hydrocarbons over zeolite H-ZSM-5: On the origin of the olefinic species. J. Catal. **249**(2), 195–207 (2007)
6. M. Bjørgen, F. Joensen, M. Spangsberg Holm, U. Olsbye, K.-P. Lillerud, S. Svelle, Methanol to gasoline over zeolite H-ZSM-5: Improved catalyst performance by treatment with NaOH. Appl. Catal. A Gen. **345**(1), 43–50 (2008)
7. C.K. Narula, Z. Li, E.M. Casbeer, R.A. Geiger, M. Moses-Debusk, M. Keller, et al., Heterobimetallic zeolite, InV-ZSM-5, enables efficient conversion of biomass derived ethanol to renewable hydrocarbons. Sci. Rep. **5**, 16039 (2015)
8. A. Wang, P. He, M. Yung, H. Zeng, H. Qian, H. Song, Catalytic co-aromatization of ethanol and methane. Appl. Catal. B Environ. **198**, 480–492 (2016)

9. D. Austin, A. Wang, P. He, H. Qian, H. Zeng, H. Song, Catalytic valorization of biomass derived glycerol under methane: Effect of catalyst synthesis method. Fuel **216**, 218–226 (2018)
10. S. Tan, Z. Zhang, J. Sun, Q. Wang, Recent progress of catalytic pyrolysis of biomass by HZSM-5. Chin. J. Catal. **34**(4), 641–650 (2013)
11. L. Chen, Y. Zhu, H. Zheng, C. Zhang, Y. Li, Aqueous-phase hydrodeoxygenation of propanoic acid over the Ru/ZrO2 and Ru–Mo/ZrO2 catalysts. Appl. Catal. A: Gen. **411–412**, 95–104 (2012)
12. L. Chen, Y. Zhu, H. Zheng, C. Zhang, B. Zhang, Y. Li, Aqueous-phase hydrodeoxygenation of carboxylic acids to alcohols or alkanes over supported Ru catalysts. J. Mol. Catal. A Chem. **351**, 217–227 (2011)
13. A.C. Basagiannis, X.E. Verykios, Catalytic steam reforming of acetic acid for hydrogen production. Int. J. Hydrog. Energy **32**(15), 3343–3355 (2007)
14. P. He, W. Shan, Y. Xiao, H. Song, Performance of Zn/ZSM-5 for in situ catalytic upgrading of pyrolysis bio-oil by methane. Top. Catal. **59**(1), 86–93 (2015)
15. A. Wang, D. Austin, A. Karmakar, G.M. Bernard, V.K. Michaelis, M.M. Yung, et al., Methane upgrading of acetic acid as a model compound for a biomass-derived liquid over a modified zeolite catalyst. ACS Catal. **7**(5), 3681–3692 (2017)
16. S. Wang, Z. Guo, Q. Cai, L. Guo, Catalytic conversion of carboxylic acids in bio-oil for liquid hydrocarbons production. Biomass Bioenergy **45**, 138–143 (2012)
17. D. Austin, A. Wang, J.H. Harrhy, X. Mao, H. Zeng, H. Song, Catalytic aromatization of acetone as a model compound for biomass-derived oil under a methane environment. Cat. Sci. Technol. **8**(19), 5104–5114 (2018)
18. A.J. Cruz-Cabeza, D. Esquivel, C. Jimenez-Sanchidrian, F.J. Romero-Salguero, Metal-exchanged beta zeolites as catalysts for the conversion of acetone to hydrocarbons. Materials (Basel) **5**(1), 121–134 (2012)
19. N. Viswanadham, S.K. Saxena, Enhanced performance of nano-crystalline ZSM-5 in acetone to gasoline (ATG) reaction. Fuel **105**, 490–495 (2013)
20. K.V. Ramanamurty, G.S. Salvapati, Catalytic cyclocondensation of acetone to isophorone. Indian J. Chem. **38B**(1), 24–28 (1999)
21. P.M. Mortensen, J.-D. Grunwaldt, P.A. Jensen, A.D. Jensen, Screening of catalysts for hydrodeoxygenation of phenol as a model compound for bio-oil. ACS Catal. **3**(8), 1774–1785 (2013)
22. Y. Hong, H. Zhang, J. Sun, K.M. Ayman, A.J.R. Hensley, M. Gu, et al., Synergistic catalysis between Pd and Fe in gas phase hydrodeoxygenation of m-cresol. ACS Catal. **4**(10), 3335–3345 (2014)
23. D. Gao, Y. Xiao, A. Varma, Guaiacol hydrodeoxygenation over platinum catalyst: Reaction pathways and kinetics. Ind. Eng. Chem. Res. **54**(43), 10638–10644 (2015)
24. J.E. Peters, J.R. Carpenter, D.C. Dayton, Anisole and guaiacol hydrodeoxygenation reaction pathways over selected catalysts. Energy Fuel **29**(2), 909–916 (2015)
25. Q. Tan, G. Wang, L. Nie, A. Dinse, C. Buda, J. Shabaker, et al., Different product distributions and mechanistic aspects of the hydrodeoxygenation of m-cresol over platinum and ruthenium catalysts. ACS Catal. **5**(11), 6271–6283 (2015)
26. G.H. Gu, C.A. Mullen, A.A. Boateng, D.G. Vlachos, Mechanism of dehydration of phenols on noble metals via first-principles microkinetic modeling. ACS Catal. **6**(5), 3047–3055 (2016)
27. A. Robinson, G.A. Ferguson, J.R. Gallagher, S. Cheah, G.T. Beckham, J.A. Schaidle, et al., Enhanced hydrodeoxygenation ofm-cresol over bimetallic Pt–Mo catalysts through an oxophilic metal-induced tautomerization pathway. ACS Catal. **6**(7), 4356–4368 (2016)
28. G. Liu, A.W. Robertson, M.M. Li, W.C.H. Kuo, M.T. Darby, M.H. Muhieddine, et al., MoS2 monolayer catalyst doped with isolated Co atoms for the hydrodeoxygenation reaction. Nat. Chem. **9**(8), 810–816 (2017)
29. Y. Xiao, A. Varma, Kinetics of guaiacol deoxygenation using methane over the Pt–Bi catalyst. React. Chem. Eng. **2**(1), 36–43 (2017)
30. M.J. Gilkey, B. Xu, Heterogeneous catalytic transfer hydrogenation as an effective pathway in biomass upgrading. ACS Catal. **6**(3), 1420–1436 (2016)
31. T.V. Choudhary, C.B. Phillips, Renewable fuels via catalytic hydrodeoxygenation. Appl. Catal. A Gen. **397**(1–2), 1–12 (2011)

32. H. Shafaghat, P. Sirous Rezaei, W.M.A.W. Daud, Catalytic hydrogenation of phenol, cresol and guaiacol over physically mixed catalysts of Pd/C and zeolite solid acids. RSC Adv. **5**(43), 33990–33998 (2015)
33. A.L. Jongerius, R. Jastrzebski, P.C.A. Bruijnincx, B.M. Weckhuysen, CoMo sulfide-catalyzed hydrodeoxygenation of lignin model compounds: An extended reaction network for the conversion of monomeric and dimeric substrates. J. Catal. **285**(1), 315–323 (2012)
34. Z. Si, X. Zhang, C. Wang, L. Ma, R. Dong, An overview on catalytic hydrodeoxygenation of pyrolysis oil and its model compounds. Catalysts **7**(6) (2017)
35. T. Prasomsri, To AT, S. Crossley, W.E. Alvarez, D.E. Resasco, Catalytic conversion of anisole over HY and HZSM-5 zeolites in the presence of different hydrocarbon mixtures. Appl. Catal. B: Environ. **106**(1–2), 204–211 (2011)
36. L.O. Alemán-Vázquez, J.L.C. Domínguez, J.L. García-Gutiérrez, Effect of Tetralin, Decalin and naphthalene as hydrogen donors in the upgrading of heavy oils. Procedia Eng. **42**, 532–539 (2012)
37. Y. Xiao, A. Varma, Catalytic deoxygenation of Guaiacol using methane. ACS Sustain. Chem. Eng. **3**(11), 2606–2610 (2015)
38. A. Wang, H. Song, Maximizing the production of aromatic hydrocarbons from lignin conversion by coupling methane activation. Bioresour. Technol. **268**, 505–513 (2018)
39. P.T.M. Do, A.J. Foster, J. Chen, R.F. Lobo, Bimetallic effects in the hydrodeoxygenation of meta-cresol on γ-Al2O3 supported Pt–Ni and Pt–Co catalysts. Green Chem., 14(5) (2012)
40. A. Popov, E. Kondratieva, L. Mariey, J.M. Goupil, J. El Fallah, J.-P. Gilson, et al., Bio-oil hydrodeoxygenation: Adsorption of phenolic compounds on sulfided (Co)Mo catalysts. J. Catal. **297**, 176–186 (2013)
41. M.S. Zanuttini, C.D. Lago, C.A. Querini, M.A. Peralta, Deoxygenation of m-cresol on Pt/γ--Al2O3 catalysts. Catal. Today **213**, 9–17 (2013)
42. P.M. de Souza, L. Nie, L.E.P. Borges, F.B. Noronha, D.E. Resasco, Role of oxophilic supports in the selective hydrodeoxygenation of m-cresol on Pd catalysts. Catal. Lett. **144**(12), 2005–2011 (2014)
43. A.J.R. Hensley, Y. Wang, J.-S. McEwen, Phenol deoxygenation mechanisms on Fe(110) and Pd(111). ACS Catal. **5**(2), 523–536 (2014)
44. L. Nie, P.M. de Souza, F.B. Noronha, W. An, T. Sooknoi, D.E. Resasco, Selective conversion of m-cresol to toluene over bimetallic Ni–Fe catalysts. J. Mol. Catal. A: Chem. **388–389**, 47–55 (2014)
45. L. Nie, D.E. Resasco, Kinetics and mechanism of m-cresol hydrodeoxygenation on a Pt/SiO2 catalyst. J. Catal. **317**, 22–29 (2014)
46. M.S. Zanuttini, B.O. Dalla Costa, C.A. Querini, M.A. Peralta, Hydrodeoxygenation of m-cresol with Pt supported over mild acid materials. Appl. Catal. A Gen. **482**, 352–361 (2014)
47. G. Liu, Y. Zhao, J. Guo, High selectively catalytic conversion of lignin-based phenols into para−/m-xylene over Pt/HZSM-5. Catalysts, 6(2) (2016)
48. P.M. de Souza, R.C. Rabelo-Neto, L.E.P. Borges, G. Jacobs, B.H. Davis, D.E. Resasco, et al., Hydrodeoxygenation of phenol over Pd catalysts. Effect of support on reaction mechanism and catalyst deactivation. ACS Catal. **7**(3), 2058–2073 (2017)
49. H.-Y.T. Chen, G. Pacchioni, Role of oxide reducibility in the deoxygenation of phenol on ruthenium clusters supported on the anatase titania (1 0 1) surface. ChemCatChem **8**(15), 2492–2499 (2016)
50. A. Wang, D. Austin, P. He, M. Ha, V.K. Michaelis, L. Liu, et al., Mechanistic investigation on catalytic deoxygenation of phenol as a model compound of biocrude under methane. ACS Sustain. Chem. Eng. **7**(1), 1512–1523 (2018)
51. S. Xiao, B. Liu, Y. Wang, Z. Fang, Z. Zhang, Efficient conversion of cellulose into biofuel precursor 5-hydroxymethylfurfural in dimethyl sulfoxide-ionic liquid mixtures. Bioresour. Technol. **151**, 361–366 (2014)
52. P. Wattanapaphawong, O. Sato, K. Sato, N. Mimura, P. Reubroycharoen, A. Yamaguchi, Conversion of cellulose to lactic acid by using ZrO2–Al2O3 catalysts. Catalysts, 7(7) (2017)
53. S. Dutta, S. De, M.I. Alam, M.M. Abu-Omar, B. Saha, Direct conversion of cellulose and lignocellulosic biomass into chemicals and biofuel with metal chloride catalysts. J. Catal. **288**, 8–15 (2012)

54. J. Akhtar, N.A.S. Amin, A review on process conditions for optimum bio-oil yield in hydrothermal liquefaction of biomass. Renew. Sust. Energ. Rev. **15**(3), 1615–1624 (2011)
55. J. Zhang, X. Liu, M. Sun, X. Ma, Y. Han, Direct conversion of cellulose to glycolic acid with a phosphomolybdic acid catalyst in a water medium. ACS Catal. **2**(8), 1698–1702 (2012)
56. M. Mascal, E.B. Nikitin, Direct, high-yield conversion of cellulose into biofuel. Angew. Chem. **47**(41), 7924–7926 (2008)
57. A. Onda, T. Ochi, K. Yanagisawa, Selective hydrolysis of cellulose into glucose over solid acid catalysts. Green Chem. **10**(10) (2008)
58. J.C. Serrano-Ruiz, D.J. Braden, R.M. West, J.A. Dumesic, Conversion of cellulose to hydrocarbon fuels by progressive removal of oxygen. Appl. Catal. B Environ. **100**(1–2), 184–189 (2010)
59. Y. Wu, Z. Fu, D. Yin, Q. Xu, F. Liu, C. Lu, et al., Microwave-assisted hydrolysis of crystalline cellulose catalyzed by biomass char sulfonic acids. Green Chem., 12(4) (2010)
60. K. Chen, M. Tamura, Z. Yuan, Y. Nakagawa, K. Tomishige, One-pot conversion of sugar and sugar polyols to n-alkanes without C-C dissociation over the Ir-ReOx /SiO2 catalyst combined with H-ZSM-5. ChemSusChem **6**(4), 613–621 (2013)
61. L. Zhou, R. Liang, Z. Ma, T. Wu, Y. Wu, Conversion of cellulose to HMF in ionic liquid catalyzed by bifunctional ionic liquids. Bioresour. Technol. **129**, 450–455 (2013)
62. Y. Wang, H. Song, L. Peng, Q. Zhang, S. Yao, Recent developments in the catalytic conversion of cellulose. Biotechnol. Biotechnol. Equip. **28**(6), 981–988 (2014)
63. G. Zhang, C. Ni, X. Huang, A. Welgamage, L.A. Lawton, P.K. Robertson, et al., Simultaneous cellulose conversion and hydrogen production assisted by cellulose decomposition under UV-light photocatalysis. Chem. Commun. (Camb.) **52**(8), 1673–1676 (2016)
64. D.A. Gunawardena, S.D. Fernando, Thermal conversion of glucose to aromatic hydrocarbons via pressurized secondary pyrolysis. Bioresour. Technol. **102**(21), 10089–10093 (2011)
65. E.L. Kunkes, D.A. Simonetti, R.M. West, J.C. Serrano-Ruiz, C.A. Gärtner, J.A. Dumesic, Catalytic conversion of biomass to Monofunctional hydrocarbons and targeted liquid-fuel classes. Science **322**, 417–421 (2008)
66. D.A. Gunawardena, S.D. Fernando, Catalytic conversion of glucose micropyrolysis vapors in methane-using isotope labeling to reveal reaction pathways. Energ. Technol. **5**(5), 708–714 (2017)
67. T.R. Carlson, J. Jae, G.W. Huber, Mechanistic insights from isotopic studies of glucose conversion to aromatics over ZSM-5. ChemCatChem **1**(1), 107–110 (2009)
68. T.R. Carlson, J. Jae, Y.-C. Lin, G.A. Tompsett, G.W. Huber, Catalytic fast pyrolysis of glucose with HZSM-5: The combined homogeneous and heterogeneous reactions. J. Catal. **270**(1), 110–124 (2010)
69. A. Zheng, Z. Zhao, S. Chang, Z. Huang, K. Zhao, H. Wu, et al., Maximum synergistic effect in the coupling conversion of bio-derived furans and methanol over ZSM-5 for enhancing aromatic production. Green Chem., 16(5) (2014)
70. Y.-T. Cheng, G.W. Huber, Chemistry of furan conversion into aromatics and olefins over HZSM-5: A model biomass conversion reaction. ACS Catal. **1**(6), 611–628 (2011)
71. X. Li, H. Zhang, J. Li, L. Su, J. Zuo, S. Komarneni, et al., Improving the aromatic production in catalytic fast pyrolysis of cellulose by co-feeding low-density polyethylene. Appl. Catal. A Gen. **455**, 114–121 (2013)
72. G. Zhou, J. Li, Y. Yu, X. Li, Y. Wang, W. Wang, et al., Optimizing the distribution of aromatic products from catalytic fast pyrolysis of cellulose by ZSM-5 modification with boron and co-feeding of low-density polyethylene. Appl. Catal. A Gen. **487**, 45–53 (2014)
73. J. Li, Y. Yu, X. Li, W. Wang, G. Yu, S. Deng, et al., Maximizing carbon efficiency of petrochemical production from catalytic co-pyrolysis of biomass and plastics using gallium-containing MFI zeolites. Appl. Catal. B: Environ **172–173**, –154–64 (2015)
74. Y. Yang, Z. Du, Y. Huang, F. Lu, F. Wang, J. Gao, et al., Conversion of furfural into cyclopentanone over Ni–Cu bimetallic catalysts. Green Chem., 15(7) (2013)
75. W. Xu, H. Wang, X. Liu, J. Ren, Y. Wang, G. Lu, Direct catalytic conversion of furfural to 1,5-pentanediol by hydrogenolysis of the furan ring under mild conditions over Pt/Co2AlO4 catalyst. Chem. Commun. (Camb.) **47**(13), 3924–3926 (2011)

76. S. Sitthisa, W. An, D.E. Resasco, Selective conversion of furfural to methylfuran over silica-supported NiFe bimetallic catalysts. J. Catal. **284**(1), 90–101 (2011)
77. Y.-T. Cheng, G.W. Huber, Production of targeted aromatics by using Diels–Alder classes of reactions with furans and olefins over ZSM-5. Green Chem. **14**(11) (2012)
78. C.L. Williams, C.-C. Chang, P. Do, N. Nikbin, S. Caratzoulas, D.G. Vlachos, et al., Cycloaddition of biomass-derived furans for catalytic production of renewable p-xylene. ACS Catal. **2**(6), 935–939 (2012)
79. A. Wang, D. Austin, H. Qian, H. Zeng, H. Song, Catalytic valorization of furfural under methane environment. ACS Sustain. Chem. Eng. **6**(7), 8891–8903 (2018)

Index

H. Song et al., *Methane Activation and Utilization in the Petrochemical and Biofuel Industries*, https://doi.org/10.1007/978-3-030-88424-6

MIX
Papier aus verantwortungsvollen Quellen
Paper from responsible sources
FSC® C105338

If you have any concerns about our products,
you can contact us on
ProductSafety@springernature.com

In case Publisher is established outside the EU,
the EU authorized representative is:
Springer Nature Customer Service Center GmbH
Europaplatz 3, 69115 Heidelberg, Germany

Printed by Libri Plureos GmbH
in Hamburg, Germany